GENETIC RESOURCES OF MEDITERRANEAN PASTURE AND FORAGE LEGUMES

Current Plant Science and Biotechnology in Agriculture

VOLUME 33

Aims and Scope
The book series is intended for readers ranging from advanced students to senior research scientists and corporate directors interested in acquiring in-depth, state-of-the-art knowledge about research findings and techniques related to all aspects of agricultural biotechnology. Although the previous volumes in the series dealt with plant science and biotechnology, the aim is now to also include volumes dealing with animals science, food science and microbiology. While the subject matter will relate more particularly to agricultural applications, timely topics in basic science and biotechnology will also be explored. Some volumes will report progress in rapidly advancing disciplines through proceedings of symposia and workshops while others will detail fundamental information of an enduring nature that will be referenced repeatedly.

The titles published in this series are listed at the end of this volume.

Genetic Resources of Mediterranean Pasture and Forage Legumes

Edited by

SARITA JANE BENNETT

Co-operative Research Centre for Legumes in Mediterranean Agriculture (CLIMA), University of Western Australia, Nedlands, Australia

and

P. S. COCKS

Faculty of Agriculture, University of Western Australia, Nedlands, Australia

KLUWER ACADEMIC PUBLISHERS

DORDRECHT / BOSTON / LONDON

Library of Congress Cataloging-in-Publication Data

International Workshop on the Genetic Resources of Pasture and Forage Legumes in Mediterranean Environments (1996 : York, W.A.)
Genetic resources of Mediterranean pasture and forage legumes : proceedingsof an International Workshop on the Genetic Resources of Plant and Forage Legumes in Mediterranean Environments, held at York, Western Australia, 14th to 18th October 1996 / edited by Sarita Jane Bennett and P.S. Cocks.
p. cm. -- (Current plant science and biotechnology in agriculture ; v. 33)
Includes index.
ISBN 0-7923-5522-9 (hardbound : alk. paper)
1. Legumes as feed--Germplasm resources--Mediterranean Region--Congresses. 2. Forage plants--Germplasm resources--Mediterranean Region--Congresses. 3. Mediterranean climate--Congresses. I. Bennett, Sarita Jane. II. Cocks, P. S. (Philip Stanley), 1939- . III. Title. IV. Series: Current plant science and biotechnology in agriculture ; 33.
SB203.3.M43I58 1996
633.3'09182'2--dc21 98-49718

ISBN 0-7923-5522-9

Published by Kluwer Academic Publishers,
P.O. Box 17, 3300 AA Dordrecht, The Netherlands.

Sold and distributed in North, Central and South America
by Kluwer Academic Publishers,
101 Philip Drive, Norwell, MA 02061, U.S.A.

In all other countries, sold and distributed
by Kluwer Academic Publishers,
P.O. Box 322, 3300 AH Dordrecht, The Netherlands.

Printed on acid-free paper

Foreword

The Co-operative Research Centre for Legumes in Mediterranean Agriculture (CLIMA) was established in July 1992 by the Australian Government. It brings together Agriculture Western Australia, The University of Western Australia, CSIRO and Murdoch University, with each group providing scientists and resources to operate the centre.

CLIMA aims to increase the profitability of Australian mediterranean agriculture by at least five times its research expenditure through basic and strategic research, education and extension programmes aimed at sustainable farming systems based on crop and pasture legumes. The main objectives of CLIMA are;

- To develop more productive pasture and grain legumes
- To enhance the benefits of legumes in increasing soil fertility while minimising soil degradation
- To achieve better adoption of appropriate new technology resulting from research, and
- To provide training for students, staff and industry representatives in legume science.

An international workshop was organised by CLIMA on the genetic resources of pasture and forage legumes from mediterranean areas and was held at York, Western Australia. The papers from that workshop are presented in this volume.

Much of the genetic material, where the sources of the genetic resources can be identified, that CLIMA is testing has been derived from overseas sources. CLIMA's policy is to utilise the material for the mutual benefit of both the originators and the Australian developers. In the many situations where ownership can not be established or when the material is held in trust by the CGIAR centres, then CLIMA will provide the material in the public domain for Australian farmers and will provide the material to any other organisation upon request.

Acknowledgements

Thanks are given to the following

Funding sources

Contributors

Nancy Longnecker for much of the initial organisation.

Sue Dodimead for keeping the accounts and booking the airfares and accommodation of many of the international delegates.

Nigel Maxted and Dirk Enneking for organising and running the one-day Vicieae workshop.

Mike Ewing for organising and co-ordinating the field day to Medina Research Station, AgWA.

John Hamblin, CLIMA Director, for assisting with the financing of the workshop

Table of Contents

List of Participants

Ailian, Ms. Yang	School of Agriculture, University of Western Sydney, Hawkesbury, Richmond 2753. AUSTRALIA
Anwar, Dr Rashid	Plant Genetic Resources Institute, National Agricultural Centre, Park Rd, Islamabad, PAKISTAN
Auricht, Mr Geoff	South Australia Research and Development Institute (SARDI), PRC, GPO Box 397, Adelaide. SA 5001. AUSTRALIA
Baker, Dr Sue	CSIRO, Private Bag, P.O. Wembley. WA 6014. AUSTRALIA
Bennett, Dr Sarita Jane	Centre for Legumes in Mediterranean Agriculture (CLIMA), University of Western Australia, Nedlands. WA 6907. AUSTALIA
Berger, Mr Jens	Plant Sciences, University of Western Australia, Nedlands WA 6907. AUSTRALIA
Bounejmate, Dr Moustapha	International Centre for Agricultural Research in Dry Areas (ICARDA), P.O. Box 5466, Aleppo, SYRIA.
Campbell, Ms. Margaret	Plant Sciences, University of Western Australia, Nedlands WA 6907. AUSTRALIA
Carr, Dr. Steve	Agriculture WA, Baron-Hay Court, South Perth, WA 6151. AUSTRALIA
Castleman, Mr Geoff H.	Victorian Institute for Dryland Agriculture, Mallee Research Station, Walpeup, VIC 3507. AUSTRALIA
Clements, Dr. Jon	CLIMA, University of Western Australia, Nedlands, WA 6907. AUSTRALIA
Cocks, Prof Phil	Plant Sciences, University of Western Australia, Nedlands, WA 6907. AUSTRALIA
Collins, Dr. Bill	Ag.WA, Baron-Hay Court, South Perth. WA 6151. AUSTRALIA
Della, Dr Athina	Agricultural Research Institute, P.O. Box 2016, Nicosia. CYPRUS
Enneking, Dr. Dirk	CLIMA, University of Western Australia, Nedlands, WA 6907. AUSTRALIA
Ewing, Dr Mike	CLIMA, University of Western Australia, Nedlands. WA 6907. AUSTALIA
Fedorenko, Ms Diana	Plant Sciences, University of Western Australia, Nedlands WA 6907. AUSTRALIA
Francis, Dr Clive	CLIMA, University of Western Australia, Nedlands. WA 6907. AUSTALIA
Hadjichristodoulou, Dr Andreas	Agricultural Research Institute, P.O. Box 2016, Nicosia. CYPRUS
Howieson, Dr John	CLIMA, University of Western Australia, Nedlands. WA 6907. AUSTALIA

Hughes, Mr. Steve	SARDI, PRC, GPO Box 397, Adelaide. SA 5001. AUSTRALIA
Loi, Mr Angelo	CLIMA, University of Western Australia, Nedlands, WA 6907. AUSTRALIA
Marshall, Prof Don	Department of Crop Sciences, A20, Faculty of Agriculture, The University of Sydney, New South Wales 2006. AUSTRALIA
Maxted, Dr Nigel	School of Biological Sciences, University of Birmingham, Edgbaston Park Road, Edgbaston, Birmingham. B15 2TT. ENGLAND
McIntyre, Mr. Allan	Private Bag 260, Horsham, VIC 3401. AUSTRALIA
Nichols, Mr Phil	Plant Sciences, University of Western Australia, Nedlands WA 6907. AUSTRALIA
Norman, Ms Hayley	Plant Sciences, University of Western Australia, Nedlands WA 6907. AUSTRALIA
Nutt, Mr. Brad	CLIMA, University of Western Australia, Nedlands, WA 6907. AUSTRALIA
Ovalle, Dr. Carlos	Institute de Investigaciones Agrpecuarias (INIA), Estacion Experimental Quilamapu, Castilla 426, CHILE
Paull, Ms Heather	Centre for Development Studies, Flinders University of South Australia, GPO Box 2100, Adelaide 5000. AUSTRALIA
Porqueddu, Dr Claudio	Centro di Studio sui Pascoli Mediterranei (CNR), Via E. Nicola, 07100 Sassari, Sardinia. ITALY
Reid, Dr Robert	Tasmanian Institute of Agricultural Research (TIAR), Mt. Pleasant Laboratories, PO Box 46, Kings Meadows, Tasmania 7249. AUSTRALIA
Revell, Dr Clinton	CLIMA, University of Western Australia, Nedlands, WA 6907. AUSTRALIA
Roberts, Prof Eric	Department of Agriculture, University of Reading, Earley Gate, P.O. Box 236, Reading, RG6 6AT, Berkshire. ENGLAND.
Robertson, Dr Larry	ICARDA, P.O. Box 5466, Aleppo, SYRIA.
Robson, Prof Alan	University of Western Australia, Nedlands. WA 6907. AUSTRALIA
Roggero, Prof Pier Paulo	Dipartimento di Biotecnologie Agrarie ed Ambientali, Universita degli Studi, Via Brecce Bianche, 60131 ANCONA. ITALY
Samaras, Mr. S	Greek Gene Bank, P.O. Box 312, Thessaloniki 57001. GREECE
Sebanci, Dr Cefar	Aegean Agricultural Research Institute (AARI), P.O. Box 9, Menemen, Izmir 35661. TURKEY.
Snowball, Mr. Richard	Agriculture Western Australia (Ag.WA), Baron-Hay Court, South Perth. WA 6151. AUSTRALIA

Sulas, Dr. Leonardo	Centro di Studio sui Pascoli Mediterranei (CNR). Via E. De Nicola, 07100 Sassari, Sardinia. ITALY
Tahtacioglu, Dr. Lutfu	East Anatolia Agricultural Research Institute, Gezkoy, Erzurum, TURKEY
Yan, Ms. Chen	C/o, Dr. Parvez, School of Agriculture, University of Western Sydney, Hawkesbury, Richmond 2753. AUSTRALIA
Zoghlami, Ms Aziza	Institut National de la Recherche Agronomique de Tunisie (INRAT), Rue Hedi Karray, 2049 Ariana, TUNISIA

CHAPTER 1

Introduction: Role of pasture and forage legumes in mediterranean farming systems

P.S. Cocks and Sarita Jane Bennett
Faculty of Agriculture and Co-operative Research Centre for Legumes in Mediterranean Agriculture, The University of Western Australia, Nedlands, WA, 6907

Abstract This paper describes the contexts in which pasture and forage legumes are used. It points out that the main use of legumes in the Mediterranean basin is on the communally-owned grasslands, and that improvement of these grasslands are constrained by socio-economic factors. However, pasture and forage legumes also have roles to play in arable land where, especially in dry areas, they have the potential to replace fallow in cereal-producing farming systems through their capacity to fix nitrogen. Examples of both grassland and fallow improvement are given.

In contrast to the Mediterranean basin annual legumes are used widely in southern Australia in a 'ley farming system'. Ley farming consists of consecutive phases of pasture and crop where the pasture regenerates spontaneously. It is a modification of the north African weedy fallow system except that annual legumes have largely replaced weeds. It is based on two groups of species: subterranean clover and the annual medics. Unfortunately ley farming has become unsustainable in southern Australia in four ways: it uses insufficient water resulting in rising water tables and dryland salinity; it results in soil acidification because of the leaching of excess nitrate and the removal of cations during cropping; an increased dependence on herbicides has resulted in herbicide resistant weeds; and finally there is widespread pasture deterioration due to many factors operating independently and together.

Several remedies are proposed. The first requires the introduction of perennial species to increase water use. While some species are available and likely to be suitable there is a need for new germplasm especially for acid soils. Phase farming is likely to replace ley farming where longer periods of pasture are required and this will require new germplasm of which the seed can be produced cheaply. Aerial seeding clovers, of which there are many, and new genera such as *Ornithopus* and *Biserrula* will be of value. Finally, to arrest pasture deterioration two remedies are proposed: firstly the use of species mixtures in which the components use a range of reproductive strategies; and secondly the selection of species that are widespread in the Mediterranean basin. In addition to being widespread the latter are often dominant in the most arid environments, those environments which, in Australia, have proven to be the most intractable.

It is concluded that new germplasm is needed in both the Mediterranean basin and in Australia and that the Mediterranean basin is a rich source of new germplasm.

Keywords Farming systems, ley farming, phase farming, annual legumes, perennial legumes, sustainability

1. INTRODUCTION

The aim of this volume is to examine the genetic resources available for improving legume pastures in mediterranean farming systems. Almost without exception these genetic resources originate from the Mediterranean basin itself. The first European settlers to southern Australia, and California and Chile accidentally introduced these species to habitats in which they were able to thrive (e.g. Gladstones 1966). Mediterranean legumes were probably introduced to the Cape region of South Africa much earlier by Arab traders, and

S. J. Bennett and P. S. Cocks (eds.), Genetic Resources of Mediterranean Pasture and Forage Legumes, 9–19.

South Africa has served as an important staging point for the colonisation of southern Australia.

Inevitably, these colonising species had an important impact on their new habitats. In particular, the non-leguminous species became important weeds. In Australia, these include annual ryegrass (*Lolium* sp.), wild oats (*Avena* sp.), wild radish (*Raphanus raphanistrum*) and capeweed (*Arctotheca calendula* - a native of South Africa). Similar species have become important weeds in Chile and California (e.g. Allard 1975). However, the legumes have also assumed great importance, especially in Australia, where they form an essential component of the ley farming system (see later). It has been estimated that pasture legumes contribute up to 150 kg/ha/year to grain-producing farming systems in the wheat belt of southern Australia, and that their contribution to meat and wool producing systems in higher rainfall areas is similar (Puckridge and French 1983; Ellington *et al.* 1979).

Strangely, pasture and forage legumes contribute much less to the farming systems of the Mediterranean basin where their impact is limited to natural grasslands, weedy fallows and a few sown forage crops, especially vetch/oat mixtures (Osman *et al.* 1991; Keatinge *et al.* 1985; Cooper *et al.* 1987). Even here, there is evidence of strong degradation - the legume component of grasslands is declining alarmingly (Cocks and Gintzburger 1993), few legumes remain in weedy fallows and vetch/oat mixtures are under pressure from more valuable cereal and food legume crops. Livestock production relies increasingly on concentrates, especially barley. Under these circumstances livestock compete with humans, which is an inefficient use of food resources for a region where food is not readily available to many people (Pellet and Young 1988).

This chapter therefore sets the scene for the main purpose of this volume - a consideration of the genetic resources available to improve the use of pasture and forage legumes in mediterranean farming systems. It begins by examining the role of legumes in the grasslands and cropping systems of west Asia and north Africa and compares these with the ley farming system of southern Australia. It then explores possibilities for introducing the ley farming system into other areas with mediterranean-type climates. Finally it discusses the sustainability of annual legume-based farming systems, how they should be modified to deal with the worst of the issues and what genetic resources are needed.

2. PASTURE AND FORAGE LEGUMES IN THE GRASSLANDS AND CROP ROTATIONS OF THE MEDITERRANEAN BASIN

The earliest recorded use of annual legumes is in the grasslands of the Mediterranean basin. From antiquity they have formed a component of the grassland flora, often a dominant component. For example, in protected grasslands at Tel Hadya in north Syria (rainfall 325 mm), some 56 separate legume species were present compared with 71 species of grasses and forbs. Provided that phosphorus is applied these legumes provided the majority of the biomass (Pagnotta *et al.* 1997). In higher rainfall areas even unprotected grasslands have high legume components, often with a great number of species (Cocks and Osman 1996). In Lebanon native grasslands failed to respond to seeding or added fertiliser, indicating that it is unlikely that their productivity can be improved by any means. One assumes that modern productivity of these particular cases differs little from their productivity in antiquity.

However, these are the exceptions. In low rainfall areas (<500 mm) the grasslands have suffered severely at the hands of farmers and sheep herders. Species diversity, seed bank size and rainfall use efficiency are all much lower than is expected (Cocks and Osman 1996). Indeed, production at one low rainfall site (270 mm) was considered insufficient to support grazing animals at any time, a situation that even the farmers realise. They believe that open grazing benefits sheep only in that it provides exercise.

Other surveys of the Mediterranean grasslands reveal similar pictures. In Morocco, Bounejmate *et al.* (1992a; 1992b) considered that several annual species of *Medicago* were at serious risk of genetic erosion. In Syria, while diversity seems to have been maintained, seed bank sizes are also severely reduced, inevitably leading to greatly reduced productivity (Ehrman and Cocks 1990). Technically, this problem is not difficult to overcome and most authors believe that use of existing germplasm accompanied by low applications of fertiliser and sensible grazing management is sufficient to increase livestock productivity. This has been achieved in Syria (Osman *et al.* 1994), Greece (Vasilios 1987; Papademetrious and Papanastasis 1993), Spain (Moreno *et al.* 1993), the stony Crau plains of southern France (Vall and Gintzburger 1991), and Italy (Bullitta and Roggero 1989).

However, the social problems are far more complex. The root of these problems is land tenure, or the ways in which control can be exercised. Traditionally, the grasslands have been communally-owned, which means that identifiable groups own and control the land. Increasingly, as population pressures increase, the land has become 'open access' to all who wish to use it (Bromley 1989). Under this regime nobody benefits from grassland improvement hence no improvement is carried out.

Nevertheless, it does seem likely that vestiges of the old land tenure systems remain. These must be identified and used to assist the uptake of new technology by farming communities. An outstanding example of this was reported by Ghassali *et al.* (1998) who used the principle of Thomson *et al.* (1990) that many legume seeds have a low digestibility when fed to sheep. Using the local Awassi sheep they were able to transfer from improved to unimproved grasslands, seeds of some ten legume species including *Trifolium campestre* and *T. purpureum*. The farmers were able to exert sufficient control over the land such that the flocks of neighbouring villages and nomadic herdsman were prevented from grazing. It was observed that the farmers needed to be able to see the land on a day to day basis: where the land was out of sight it was not possible to eliminate grazing by flocks other than by those who shared ownership of the land.

While annual legumes continue to be a component of grassland ecosystems their role in arable agriculture is confined to the growing of forage crops. Forages are defined as crops sown annually to be grazed or fed to livestock: in contrast pastures are expected to regenerate spontaneously. Recent efforts have been made to replace fallows with forage legumes in the barley/livestock zone of west Asia (see Cooper *et al.* 1987), but this is less attractive to farmers than continuous cropping with barley (though possibly more sustainable). On-farm experiments have demonstrated that inclusion of forage legumes (*Vicia sativa* and *Lathyrus sativus*) increases both the milk and grain production of low rainfall north Syrian farming systems compared with continuous barley or barley/fallow (Thomson *et al.* 1992). Experimentation with vetch based farming systems is proceeding in Turkey (Elicin *et al.* 1993) where similar results have been obtained. While there is little data available from north Africa, vetch/oat mixtures are both widely grown and valued by farmers. Nevertheless it seems likely that competition from food legumes and cereals will limit the use and value of forage legumes in the region.

3. LEGUMES IN THE LEY FARMING SYSTEM OF SOUTHERN AUSTRALIA

Ley farming is a system where cereals are rotated with one or more years of spontaneously regenerating pasture. It is a modification of the weedy fallow system of north Africa, except that the weeds are replaced by legumes. In both cases the 'fallow' is used for livestock production and sheep are used to control weeds. In southern Australia the system was developed using introduced annual legumes, mainly subterranean clover (*Trifolium subterraneum*) and a few medic species (e.g. *M. truncatula, M. polymorpha,* and *M. littoralis*). Ley farming is considered to be largely responsible for the 50% increase in wheat yields in Australia between 1950 and 1980 (Donald 1981).

The key to the success of the ley farming system is the capacity of the annual legumes to survive one or more years of crop without setting fresh seed. Like most mediterranean annual legumes, medics and subterranean clover produce hard seeds, that is seeds that are impermeable to water. In the case of medics (e.g. *M. noeana*) the seeds may remain hard for five years or more (Cocks 1993), while unburied subterranean clover seed loses most of its hard seed in one year. Seed burial has little effect on medics but does increase the longevity of subterranean clover seed (Taylor and Ewing 1996), possibly to as long as the medics.

It is worth spending a little more time on this important process. Seed softening appears to be governed by a two stage process, the first of which involves heat inputs and the second a much shorter period of alternating temperatures (Taylor 1981). The first stage is responsible for the longevity of the seed bank, that is its survival from year to year. The second controls the pattern of softening within a year, that is whether softening occurs in late summer, or early or late autumn. This is extremely important because many mediterranean climates experience 'false breaks'; periods of rainfall followed by long drought during which newly germinated seeds are likely to die. For this reason, in most mediterranean environments seeds that soften late in the autumn are more likely to produce seedlings capable of reproduction than seeds germinating earlier. Smith *et al.* (1996) refer to the time at which seeds soften as the half life of softening within any year. They present evidence showing that the cultivars of subterranean clover have a half life of just over 40 days, compared with medics and some naturalised clovers with half lives of more than 120 days. More recently the issue of seed softening has become even more complex with most species having distinctive patterns (Norman *et al.* 1998), often with variation between ecotypes within species (Fresnillo-Fedorenko pers. comm.), and possibly involving light as a another control mechanism (Revell *et al.* 1998).

Ley farming integrates the livestock and cereal components of the farm enterprise in the following ways: 1) the pasture produces nutritious feed for livestock. Biomass yields of more than 10 t/ha have been recorded. The quality and quantity of feed varies seasonally, with highest quality in winter, lowest in summer, and the greatest quantity in spring (Allden 1980); 2) the pasture adds nitrogen to the soil, improves the soil structure and builds up soil organic matter (Greenland 1971); 3) provided that grasses are removed, the pastures control root diseases of cereals (King *et al.* 1982; MacNish 1989); and 4) weeds are controlled, especially grassy weeds, by grazing (MacNish and Nicholas 1987) and by using herbicides on pastures that are toxic to cereals (Thorn and Perry 1987).

Although ley farming has been widely adopted in southern Australia it is not used in other regions with mediterranean-type climates, and recently it has proven to be unsustainable in southern Australia. The implications of both these issues are discussed in the following sections.

4. ATTEMPTS TO INTRODUCE LEY FARMING TO THE MEDITERRANEAN BASIN

On the face of it, the only technical reasons why ley farming might fail in the Mediterranean basin are lack of adapted cultivars and rhizobia, and, where cultivars are available, cheap sources of pasture seed. Soils and climate are similar to southern Australia and, while access to machinery is often limited, Mediterranean farmers are as capable as Australians in modifying existing equipment.

In a thoughtful paper, Springborg (1986) classifies the proponents of ley farming into two schools: those who believe that a simple transfer of technology, requiring little basic or adaptive research, is involved, and those who believe that technical problems loom so large there is little point in working with farmers until they are resolved. The first school, to which Chatterton and Chatterton (1984) belong, argues that there is no time for research, the need for extension is urgent, and seeds and machinery must be imported as quickly as possible.

This school has resulted in the implementation of large scale government-run programmes in countries as diverse as Iraq, Algeria and Morocco (Halse 1993). The second argues that ley farming as practised in Australia is simply not suitable for the Mediterranean basin. It was developed in and for Australia where climate and soils may differ, farmers are comparatively highly educated, and resources for development are readily available. Extremists contend that even research is a waste of time, but most agree that the problems are soluble.

We now know that *Medicago rigidula* and *M. rotata*, neither of which are used in Australia, are suitable for ley farming in areas subject to severe frost (Cocks and Ehrman 1987). The Australian cultivars are more likely to succeed in north Africa, where *M. truncatula* is widespread (Bounejmate *et al.* 1992a), than in west Asia, where it is rare (Ehrman and Cocks 1990). Research involving farmer participation has revealed that wheat yields after medic are higher than in traditional rotations (lentil/wheat), supervised farmers are able to maintain seed banks at more than 200 kg/ha, and that profit from the pasture phase can be three times that of the wheat phase (Cocks *et al.* 1993). Still, the ley farming system has not been adopted.

It appears to us that there are three constraints to widespread adoption of ley farming. 1) Farmers find it difficult to manage grazing to maintain the seed bank. Even where they assume control of the land, grazing management is difficult because sheep grazing medic pods are able to increase or maintain liveweight only at unsustainable seed bank levels (Cocks 1988). As a result seed bank levels can fall to as low as 10 kg/ha before sheep liveweight gain is adversely affected. 2) Lack of local commercial seed production. Few of the countries are prepared to import Australian seed and in most cases even if they did the cultivars would be unsuitable. Throughout the Mediterranean basin there is a need for local seed industries to link with local plant breeding programmes. 3) Farmers need more information. Extension systems can barely deal with the simplest enquiries, partly because they themselves lack information but mainly because they are badly under-resourced.

In some parts of north Africa tillage systems also prevent successful adoption of ley farming. For example, Algeria continues to rely on the European mouldboard plough. Under these conditions seeds are buried far too deeply and few manage to produce seedlings (Cocks 1994).

In recent years the effort to introduce ley farming has largely abated, partly because of political instability in the countries concerned (e.g. Algeria and Iraq), partly because of serious under-funding in the international centres, but mainly because of disillusionment by both scientists and farmers. This is a pity because the ley farming system has much to offer a region where livestock products are highly valued and soil fertility is low.

5. SUSTAINABILITY OF FARMING SYSTEMS BASED ON ANNUAL LEGUMES

In this section we wish to touch on four sustainability issues brought about by our dependence on annual pastures. These are 1) soil acidification; 2) rising water tables and soil salinisation; 3) pasture deterioration; and 4) herbicide resistance.

Farming accelerates two processes that produce acidity as a side effect; the amount of nitrogen that flows through the soil and the amount of nutrients absorbed from the soil and exported as produce. Legumes strongly influence the first of these processes by contributing fixed atmospheric nitrogen. Many annual pasture and crop plants further contribute to the problem by their ability to explore only the surface soil, thereby decreasing their capacity to absorb nitrogen deeper in the soil profile. Soil acidification has been most serious on soils that were initially acid and have high aluminium levels (Carr *et al.* 1991), the situation in much of south-east and south-west Australia.

Soil acidification can be overcome by the application of lime (Coventry *et al.* 1989; Dolling *et al.* 1991). However, lime is applied to the surface, especially now that farmers prefer to use minimum and zero tillage techniques. The result is that lime increases surface pH but has little, if any, effect on subsurface pH. Since this subsurface layer is often only 10 cm below the surface, plant roots can be severely inhibited, reducing water and nutrient uptake, especially in spring.

In addition to reducing root growth, low pH has a severe effect on the survival of rhizobia (Evans *et al.* 1988). This is particularly so for annual medics, many of whose rhizobia are poorly adapted to even mildly acid soils (Howieson 1995). Koopman *et al.* (1995) suggest that low pH appears to be one of the reasons for the poor performance of lucerne in the mid-north of South Australia. They were able to show that some strains of rhizobia (e.g. NA39) are more tolerant of low pH than others (e.g. CC169). Similar results have been obtained for *M. polymorpha* (Howieson and Ewing 1986). Even *Rhizobium trifolii*, normally considered acid tolerant, survives better in limed soil over summer than it does in unlimed soil (Evans *et al.* 1988).

Rates of acidification are frightening and vary between the equivalent of 8 kg lime/ha/year (Dolling *et al.* 1994) and 95 kg lime/ha/year (Conyers *et al.* 1996). Most measurements of rate are nearer the higher than the lower figure.

An even more intractable problem is caused by rising water tables leading to salinisation. In fact, salinisation is related to acidification in that both have roots in the inadequate use of water by annual crops and pastures. In pre-European times there was a balance between water intake and its use by the perennial native vegetation. When the latter was removed, the newly introduced Mediterranean annuals were unable to use summer rainfall or rainfall in early autumn. For example, in the Lemon catchment of Western Australia, water tables in the valley floors have risen from 17 m below the surface to the surface in only 20 years (Anon. 1996). In the Darling peneplain up to 40% of rainfall may be lost by pathways other than evaporation (Gregory *et al.* 1992). As a result yields of wheat are poorly related to evapotranspiration. This water eventually reaches the water table. The water is salty because sea salt has been carried inland for millenia by the prevailing winds and deposited on the land at rates of 20-200 kg/ha/year (Hingston and Gailitis 1976). It is this salt that is returning to the surface as water tables rise.

Of the 25 million ha of land in south-western Australia previously covered by native perennial vegetation, almost 18 million ha have been cleared. Currently some 1.8 million ha are affected by salt and it is expected that, if present trends continue, 6 million ha will be affected within 30 years. Even at the present rate of salinisation the loss of productivity due to salinity almost equals the gains in productivity due to plant breeding and other new technologies.

Pasture deterioration is loss of the legume component. It results in a loss of productivity because of the high risk of nitrogen deficiency in the non-legumes. It affects both the livestock phase of rotations and the cereal phase, the former through reduced feed supplies and the latter through reduced nitrogen, and increased diseases and weeds. It is thought to be the result of many factors, some of them the result of neglect by farmers, others the inevitable result of an ecosystem that is heavily dependent on one or a few species.

The following is a list of the causes of pasture deterioration (Carter 1982):

- increasing problems with insects, especially red-legged earth mites and several aphid species;
- reduced use of superphosphate associated with the reduced value of livestock products since 1990;
- increased grazing pressure, especially in summer when seed banks are vulnerable to overgrazing;
- increased use of herbicides during seed set;
- reduced sowings of medics and clovers;
- increased use of nitrogen fertilisers;
- increased frequency of cropping, which places extreme pressure on the pasture seed banks;

- increased incidence of root and foliar diseases; and
- lack of diversity in legume germplasm available to farmers.

Herbicide resistance is defined as the 'inherited ability of a weed population to survive a herbicide application that is normally lethal' (Powles *et al.* 1997). It is essentially a genetic and evolutionary problem. Given time, some grasses will even evolve resistance to heavy metals (e.g. Bradshaw 1960).

Herbicide resistance looms as the major short term sustainability issue in south-western Australia. At least four species have developed herbicide resistance, wild oats (*Avena* sp.), ryegrass (*Lolium* sp.), wild radish (*Raphanus raphanistrum*) and wild mustard (*Synapis arvensis*). In all, herbicide resistance affects 2 million ha in Western Australia, with a further 6 million ha at risk. Ineffective herbicide application is estimated to cost \$15-25/ha/yr, and yield losses from ineffective weed control are of the order of \$20-200/ha/yr. Conservatively, herbicide resistance costs Western Australian grain growers some \$100 million/year. The costs are also high in south-eastern Australia and, in other parts of the world (Powles *et al.* 1997).

The selection pressures imposed by herbicides are extremely high. Gene frequencies in an untreated population range from 3×10^{-3} (Darmency and Gasquez 1990) to 1×10^{-42}, although more often they are around 1×10^{-6}. In the latter case, the selection pressure imposed by routine herbicide use might result in herbicide resistance in 3-5 years (Gill 1995), depending on its cost to fitness, rates of herbicide application, herbicide efficacy, whether the gene is dominant or recessive (Darmency 1994), and whether multiple genes are involved. Herbicides with residual effects apply more pressure than those without.

In the next section we discuss how new pasture and forage germplasm may address these sustainability issues.

6. NEW PASTURE LEGUME-BASED FARMING SYSTEMS

Acidification and salinisation can be discussed together since they are different aspects of the same problem: the inadequate use of rainfall.

The use of perennial plants in farming systems where water use is inadequate is being advocated in most of southern Australia. For example, the salinity action plan in Western Australia (Anon. 1997) puts forward the case that some 3 million ha of trees and shrubs need to be planted across the agricultural area of the state. However, such a plan is unlikely to be welcomed by grain growers who would not only face substantial costs in establishing the trees but would also lose the capacity to derive income from grain growing.

An alternative is to use perennial pasture species, especially perennial legumes. For example, lucerne is estimated to use 70% of the water used by complete tree cover and provide 30 times the return to farmers. Where farmers currently use the ley farming system, 'phase farming', in which several years of crop are followed by several years of lucerne pasture, may increase water use substantially while maintaining the farmer's capacity to produce grain. Evidence to this effect is provided by Dear (pers. comm.), who found that perennials (lucerne, *Phalaris aquatica*, *Dactylis glomerata* and *Danthonia richardsonii*) dried the soil to 140 cm depth, where annual subterranean clover had no effect. Furthermore, while the profile partially refilled at wetter sites (550 mm), the soil remained dry at drier sites (440 mm). A particularly interesting aspect of Dear's results is that all perennials, including lucerne, were equally effective at using nitrate and ammonium nitrogen. It is therefore likely that using perennials will slow the rate of acidification.

Nevertheless, there are problems associated with using perennials. This is especially so with the perennial grasses where no nitrogen is fixed. Dear and Cocks (1997) demonstrated that perennials compete strongly with subterranean clover for water in autumn, probably because the subterranean clover is unable to regulate its germination in response to out-of-season rainfall. What is needed are annual legumes with more appropriate within season seed softening patterns. Thus, in addition to the need for perennial legumes, new annuals are needed specifically adapted to an environment strongly influenced by the presence of perennials.

Phase farming is also likely to play a role in the control of herbicide resistance in weeds. During the long pasture phase grazing is manipulated to minimise seed set of the weeds. In the initial years, grazing soon after the opening rains when most grasses are erect and legumes prostrate will favour the legumes at the expense of grasses. Grazing can again be used during seed set, when most grasses produce seed heads that rise above the pasture canopy. Later in the rotation, after the necessity for spontaneous regeneration of the legumes has passed, weeds can be controlled with complete herbicides (e.g. glyphosate).

Phase farming depends for its success on a supply of cheap seed. Because subterranean clover buries its seeds and the annual medics shed their seeds inside hard, woody pods, highly specialised harvesting techniques are required in both cases. In contrast, aerial seeding pasture legumes can be harvested with conventional machinery using skills available to most farmers. Even if farmers prefer to buy seed the price will be maintained at a low level by competition from other suppliers. Some germplasm is currently available, especially for the deep sandy soils of Western Australia where serradella (*Ornithopus sativus*) cv. Cadiz is proving popular. However, for many soil and climatic types aerial seeders with the right hard seed breakdown patterns are still required.

Although pasture deterioration is caused in part by poor management there is also a need for new germplasm. In their native Mediterranean basin pasture legumes have developed a number of reproductive strategies to overcome problems associated with mediterranean climates. For example, Ehrman and Cocks (1996) found that the range of reproductive strategies show the greatest diversity under the mild, wet conditions of the west Asian coast and least diversity in the arid interior. Increasing aridity leads to a subset of reproductive patterns that includes high levels of seed dormancy, high seed to pod ratios, restricted dispersal capabilities and early flowering - in other words a set of conservative strategies. However, even here, diversity of strategies is important.

It is possible that pasture deterioration can be addressed using one of two approaches involving new germplasm. The first is to take account of the most widespread reproductive strategies used in the mediterranean environment and focus on them. Such strategies are similar to those used by species adapted to arid areas. The second is to use more diverse mixtures, with the focus on diversity of strategies. In this way, productivity will be maintained by species using more radical strategies, and persistence by those using conservative strategies. Such an approach is similar to the result achieved through natural selection in the Mediterranean basin itself.

7. CONCLUSIONS

In both southern Australia and the Mediterranean basin legumes form the basis of grassland production. In both environments there are serious problems associated with the sustainability of farming systems, and both these problems can be partly addressed through the use of unexplored germplasm. Although bad management is largely to blame for the present problems, at least in southern Australia, the narrow genetic base is also responsible. Broadening the genetic base will go far towards overcoming the issues of sustainability.

In particular Australian farming systems need perennial legumes. These are widely available although few species are presently in commercial use. Of these only lucerne shows promise in southern Australia. However, even this species has many wild relatives, and these should be examined as soon as possible. Many other genera are available, including *Trifolium, Astragalus, Hedysarum* and *Coronilla*. In the Mediterranean basin diversity is present and the main requirement is better management, including a clear system of land tenure. Once this is available production from the mediterranean grasslands will increase several fold (Osman *et al.* 1994).

The remainder of this book examines the legume germplasm available to improve farming systems in the mediterranean areas of the world and how it might be accessed. Included are the principles by which legumes survive in their native habitats (chapters 2 and 12), an outline of the genetic resources available in international collections (chapters 3 and 11), some of the principles by which plant collectors might approach their task (chapters 4, 5 and 7), some of the reasons why plants need to be collected (chapters 6 and 8), an outline of the importance of including rhizobia (chapter 9) and some of the political dimensions in plant collection (chapter 10). Evaluation is discussed in chapters 13, 14 and 15 (the latter examines evaluation from the point of view of grazing animals). The volume concludes with a consideration of seed storage (chapter 16) and three case studies based in Cyprus, Greece and Tunisia.

REFERENCES

Allard, R.W. (1975). The mating system and microevolution. *Genetics* **79**, 115-26.

Allden, W.G. (1980). Integration of animals into dryland farming systems. *International Congress on Dryland farming, Adelaide, August 1980.*

Anon. (1996). Salinity: a situation statement for Western Australia. A Report to the Government of Western Australia.

Bounejmate, M., Beale, P.E. and Robson, A.D. (1992a). Annual *Medicago* species in Morocco. I. Species and their abundance. *Australian Journal of Agricultural Research* **43**, 739-49.

Bounejmate, M., Robson, A.D. and Beale, P.E. (1992b). Annual *Medicago* species in Morocco. II. Distribution in relation to soil and climate. *Australian Journal of Agricultural Research* **43**, 751-63.

Bradshaw, A.D. (1960). Population differentiation in *Agrostis tenuis* Sibth. III. Populations in varied environments. *New Phytologist* **59**, 92-103.

Bromley, D.W. (1989). The management of common property natural resources: some conceptual and operational fallacies. *World Bank 9th Agricultural Symposium.* Washington DC, USA.

Bullitta, P.S. and Roggero, P.P. (1989). Agronomic methods to increase pasture land production in Mediterranean marginal areas. *Proceedings of the XVIth International Grasslands Congress*, pp. 1591-92. Nice, France.

Carr, S.J., Ritchie, G.S.P. and Porter, W.M. (1991). A soil test for aluminium toxicity in acidic subsoils of yellow earths in Western Australia. *Australian Journal of Agricultural Research* **42**, 875-92.

Carter, E.D. (1982). The need for change in making the best use of medics in the cereal/livestock farming systems of South Australia. In *Proceedings of the 2nd Australian Agronomy Conference, Wagga Wagga, NSW*, page 180.

Chatterton, B. and Chatterton, L. (1984). Alleviating land degradation and increasing cereal and livestock production in north Africa and the Middle East using annual *Medicago* pasture. *Agriculture, Ecosystems and Environment* **11**, 117-29.

Cocks, P.S. (1988). Seed production and seed survival under grazing of annual medics (*Medicago* spp.) in north Syria. *Journal of Agricultural Science (Cambridge)* **110**, 455-63.

Cocks, P.S. (1993). Seed and seedling dynamics over four consecutive years from a single seed set of six annual medics (*Medicago* spp.) in north Syria. *Experimental Agriculture* **29**, 461-72.

Cocks, P.S. (1994). Effect of tillage system on the spontaneous regeneration of two annual medics (*Medicago* spp.) after wheat in north Syria. *Experimental Agriculture* **30**, 237-48.

Cocks, P.S. and Ehrman, T.A.M. (1987). The geographic origin of frost tolerance in Syrian pasture legumes. *Journal of Applied Ecology* **24**, 678-83.

Cocks, P.S. and Gintzburger, G. Long term sustainability of livestock-producing farming systems in contrasting regions with mediterranean-type climates. *Proceedings of the XVIIth International Grasslands Congress*, pages 247-51. New Zealand Grassland Association, Palmerston North, NZ.

Cocks, P.S., Malawi, B. and Sawmy, H. (1993). Introduction of the ley farming system to Syrian villages. In *Introducing Ley Farming to the Mediterranean Basin* (eds S. Christiansen, L. Materon, M. Falcinelli and P. Cocks), pages 52-64. ICARDA, Aleppo, Syria.

Cocks, P.S. and Osman, A.E. (1996). Productivity and botanical composition of communally-owned grasslands in the marginal farming areas of north Syria. *Journal of Arid Environments* **33**, 389-98.

Conyers, M.K., Heenan, D.P., Poile, G.J., Cullis, B.R. and Helyar, K.R. (1996). Influence of dryland agricultural management practices on the acidification of a soil profile. *Soil and Tillage Research* **37**, 127-41.

Cooper, P.J.M., Gregory, P.J., Tully, D. and Harris, H.C. (1987). Improving water use efficiency of annual crops in the rainfed farming systems of west Asia and north Africa. *Experimental Agriculture* **23**, 113-58.

Coventry, D.R., Walker, B.R., Morrison, G.R., Hyland, M.T., Avery, J.C., Maden, J.J.L. and Bartram, D.C. (1989). Yield responses to lime of wheat and barley on acid soils in north-eastern Victoria, Australia. *Australian Journal of Experimental Agriculture* **29**, 209-14.

Darmency, H. (1994). Genetics of herbicide resistance in weeds and crops, In *Herbicide Resistance in Plants: Biology and Biochemistry* (eds S. B. Powles and J.A.M. Holtum), pages 263-97. Lewis, Boca Raton, Florida.

Darmency, H. and Gasquez, J. (1990). Appearance and spread of triazine-resistance in common lambsquarters (*Chenopodium album*). *Weed Technology* **4**, 173-77.

Dear, B.S. and Cocks, P.S. (1997). Effect of perennial pasture species on surface soil moisture and early growth and survival of subterranean clover (*Trifolium subterraneum* L.) seedlings. *Australian Journal of Agricultural Research* **48**, 683-93.

Dolling, P.J., Porter, W.M. and Robson, A.D. (1991). Effect of soil acidity on barley production in the south-west of Western Australia. 1. The interaction between lime and nutrient application. *Australian Journal of Experimental Agriculture* **31**, 803-10.

Dolling, P.J., Porter, W.M. and Rowland, I.C. (1994). Acidification rates in the central wheatbelt of Western Australia. *Australian Journal of Experimental Agriculture* **34**, 1165-72.

Donald, C.M. (1981). Innovation in Australian agriculture. In *Agriculture in the Australian Economy (Ed. D.B. Williams)*, pages 57-86. Sydney University Press, Sydney, Australia.

Elicin, A., Ilaslan, M., Munzur, M., Cangir, S. and Karabulut, A. (1993). Fattening performance of weaned lambs on common vetch-barley mixtures sown on fallow land. *Proceedings of the XVIIth International Grasslands Congress*, page 253. Palmerston North, New Zealand.

Ellington. A., Reeves, T.G., Boundy, K.A. and Brooke, H.D. (1979). Increasing yields and soil fertility with pasture/wheat/grain legume rotations and direct drilling. *Paper delivered to 49th ANZAAS Congress, Auckland, NZ.*

Ehrman, T. and Cocks, P.S. (1990). Ecogeographic survey of wild annual legumes in Syria: distribution patterns. *Journal of Applied Ecology* **27**, 578-91.

Ehrman, T. and Cocks, P.S. (1996). Reproductive patterns in annual legume species on an aridity gradient. *Vegetatio* **122**, 47-59.

Evans, J., Hochman, Z., O'Connor, G.E. and Osborne, G.J. (1988). Soil acidity and *Rhizobium*: their effects on nodulation of subterranean clover on the slopes of southern New South Wales. *Australian Journal of Agricultural Research* **39**, 605-18.

Ghassali, F., Osman, A.E. and Cocks, P.S. (1998). Rehabilitation of degraded grasslands in north Syria: the use of Awassi sheep to disperse the seeds of annual pasture legumes. *Experimental Agriculture* **34** (in press).

Gill, G. (1995). Development of herbicide resistance in annual ryegrass populations (*Lolium rigidum* Gaud.) in the cropping belt of Western Australia. *Australian Journal of Experimental Agriculture* **35**, 67-72.

Gladstones, J.S. (1966). Naturalized subterranean clover (*Trifolium subterraneum* L.) in Western Australia: the strains, their distributions and possible origins. *Australian Journal of Botany* **14**, 329-54.

Greenland, D.J. (1971). Changes in the nitrogen status and physical condition of soil under pastures, with special reference to the maintenance of the fertility of Australian soils. *Soils and Fertilizers* **34**, 237-51.

Gregory, P.J., Tennant, D., Hamblin, A.P. and Eastham, J. (1992). Components of the water balance on duplex soils in Western Australia. *Australian Journal of Experimental Agriculture* **32**, 845-55.

Halse, N.J. (1993). Australian attempts to introduce ley farming to west Asia and north Africa. In *Introducing Ley Farming to the Mediterranean Basin* (eds S. Christiansen, L. Materon, M. Falcinelli and P.S. Cocks), 1-14. ICARDA, Aleppo, Syria.

Hingston, F.J. and Gailitis, V. (1976). The geographic variation of salt precipitated over Western Australia. *Australian Journal of Soil Research* **14**, 319-35.

Howieson, J.G. (1995). Characteristics of an ideotype acid tolerant pasture legume symbiosis in Mediterranean agriculture. *Plant and Soil* **171**, 71-76.

Howieson, J.G. and Ewing, M.A. (1986). Acid tolerance in the *Rhizobium meliloti-Medicago* symbiosis. *Australian Journal of Agricultural Research* **36**, 55-64.

Keatinge, J.D.H., Cooper, P.J.M. and Hughes, G. (1985). The potential of peas as a forage in the dryland cropping systems of western Asia. In *Proceedings of the 1984 Nottingham School of Agriculture, The Pea Crop - The Basis for Improvement*, pages 185-91. Butterworths, London, UK.

King, P.M, Rovira, A.D., Brisbane, P.G., Simon, A. and Brown, R.H. (1982). Population estimates of cereal cyst nematode and response of wheat to granular nematocides. *Australian Journal of Experimental Agriculture and Animal Husbandry* **22**, 209-20.

Koopman, D.J., Tow, P.G., Reeves, T.G. and Gibson, A.H. (1995). Soil acidification, chlorsulfuron application, and *Rhizobium meliloti* as factors in lucerne yield decline. *Soil Biology and Biochemistry* **27**, 673-77.

MacNish, G.C. (1989). Root disease as a major constraint in high rainfall cropping systems. *Proceedings of the 5th Australian Agronomy Conference, Perth*, pages 160-64. Australian Society of Agronomy, Melbourne.

MacNish, G.C. and Nicholas, D. (1987). Some effects of field history on the relationship between grass production in subterranean clover pasture, grain yield and take-all (*Gaeumannomyces graminis* var. *tritici*) in a subsequent crop of wheat at Bannister, Western Australia. *Australian Journal of Agricultural Research* **38**, 1011-18.

Moreno, V., Bueno, C. and Santos, A. (1993). Natural pasture fertilization with superphosphate in the Dehesa ecosystem. *Proceedings of the 7th Meeting of the FAO-European Sub-network on Mediterranean Pastures and Forage Crops*, pages 80-84. Crete, Greece.

Norman, H.C., Cocks, P.S., Smith, F.P. & Nutt, B. (1998). Reproductive strategies in Mediterranean annual clovers: germination and hardseededness. *Australian Journal of Agricultural Research* **49**, 973-82.

Osman, A.E. and Cocks, P.S. (1996). Prospects for improving Mediterranean grasslands in Lebanon through seeding, fertilization and protection from grazing. *Experimental Agriculture* **28**, 461-71.

Osman, A.E., Cocks, P.S., Russi, L. and Pagnotta, M.A. (1991). Response of Mediterranean grassland tp phosphate and stocking rates: biomass and botanical composition. *Journal of Agricultural Science (Cambridge)* **116**, 37-46.

Osman, A.E., Cocks, P.S. and Bahhady, F. (1994). Response of Mediterranean grassland to phosphate and stocking rate: livestock production. *Journal of Agricultural Science (Cambridge)* **123**, 319-26.

Pagnotta, M.A., Snaydon, R.W. and Cocks, P.S. (1997). The effects of environmental factors on components and attributes of a Mediterranean grassland. *Journal of Applied Ecology* **34**, 29-42.

Papademetrious, K. and Papanastasis, V. (1993). Herbage and seed yield of annual legumes in Macedonia, Greece. *Proceedings of the 7th meeting of the FAO-European Sub-network on Mediterranean Pasture and Forage Crops*, pages 28-31. Crete, Greece.

Pellet, P.L. and Young, V.R. (1988). The contribution of livestock products to human dietary needs with special reference to north Africa and west Asia. In *Increasing Small Ruminant Productivity in Semi-arid Areas*, pages 3-25. Kluwer, Dordrecht, The Netherlands.

Powles, S. B., Preston, C., Bryan, I. B. and Jutsum, A. R. (1997). Herbicide resistance: impact and management. *Advances in Agronomy* **58**, 57-92.

Puckridge, D.W. and French, R.J. (1983). He annual legume pasture in cereal ley farming systems of southern Australia: a review. *Agriculture, Ecosystems and Environment* **9**, 229-67.

Revell, C.K., Taylor, G.B. and Cocks, P.S. (1998). Long-term softening of surface and buried seeds of yellow serradela grown in a range of environments. *Australian Journal of Agricultural Research* **49**, 673-85.

Smith, F.P., Cocks, P.S. and Ewing, M.A. (1996). Short term patterns of seed softening in *Trifolium subterraneum*, *T. glomeratum* and *Medicago polymorpha*. *Australian Journal of Agricultural Research* **47**, 775-85.

Springborg, R. (1986). Impediments to the transfer of Australian dryland agricultural technology to the Middle East. *Agriculture, Ecosystems and Environment* **17**, 229-51.

Taylor, G.B. (1981). Effect of constant temperature treatments followed by fluctuating temperatures on the softening of hard seeds of *Trifolium subterraneum* L. *Australian Journal of Plant Physiology* **8**, 547-58.

Taylor, G.B. and Ewing, M.A. (1996). Effected of extended (4-12 years) burial on seed softening of subterranean clover and annual medics. *Australian Journal of Experimental Agriculture* **36**, 145-50.

Thomson, E.F, Jaubert, R. and Oglah, M. (1992). Using on-farm trials to study the benefits of feed legumes in barley-based rotations of north-west Syria. *Experimental Agriculture* **28**, 143-54.

Thomson, E.F., Rihawi, S., Cocks, P.S., Osman, A.E. and Russi, L. (1990).Recovery and germination rates of Mediterranean medics and clovers offered to sheep at a single meal or continuously. *Journal of Agricultural Science (Cambridge)* **114**, 295-99.

Thorn, C.W. and Perry, M.W. (1987). Effect of chemical removal of grasses from pasture leys on pasture and sheep production. *Australian Journal of Experimental Agriculture* **27**, 349-57.

Vall, E. and Gintzburger, G. (1991). Sub clover and annual medics: spring pasture for rangeland sheep in the Crau plain of southern France. *Proceedings of the 4th International Rangeland Congress*, page 372. Montpellier, France.

Vasilios, P. (1987). Potential of indigenous and exotic cultivars of *Dactylis glomerata* for developing dry pastures in northern Greece. *FAO-European Cooperative Network on Pasture and Forage Crop Production, Bulletin No. 5*, pages 143-47. Montpellier, France.

CHAPTER 2

Reproductive strategies and genetic structure of wild and naturalised legume populations

P.S. Cocks

Faculty of Agriculture and Co-operative Research Centre for Legumes in Mediterranean Agriculture, The University of Western Australia, Nedlands, WA, 6907

Abstract

The genetic diversity of legumes in the Mediterranean basin far exceeds that of Australia. The one exception is subterranean clover, where Australian variation approaches that of the Mediterranean. Evidence from weed populations suggests that capacity to outcross is important for the widespread establishment of volunteer species, including pasture legumes. Subterranean clover outcrosses occasionally, and this may be important in producing the genetic variation necessary to cope with environmental variation. Based on pollen/ovule ratios most other naturalised legumes appear to outcross less frequently than subterranean clover.

Mediterranean species employ many reproductive strategies. In high rainfall areas these appear to be substitutable, early flowering, for example, being replaced by high seed dormancy in some species. Reproductive strategies narrow substantially in arid areas, and include high seed dormancy, high seed to pod ratio, restricted dispersal capabilities, low flower abortion and early flowering. Most of these arid-type strategies are also found in widespread species.

The control of germination, both within and between seasons, is a strategy of special significance. The within season pattern of hard seed breakdown is seen to be important in a number of species, including some that are widely naturalised in Australia. This strategy controls germination in summer and early autumn, restricting it to a period when following rains are likely. A slow rate of germination and dormancy at high temperatures may serve similar purposes. Dispersal mechanisms that assist seeds escape or tolerate grazing are also highly desirable traits to which we pay insufficient attention. These include the capacity of small seeds (< 1 mg) to survive ingestion by sheep, and the burial of larger seeds (5-60 mg) at depths where they become unavailable.

It is concluded that we must continue to explore the genetic resources of the Mediterranean basin, especially of species other than subterranean clover and the annual medics. The diversity of species and genotypes available, and the reproductive strategies and genetic variability that they possess, bode well for increasing the range of pasture legumes available to Australian farmers.

Keywords

1. INTRODUCTION

Naturalised Mediterranean legumes in southern Australia include subterranean clover (*Trifolium subterraneum* L.), cluster clover (*T. glomeratum* L.), barrel medic**Error! Bookmark not defined.** (*Medicago truncatula* Gaertn.), and woolly bur medic (*Medicago minima* (L.) Bart.) plus several others. Black's Flora of South Australia lists 66 exotic legumes (Black 1948), mainly from the Mediterranean basin. In the Perth region (from Gingin to Bunbury, and inland to Mt Cooke) Marchant *et al.* (1987) lists 55 exotic legumes, again mostly from the Mediterranean basin. Of these, only a handful are deliberately sown, and formal breeding programmes are centred on a few medics and subterranean clover.

In contrast, Davis (1970) lists 900 legumes from Turkey alone. There are nearly 400 species of *Astragalus*, a genus that is almost unrepresented in Australia, 58 *Lathyrus*, 59 *Vicia*, 30 *Medicago*, and 94 *Trifolium* (Table 1). In neighbouring Syria, Ehrman and Cocks

S. J. Bennett and P. S. Cocks (eds.), Genetic Resources of Mediterranean Pasture and Forage Legumes, 20–31.

(1990) found large populations of 19 *Medicago* and 24 *Trifolium* species, and a wide range of *Lotus, Trigonella, Anthyllis, Securigera, Ornithopus, Onobrychis, Coronilla, Biserrula, Hymenocarpus, Scorpiurus, Astragalus* and *Hippocrepis*. Compared with the naturalised Australian legumes, genetic diversity in the Mediterranean basin is indeed immense.

Table 1. Some important legume (*Leguminosae*) genera native to Turkey, species of existing or potential economic importance, and the number of species present (source: Davis (1970)). The total number of species in 60 genera is 925 (see also Cocks 1993).

Genus	Examples of economically-important species	Number.
Astragalus	*A. hamosus* L; *A. asterias* Stev. ex Ledeb.	372
Cicer	*C. arietinum* L. (chickpea)	9
Coronilla	*C. varia* L.	8
Hedysarum	*H. hedysaroides* (L.) Schinz and Thell.	21
Lathyrus	*L. sativus* L.; *L. ochrus*; *L. cicera* L. (grasspeas)	58
Lens	*L. culinaris* Medik. (lentils)	5
Lotus	*L. corniculatis* L. (birdsfoot trefoil)	17
Lupinus	*L. albus* L.; *L. angustifolius* L. (lupins)	5
Medicago	*M. sativa* L. (lucerne); *M. truncatula* Gaertn. (barrel medic)	30
Melilotus	*M. alba* Desr. (sweet clover)	10
Onobrychis	*O. viciifolia* Scop.	46
Ononis	*O. viscosa* L.	18
Ornithopus	*O. compressus* L. (serradella)	3
Pisum	*P. sativum* L. (garden and dry pea)	2
Trifolium	*T. subterraneum* L. (subterranean clover); *T. purpureum* Lois.	94
Trigonella	*T. foenum-graecum* L. (fenugreek)	49
Vicia	*V. narbonensis* L.; *V. sativa* L.; *V. ervilia* (L.) Willd.	59

Most of the species used in Australia were selected from accidental introductions. We therefore use them as models of what we expect from pasture legumes. Only recently have selection programmes looked at newly introduced species, and few have reached the market place. It is likely that this will change however, although the inertia associated with "traditional" species is remarkably high. For example, until recently we knew little of the reproductive and adaptive strategies of Mediterranean annual legumes in their native habitat, so it made sense to continue with the species that we knew succeeded in Australia. With new insights into the wider Mediterranean flora, it is increasingly important that we open our eyes to widespread and successful species that employ different adaptive strategies.

This paper examines the genetic resources available in the Mediterranean basin, and the traits that influence their ecological success. It focuses on the reproductive strategies, genetic diversity and structure of wild and colonising populations, and their relevance to Australian farming systems.

2. GENETIC DIVERSITY OF MEDITERRANEAN GRASSLANDS

I have already referred to the diversity of legumes in Turkey and Syria. However, one of the major differences between Mediterranean and southern Australian grasslands is the genetic diversity that occurs within any ecosystem. This applies both within and between species.

In Australia, Cocks (1994) surveyed a volunteer pasture near Mt Gambier, South Australia (800 mm rainfall). He found seven legumes, the most common being subterranean clover. There were also 17 grasses and 10 other species, a total of 34 species, all but two of which came from the Mediterranean basin. In comparison, in the much drier Mediterranean environment of Tel Hadya, near Aleppo, Syria (320 mm rainfall) there were more than four times as many legumes (44). These included species such as *Lathyrus sativus*, *Pisum sativum*, and *Vicia narbonensis* (Osman *et al.* 1991). The most frequent were the small-seeded *T. campestre* and *T. tomentosum* and the very early flowering *T. stellatum*. Medics (mainly *M.*

rigidula), though present, were rare. The difference between the Australian and west Asian sites is striking, all the more so since, on first principles, we would expect greater diversity at the wetter, Australian site.

Within species, diversity in Australia is also restricted. For example, *M. laciniata* from eastern Australia shows little, if any, variation (Young *et al.* 1992; Young and Lagudah 1994), while accessions from the Mediterranean vary markedly (Lesins and Lesins 1979). *M. minima* within Western Australia seems not to vary at all (D. Fresnillo-Fedorenko pers. comm.), although it does in South Australia (G. Sweeney pers. comm.). On the other hand, subterranean clover has a genetic diversity that is similar to that in the Mediterranean (Cocks and Phillips 1979; Rossiter and Collins 1988; Cocks 1992). This is probably the result of continuous and widespread sowing for more than 50 years and limited outcrossing (see later section of this paper). Most Mediterranean legume populations vary from site to site and within sites (Ehrman and Cocks 1990). Their genetic diversity seems infinite.

There are sound ecological reasons for increasing the diversity of Australian grasslands (Marshall 1977), even within single sites. Specific and genetic mixtures are tolerant of epidemic diseases, are more likely to colonise soil mosaics, and have the potential to survive environmental disasters. They also maintain production and botanical composition at stable levels (Tilman and Downing 1984), and are more likely to provide a balanced diet for grazing animals. The use of pure stands in pastures are partly the result of convenience, but mainly the result of our experience with crops, where quality often dictates genetic purity. This is rarely, if ever, the case in pastures.

3. THE GENETIC STRUCTURE OF WILD AND COLONISING POPULATIONS

Annual pasture species and weeds have much in common. Both are self-perpetuating, colonising groups of species. An examination of the genetic structure of successful weed populations may have much to offer those seeking to understand the success or failure of pasture legumes.

Firstly, there is evidence that successful weeds in the early stages of colonisation are less variable than those species that succeed them (Hamrick *et al.* 1979). This seems also to be true of weeds with restricted distribution. The reason is likely to be associated with the isolated nature of individual plants during the colonising process. In these circumstances a high degree of self fertility results in a greater number of individuals able to establish themselves. The fact that the population is less variable as a result is probably incidental.

Widespread and established weed populations are often variable, the most successful being outbreeders. The best example is annual ryegrass (*Lolium rigidum* Gaud.), a widespread weed in the cereal belt of southern Australia that outcrosses freely. Not only has it colonised many different habitats but it also adapts rapidly to changed circumstances. For example, it has developed resistance to some 20 herbicides, and can do so within a few years of the first application of the spray (Hall *et al.* 1994; Gill 1995; McAllister *et al.* 1995). Similarly in Canada, wild radish (*Sinapsis arvensis* L.) has developed resistance to several herbicides, including dicamba and MCPA (Webb and Hall 1995). There are numerous other examples throughout the world. Other widespread outcrossing weeds that have successfully colonised huge areas of southern Australia are salvation jane (*Echium plantagineum* L.) (Brown and Burdon 1983) and capeweed (*Arctotheca calendula*) (for further discussion of this topic see the review by Barrett and Richardson 1986).

Nevertheless not all successful widespread weeds are outbreeders. Indeed, most annual grasses are self-pollinating and, as will be seen, so are most annual legumes. However, self pollination does not necessarily imply cleistogamy (where cross-pollination is impossible), and the degree of outcrossing in a species, even if only a small amount, may be all important in its success or otherwise.

Most annual legumes are predominantly self-pollinating, although there are some exceptions. For example, although *Vicia sativa*, *V. ervilia*, and *V. lutea* are self-pollinating, *V. villosa* is largely outcrossing (Zhang and Mosjidis 1995). *Trifolium stellatum*, *T. tomentosum* and *T. campestre* are all predominantly self-pollinating, although field variation suggests that they occasionally outcross (Pagnotta *et al.* 1995). In contrast, most perennials are outcrossing (Kongkiatingam *et al.* 1995) as are a few annuals (e.g. *T. michelianum*). The paucity of variation in *M. laciniata*, mentioned earlier, suggests that this species, at least, rarely, if ever, outcrosses. Another explanation is that only one genotype was ever introduced to Australia. The evidence that subterranean clover, often considered to be cleistogamous, outcrosses frequently is mounting. Cocks and Phillips (1979) discovered large numbers of subterranean clover strains in South Australia that had not been sown. These populations were too frequent to be the result of accidental introductions as described by Gladstones (1966). Rossiter and Collins (1988) found that old Dwalganup pastures have been almost completely replaced by what they termed 'divergent strains', the result of hybridisation. Cocks (1992) confirmed their findings in old Mt Barker/Dwalganup stands where up to 67% of the original population had been replaced by hybrids. These results indicated that diversity within a single South Australian field almost equalled that of a collection of 80 strains from throughout the Mediterranean basin. Clearly, in the case of subterranean clover, limited outcrossing has had a profound effect on the genetic variation.

The ability of legumes to naturally increase their variation in this way may be of considerable ecological significance. The idea is that no sown cultivar will be ideal for all environments in a field, and indeed may not be ideal for any. Limited outcrossing and subsequent natural selection may provide the mechanism that supplies the required variation.

Indeed, the ability to outcross in this way has been shown to be crucial in the colonisation of Californian grasslands by the annual grass *Avena barbata* (Allard 1975). *A. barbata* has an outcrossing frequency of 2-5%, which means that colonising populations are able to outcross relatively frequently. Allard (1975) demonstrated that initial outcrossing creates adapted genotypes, which are consolidated by self pollination. Self pollination maintains what Allard called a 'super gene', a term he uses for a mainly homozygous individual adapted to its own ecological niche. *A. barbata* is enormously successful, and should provide a genetic model for colonising legumes.

We are yet to determine the low rates of outcrossing that are needed to establish and maintain the required amount of genetic diversity. Using conventional techniques this is difficult, if not impossible. One possible technique is to examine pollen/ovule ratios (Cruden 1977). Where there is a great deal of pollen produced for each ovule the species is likely to be outcrossing, whereas a low amount indicates self-pollination. A method such as this could be used to screen for ability to outcross infrequently, if this proves to be important in the colonisation process. Such a survey of naturalised legumes in the southern Flinders Ranges of South Australia has revealed that, for most annual legumes, the pollen/ovule ratio is indeed low, although there is considerable variation between species; e.g. subterranean clover has a ratio of 800/1, while those of *M. minima* and *T. glomeratum* are 250/1 or less (Table 2). All three species could be classified as successful colonisers.

Table 2. Ovule number, pollen grains/sac (five sacs/flower in *Medicago* and 10/flower in *Trifolium*) and the pollen/ovule ratio for five annual legume species naturalised in southern Australia (source: G. C. Sweeney, pers. comm.)

Species	Ovul es	±S.E.	Pollen/sac	±S.E.	Pollen/ovule ratio	±S .E.
M. minima	5.3	0.098	106	1.46	205	3.
T.	2.0	0	45	0.79	227	98
glomeratum	2.0	0	187	4.06	935	3.
T. hirtum	2.0	0	159	6.98	794	98
T.	2.0	0	61	12.4	309	20
subterraneum						.9

T. tomentosum			34.9 12.8

4. REPRODUCTIVE STRATEGIES OF MEDITERRANEAN ANNUAL LEGUMES

4.1 Substitutability

Subterranean clover is well adapted to much of the high rainfall acid soils of southern Australia. Its adaptation depends on appropriate flowering time, ability to produce a long-lived seed bank, protection of seeds from grazing by seed burial and prostrate growth habit. Its adaptive strategies result in adaptation to intensive grazing. It also grows well in crop rotations, although many would consider it inadequately adapted to low rainfall sites.

That subterranean clover is successful leads us to believe that its strategies provide us with an ideotype. However, recent evidence from ecological modellers suggests that this may not be so, and that there may be many other paths to ecological success (Venable 1985; Venable and Brown 1988). For example, early flowering and high hardseededness may substitute one for another. Early flowering is necessary if long term seed dormancy is absent. In its presence, a species may flower much later and use rain that falls in good seasons to build up a seed bank. In their model, Venable and Brown (1988) considered that seed size, dispersal and seed dormancy are interchangeable. We refer to this interchangeability as "substitutability of reproductive traits".

Ehrman and Cocks (1996) used a different approach to the models of Venable and Brown (1985). They collected legumes from throughout Syria (Ehrman and Cocks 1990) and examined reproductive strategies, relating them to site of collection. They found that in high rainfall areas reproductive strategies were indeed substitutable, but in more arid areas the range of strategies narrowed substantially. "Arid-type" strategies included a high level of seed dormancy, high seed to pod ratios, restricted dispersal capabilities and early flowering. Other evidence suggests (Cocks 1990) that flower abortion is less and that, although the total number of flowers produced is less, the number of seeds/flower initiated is greater than in plants from wet sites. Medics, as we might expect, fit this model better than subterranean clover. Most widespread species, of which many medics are examples, have "arid-type" strategies.

Norman *et al.* (1998) have confirmed the results of Ehrman and Cocks (1996) and Venable and Brown (1985). They grew pairs of species collected at high rainfall sites in the east and west Mediterranean. Species from the same site differed in their adaptation to the site in all cases. Seed size, seed dormancy, germination rate and flowering time were interchangeable.

4.2 Flowering time

Most species from arid environments flower early. In Syria there is a well-defined shift from late to early flowering from the high rainfall coast to the arid interior (Ehrman and Cocks 1996). Similarly in Australia, flowering time of *T. glomeratum* increases with increasing rainfall (Woodward and Morley 1974; Smith *et al.* 1995). Species from predictable environments also flower later than those from unpredictable environments (Cohen 1976). Flowering time is the most responsive of all reproductive strategies to environmental change, most species showing ecotypic differentiation (Ehrman and Cocks 1996).

Besides substituting early flowering for high seed dormancy, it is likely that small seed size can substitute for early flowering. Thus, because small seeds fill quicker than large seeds, the flowers are initiated closer to the end of the growing season. *T. clusii* and *T. glanduliferum* are good examples. In northern Israel their small seed size makes them among the last-flowering native clovers, and much later than larger seeded *T. purpureum* from the same sites (Norman *et al.* 1998).

The importance our breeding programmes place on flowering time is justified. However, in some circumstances, a better strategy may be to select for small seed size and high seed dormancy. There is more than one way to achieve adaptation.

4.3 Hardseededness

Most authors believe that hardseededness is the main dormancy mechanism in Mediterranean legumes. It is clearly the most important "long term" dormancy mechanism, that is the mechanism that maintains the seed bank from year to year. Taylor and Downes (1995) calculated that a hardseededness of more than 40% is necessary for a single species to maintain its seed bank. However, among Mediterranean legumes it varies widely, from almost non-existent in ecotypes of *Trifolium isthmocarpum* (Norman *et al.* 1998) and *Ornithopus sativus* (Davies 1995), to greater than 90% of the original seed produced in *Medicago noeana* (Cocks 1995) and *Biserrula pelecinus* (A. Loi, pers. comm.). Mediterranean grasslands themselves are usually complexes of legumes, such that hard-seeded and soft-seeded species occur together, maintaining legume stability. For example, Cocks (1993) demonstrated that a mixture of *M. noeana, M. polymorpha* and *M. rigidula* steadily released hard seeds over a five year period. A similar concept, using mixtures of *Ornithopus sativus* (soft-seeded) and *O. compressus* (hard-seeded) is proposed for south-western Australia.

'Short term' hardseededness has recently emerged as an important reproductive strategy. Taylor (1981) developed the idea of two-stage softening in subterranean clover. In doing so he opened the door to selecting for a mechanism that takes into account the events of the first few weeks of the growing season. Taylor's idea was that a preconditioning period of high temperature is required before a short period of alternating temperatures (60/15°C) finally softens the seed. In the case of subterranean clover, the alternating temperature requirement appears to be met at the same time as the preconditioning, that is, during summer. For this reason most subterranean clover cultivars become soft by early autumn, and when false breaks occur, the seeds run a high risk of death following germination. Subterranean clover is said to have a short half life, indicating that its hard seeds break down early (Table 3).

Table 3. Calculated half lives (time for half the hard seed population to soften in any year) and initial and final soft seed levels of subterranean clover cv. Nungarin, Medicago polymorpha cv. Serena and two accessions of *T. glomeratum*. Standard errors are in parentheses. Source: Smith *et al.* (1996)

Strain	Half life (days)	Initial soft seed (%)	Final soft seed (%)
Subterranean clover cv. Nungarin	46 (28)	10 (15)	47 (17)
M. polymorpha cv. Serena	126 (3.6)	5 (2.1)	37 (3.3)
T. glomeratum accn 30	136 (6.7)	-1 (3.3)	54 (7.5)
T. glomeratum accn 42	104 (2.5)	3 (3.7)	95 (5.1)

In contrast, *M. polymorpha* requires a lower alternating temperature range (30/15 °C) before its seeds will soften. Consequently, if the seeds do soften in any year (that is, if the requirements of the first stage are met), they will not soften until the likelihood of following rains is high. Other species with long half lives are *T. glomeratum* (Smith *et al.* 1996), *Vicia sativa* subsp. *amphicarpa* (Christiansen *et al.* 1996), *T. lappaceum*, *T. purpureum* and *T. glanduliferum* (Norman *et al.* 1998). Some also have high "long term hardseededness" (*M. polymorpha, T. glanduliferum, T. lappaceum*), while others have extremely low levels of long term hardseededness (*T. purpureum, V. sativa* subsp. *amphicarpa*). Short term hardseededness is a highly desirable characteristic for all but the coolest and wettest of southern Australian mediterranean environments.

While the short term hard-seed patterns of many clovers (e.g. *T. subterraneum*, *T. isthmocarpum*, *T. cherleri*, and *T. angustifolium*) are undesirable, there is evidence that more desirable patterns may exist in some genotypes within each species. For example, Smith *et al.* (1996) compared short term seed softening patterns in several ecotypes of subterranean clover. Hard-seeded subterranean clover ecotypes from Sardinia and Turkey did not soften as late as cv. Serena or *T. glomeratum*, but they softened much later than cv. Nungarin (compare Tables 3 and 4). Since they also have much higher long term hardseededness than cv. Nungarin, it appears that these ecotypes possess the potential to go some way towards matching *T. glomeratum* and *M. polymorpha* in dry areas.

4.4 Other germination strategies

Although hardseededness is almost certainly the major mechanism of germination control in Mediterranean legumes, it seems possible that we have neglected other mechanisms. For example, in their study of germination and hardseededness in Mediterranean legumes, Norman *et al.* (1998) demonstrated that, at any site, different species employ different hard-seed or germination strategies, that species tend to employ the same strategy at different sites, and that all species employ some means of protecting seeds from out-of-season rainfall. Apart from short term hardseededness, slow rate of germination and high temperature dormancy also control germination during this period.

Table 4. Calculated half lives of subterranean clover cv. Nungarin and four recent introductions from Turkey and Sardinia

Cultivar or accession	Half life (days)
Nungarin	44 (15)
Sardinia-B	77 (4.8)
Sardinia-J	79 (4.1)
Turkey-A	64 (5.0)
Turkey-B	72 (8.8)

Inhibition of germination at high temperatures is a widely recognised phenomenon. It is reversed by placing the seeds in cooler temperatures where germination continues normally. Not all species react similarly. For example, Espigares and Peco (1993) found that of the clovers, *T. striatum*, *T. gemellum* and *T. suffocatum* were inhibited by temperatures as low as 20/15°C. Jansen and Ison (1994) considered that temperatures of 30-40°C enforced dormancy in *T. resupinatum*, *T. subterraneum*and *T. michelianum*. Brar *et al.* (1991) observed that subterranean clover cv. Mt Barker germinated slowly at temperatures of 25°C or more. Gladstones (1987) noticed that subterranean clover cultivars Dinninup and Geraldton resisted germination at 24°C while cvs. Clare and Woogenellup did not. Norman *et al.* (1998) detected dormancy at 28°C in *T. cherleri*, *T. isthmocarpum*, *T. angustifolium*, and *T. glanduliferum*, while *T. argutum*, *T. clypeatum*, *T. purpureum* and *T. obscurum* germinated freely at 30°C.

Slow germination is another mechanism whereby seeds can resist out-of-season rains. In contrast to seeds that germinate rapidly and thereby respond to short periods of high moisture, species that germinate slowly will safely dehydrate unless soil moisture is maintained. In legumes, slow germination is often associated with large seeds, which are produced in smaller numbers by the mother plant (Table 5). Examples are *T. obscurum* and *T. clypeatum* (Norman *et al.* 1998). It is probably the least reliable of the germination regulating mechanisms, but its flexibility allows for germination at any time in summer or autumn when conditions are favourable.

4.5 Seed dispersal

Dispersal is the mechanism whereby seeds are distributed by the mother plant. Some species bury their seeds (e.g. subterranean clover, *V. sativa* subsp *amphicarpa*, *Lathyru ciliolatus*), in which case dispersal is limited to the base of the mother plant (Christiansen *et al.* 1996). Others produce inflorescences that are light enough to be dispersed by wind (e.g. *T. tomentosum*). Yet others are dispersed by animals (e.g. most medics, in wool and fur).

Table 5. Calculated half lives of germination (degree-days) and high temperature dormancy at 28°C for 12 annual clovers collected in the Mediterranean basin (source: Norman *et al.* 1998)

Species	Half life	Dormancy
T. obscurum	30.0	0
T. isthmocarpum	14.3	24
T. cherleri	19.9	61
T. nigrescens	12.4	11
T. angustifolium	14.1	65
T. lappaceum	14.5	0
T. argutum	15.9	0
T. glanduliferum	11.8	31
T. purpureum	15.4	0
T. clusii	9.7	7
T. clypeatum	33.5	0

Several plant species exhibit amphicarpy (flower above and below the ground). They include *Cyperaceae* (Bruhl 1994), *Poaceae* (Cheplick 1995), *Asteraceae* (Declavijo 1995) and, of course, legumes. In addition to the Mediterranean legumes mentioned above, the tropical legumes *Macroptilium gracile* (Pengelly and Eagles 1995), *Glycine pindanica* and *G. hirticaulis* (Tindale and Craven 1993) exhibit amphicarpy. Amphicarpy encourages retention of the seed in a proven environment, one where the mother plant flourished, and protects seeds from predation, fire and desiccation. However, it restricts gene exchange, has a high energy cost, reduces dispersal and increases sibling competition. The interaction of these factors will determine the success or otherwise of amphicarpic legumes in pastures.

An example of interactions which ensure survival and dispersal is described by Declavijo (1995). He found that *Catananche lutea* (*Asteraceae*) has two underground morphs, which are produced in February to April (northern hemisphere), and at least two above ground morphs, produced in April and May. One of the underground morphs germinates immediately, while the other remains dormant for several months. One aerial morph produces achenes with no or little pappus while a second produces achenes with a long pappus (there are intermediates). The first has limited dispersal, while the second is carried by the wind for long distances. The germination of all seeds is inhibited by light, which means that buried non-dormant seeds produced in early spring germinate most promptly. The whole strategy is aimed at spreading germination in both space and time.

It is becoming increasingly evident that dispersal interacts with grazing. Thus, if seeds are dispersed only to the base of plants there needs to be some mechanism by which they are protected from grazing. Unfortunately, even seeds that have the potential to be widely dispersed (e.g. medics) are vulnerable to grazing. For example, Cocks (1988) demonstrated that sheep will eat all but the final 10 kg/ha of seeds of *M. rigidula*, *M. rotata* and *M. noeana*, even when the medics produced more than 500 kg/ha. Where medic seeds are ingested by sheep they rarely survive (Thomson *et al.* 1991). It is fair to say that vulnerability to grazing is the single most significant weakness in the annual medics.

Thomson *et al.* (1991) established a relationship between seed size and survival after ingestion. In this relationship seeds larger than 2 mg (most medics are larger than 2 mg, and subterranean clover is larger than 5 mg) are unlikely to survive in large numbers, while up to 60% of seeds smaller than 1 mg will survive. Small seed size is therefore a potent method of protection from grazing and an efficient means of seed dispersal. It is significant that two of the three most widely distributed legumes in the Western Australian wheatbelt (*T. glomeratum* and *T. tomentosum*) have seeds of about 0.5 mg. It is also significant that these species often dominate the degraded grasslands of west Asia (Osman *et al.* 1991).

As long ago as 1952, Suckling (1952) advocated using sheep to distribute the small seeds of white clover (*T. repens*) to inaccessible hills in New Zealand. More recently the role of sheep in distributing seeds to restore west Asian grasslands has been investigated (F. Ghassali, pers. comm.). Seeds of *T. campestre*, *T. tomentosum*, *T. resupinatum* and *T. noeana*, amongst others, are fed to penned sheep that are then sent out to graze. In the following 24-48 hours the small seeds are deposited on the grassland by the sheep in faeces. In the short term, hard seed levels are high (soft seeds are digested), so it is important that the process takes place in early summer to maximise hard seed breakdown. The results are most promising.

Seeds are more likely to survive ingestion by cattle than sheep. For example, in Spain, up to 4 seeds/g of cattle faeces explains up to 20% of variation in natural *Biserrula pelecinus* populations (Malo and Suarez 1995). In Australia, incursion of cattle into natural grasslands is often responsible for the establishment of naturalised subterranean clover populations (e.g. Cocks 1994). Cicer milkvetch (*Astragalus cicer*) also survives ingestion by cattle, although attempts to use cattle to spread the species have not been successful (Williams *et al.* 1995). Even rabbits and foxes are involved in the dispersal of seeds, where they are known to be important in the recolonisation of small disturbances in Spanish grasslands (Malo *et al.* 1995; Ortiz *et al.* 1995)

Dispersal is difficult to work with experimentally, and has therefore been neglected. Most information is descriptive, e.g. Zohary and Heller (1984) for *Trifolium*. We urgently need to approach this problem experimentally, because it holds the key to understanding the survival of pasture legumes.

5. CONCLUSIONS

While recent research has revealed some of the reasons why species succeed or fail, there remains a need to better understand the adaptive strategies of Mediterranean annual legumes. Once discovered, these will lead to more efficient and appropriate approaches to the problem of plant adaptation and pasture improvement. I suggest that the most rewarding investigations include how seeds are dispersed by grazing animals, what is the appropriate level of genetic diversity, and what are the most efficient means of escaping germination after false breaks in autumn. The idea that one reproductive strategy can substitute for another leads us to conclude that we should not base selection programmes on one or a few traits. Hardseededness and early flowering, valuable though they are, need to be considered in the context of all plant traits. This approach is likely to lead to the introduction of a number of new species, markedly different from subterranean clover or the commercial medics.

To facilitate plant introduction and selection, plant collection needs to be more systematic. Collection site characteristics have been shown to be of great value in the selection process (e.g. Ehrman and Cocks 1996). However, more accurate measurements of soil and climate at the collection sites are needed. With enormous collections of medics and subterranean clover already at hand, we need more germplasm of these species only to help us understand their ecology. This is not so for other genera where we badly need more germplasm. For example, *Trigonella*, *Astragalus*, *Ornithopus*, *Biserrula* and *Trifolium* other than subterranean clover, plus a number of others.

I believe that pasture plant improvement in Australia is at a turning point. New species and new approaches to their selection will lead to significant rewards for the drier wheatbelt. While we must continue with medics and subterranean clover, and apply more rigorous selection procedures to them, the breakthroughs for this difficult agroecosystem are likely to come from other species. These will occupy ecological niches that currently have no legumes. They will be species in which grazing assists instead of hinders persistence and seed dispersal. They will be capable of producing long-lived and responsive seed banks instead of short-lived seed banks that germinate unseasonably. Productivity will not necessarily be greater in the short run, but it will be in the longer run, when greater persistence and stability will eventually lead to higher productivity.

ACKNOWLEDGEMENTS

Much of the material in the paper has been collected by students at the University of Western Australia and by my former colleagues at ICARDA. I acknowledge their ideas and hard work.

REFERENCES

Allard, R. W. (1975). The mating system and microevolution. *Genetics* **79**, 115-26.

Barrett, S. C. H. and Richardson, B. J. (1986). Genetic attributes of invading species. In Ecology of Biological Invasions (eds R. H. Groves and J. J. Burdon) pages 21-33. Canberra: Australian Academy of Science.

Black, J. M. (1948). Flora of South Australia, Part II. Adelaide: Government Printing Office.

Brar, G. S., Gomez, J. F., McMichael, B. L., Matches, A. G. and Taylor, H. M. (1991). Germination of twenty forage legumes as influenced by temperature. *Agronomy Journal* **83**, 173-75.

Brown, A. H. D. and Burdon, J. J. (1983). Multilocus diversity in an outbreeding weed, *Echium plantagineum* L. *Australian Journal of Biological Sciences* **36**, 503-9.

Bruhl, J. J. (1994). Amphicarpy in the Cyperaceae, with novel variation in the wetland sedge, Eleocharis caespitosissima Baker. *Australian Journal of Botany* **42**, 441-48.

Cheplick, G. P. (1995). Life history evolution in amphicarpic plants. *Plant Species Biology* **9**, 119-31.

Christiansen, S., Abd El Moneim, A. M., Cocks, P. S. and Singh, M. (1996). A comparison of seed yield, hardseededness and seed dormancy between two amphicarpic pasture legumes (*Vicia sativa* subsp. *amphicarpa* and *Lathyrus ciliolatus*) and two annual medics (*Medicago rigidula* and *M. noeana*). *Journal of Agricultural Science (Cambridge)* (in press).

Cocks, P. S. (1988). Seed production and seed survival under grazing of annual medics (*Medicago* sp.) in north Syria. *Journal of Agricultural Science (Cambridge)* **110**, 455-63.

Cocks, P. S. (1990). Dynamics of flower and pod production in annual medics. 1. In spaced plants. *Australian Journal of Agricultural Research* **41**, 911-21.

Cocks, P. S. (1992). Evolution in sown populations of subterranean clover (*Trifolium subterraneum* L.) in South Australia. *Australian Journal of Agricultural Research* **43**, 1583-95.

Cocks, P. S. (1993). Legumes from the Mediterranean basin: a continuing source of agricultural wealth for southern Australia. Perth: Co-operative Research Centre for Legumes in Mediterranean Agriculture: Technical Paper No. 1.

Cocks, P. S. (1993). Seed and seedling dynamics over four consecutive years from a single seed set of six annual medics (Medicago spp.) in north Syria. *Experimental Agriculture* **29**, 461-72.

Cocks, P. S. (1994). Colonisation of a South Australian grassland by invading Mediterranean annual and perennial pasture species. *Australian Journal of Agricultural Research* **45**, 1063-76.

Cocks, P. S. and Phillips, J. R. (1979). Evolution of subterranean clover in South Australia. 1. The strains and their distribution. *Australian Journal of Agricultural Research* **30**, 1035-52.

Cohen, D. (1976). The optimal timing of reproduction. *American Naturalist* **110**, 801-7.

Cruden, R. W. (1977). Pollen-ovule ratios: a conservative indicator of breeding systems in flowering plants. *Evolution* **31**, 32-46.

Davis, P. H. (ed.) (1970). Flora of Turkey and the East Aegean Islands. Edinburgh: University Press.

Davies, K. S. (1995). Germination strategies and hard seed breakdown of five lines of serradella (*Ornithopus* spp.). Dissertation for the BSc(Agric), The University of Western Australia.

Declavijo, E. R. (1995). The ecological significance of fruit heteromorphism in the amphicarpic species Catananche lutea (Asteraceae). *International Journal of Plant Sciences* **156**, 824-33.

Ehrman, T. and Cocks, P. S. (1990). Ecogeography of annual legumes in Syria: distribution patterns. *Journal of Applied Ecology* **27**, 578-91.

Ehrman, T. and Cocks, P. S. (1996). Reproductive patterns in annual legume species on an aridity gradient. *Vegetatio* **122**, 47-59.

Espigares, T. and Peco, B. (1993). Mediterranean pasture dynamics: the role of germination. *Journal of Vegetation Science* **4**, 189-94.

Gill, G. S. (1995). Development of herbicide resistance in annual ryegrass populations (Lolium rigidum Gaud.) in the cropping belt of Western Australia. *Australian Journal of Experimental Agriculture* **35**, 67-72.
Gladstones, J. S. (1966). Naturalised subterranean clover (*Trifolium subterraneum* L.) in Western Australia: the strains, their distribution, characteristics and possible origins. *Australian Journal of Botany* **14**, 329-354.
Gladstones, J. S. (1987). Measurement of high temperature seed dormancy in subterranean clover. *Journal of the Australian Institute of Agricultural Science* **53**, 99-101.
Hall, L. M., Tardif, F. J. and Powles, S. B. (1994). Mechanisms of cross and multiple resistance in *Alopecurus myosuroides* and *Lolium rigidum*. *Phytoproduction* **75**, 17-23.
Hamrick, J. L., Linhart, Y. B. and Mitton, J. B. (1979). Relationships between life history characteristics and electrophoretically detectable genetic variation in plants. *Annual Review of Ecology and Systematics* **10**, 173-200.
Jansen, P. I. and Ison, R. L. (1994). Temperature effects on germination of Trifolium balansae and T. resupinatum with special reference to high temperature dormancy. *Australian Journal of Agricultural Research* **45**, 689-701.
Kongkiatingam, P., Waterway, M. J., Fortin, M. G. and Coulman, B. E. (1995). Genetic variation within and between two cultivars of red clover (*Trifolium pratense* L.): a comparison of morphological, isozyme and RAPD markers. *Euphytica* **84**, 237-46.
Lesins, K. A. and Lesins, I. (1979). Genus *Medicago* (Leguminosae): a Taxogenetic Study. The Hague: Dr W. Junk.
Malo, J. E. and Suarez, F. (1995). Cattle dung and the fate of *Biserrula pelecinus* L. in a Mediterranean pasture: Seed dispersal, germination and recruitment. *Botanical Journal of the Linneaen Society* **118**, 139-48.
Malo, J. E., Jiminez, B. and Suarez, F. (1995). Seed bank build-up in small disturbances in a Mediterranean pasture: The contribution of endozoochorous dispersal by rabbits. *Ecogeography* **18**, 73-82.
Marchant, N. G., Wheeler, J. R., Rye, B. L., Bennett, E. M., Lander, N. S., and McFarlane, T. D. (1987). Flora of the Perth Region, Part I. Perth: Western Australian Herbarium, Department of Agriculture.
Marshall, D. R. (1977). The advantages or hazards of genetic homogeneity. *Annals of the New York Academy of Science* **287**, 1-20.
McAllister, F. M., Holtum, J. A. M. and Powles, S. B. (1995). Dinitroaniline herbicide resistance in rigid ryegrass (Lolium rigidum). *Weed Science* **43**, 55-62.
Norman, H. C., Cocks, P. S. and Smith, F. P. (1998). Reproductive strategies in Mediterranean annual clovers. 1. Germination and hardseededness. *Australian Journal of Agricultural Research* (in press).
Ortiz, P. L., Arista, M. and Tolavera, S. (1995). Germination ecology of *Cerastonia siliqua* L. (Caesalpiniaceae), a Mediterranean tree. *Flora (Jena)* **190**, 89-95.
Osman, A. E., Cocks, P. S., Russi, L. and Pagnotta, M. (1991). Response of Mediterranean grassland to phosphate and stocking rates: biomass production and chemical composition. *Journal of Agricultural Science (Cambridge)* **116**, 37-46.
Pagnotta, M. A., Cocks, P. S. and Snaydon, R. W. (1995). The breeding system of three annual clovers native to north Syria. *Israel Journal of Plant Science* **43**, 347-358.
Pengelly, B. C. and Eagles, D. A. (1995). Geographical distribution and diversity in a collection of the tropical legume *Macroptilium gracile* (Poeppiga ex Bentham) Urban. *Australian Journal of Agricultural Research* **46**, 569-80.
Rossiter, R. C. and Collins, W. J. (1988). Genetic diversity in old subterranean clover (*Trifolium subterraneum* L.) populations in Western Australia. I. Pasture sown initially to the Dwalganup strain. *Australian Journal of Agricultural Research* **39**, 1051-62.
Smith, F. P., Cocks, P. S. and Ewing, M. A. (1995). Variation in the morphology and flowering time of cluster clover (*Trifolium glomeratum* L.) and its relationship to distribution in southern Australia. *Australian Journal of Agricultural Research* **46**, 1027-38.
Smith, F. P., Cocks, P. S. and Ewing, M. A. (1996). Short term patterns of seed softening in Trifolium subterranean, T. glomeratum and Medicago polymorpha. *Australian Journal of Agricultural Research* **47** (in press).
Suckling, F. E. T. (1952). Dissemination of white clover (Trifolium repens) by sheep. *New Zealand Journal of Science and Technology* **35**, 64-77.
Taylor, G. B. (1981). Effect of constant temperature treatments followed by fluctuating temperatures on the softening of hard seeds of Trifolium subterranean L. *Australian Journal of Plant Physiology* **8**, 547-58.
Taylor, G. B. and Downes, P. A. (1995). MEDSIST: A model for predicting the effect of hardseededness on the persistence of annual medics in ley farming systems in a mediterranean-type environment. CLIMA Technical Paper No. 4.
Thomson, E. F., Rihawi, S., Cocks, P. S., Osman, A. E. and Russi, L. (1990). Recovery and germination rates of seed of Mediterranean medics and clovers offered to sheep at a single meal or continuously. *Journal of Agricultural Science (Cambridge)* **114**, 295-99.
Tilman, D. and Downing, T. A. (1994). Biodiversity and stability in grasslands. *Nature (London)* **367**, 363-65.
Tindale, M. D. and Craven, L. A. (1993). Glycine pindanica, a new species from west Kimberley, Western Australia. *Australian Systematic Botany* **6**, 371-76.

Venable, D. L. (1985). Ecology of achene dimorphism in Heterotheca latifolia. 3. Consequences of varied water availability. *Journal of Ecology* **73**, 757-63.
Venable, D. L. and Brown, J. S. (1988). Selective interactions of dispersal, dormancy and seed size as adaptations for reducing risk in variable environments. *American Naturalist* **131**, 360-84.
Webb, S. R. and Hall, J. C. (1995). Auxinic herbicide resistant and susceptible wild mustard (*Sinapsis arvensis* L.) biotypes: Effect of auxinic herbicides on seedling growth and auxin binding activity. *Pesticide Biochemistry and Physiology* **52**, 137-48.
Williams, W. D., Acharya, S. N. and Rode, L. M. (1995). Feasibility of using cattle to disperse cicer milkvetch (*Astragalus cicer* L.) seed in pastures. *Canadian Journal of Animal Science* **75**, 173-75.
Woodward, R. G. and Morley, F. H. W. (1974). Variation in Australian and European collections of *Trifolium glomeratum* L. and the provisional distribution of the species in southern Australia. *Australian Journal of Agricultural Research* **25**, 73-78.
Young, R. R., Croft. P. H. and Sandral, G. A. (1992). Variation in flowering times and agronomic characteristics of *Medicago laciniata* (L.) Miller collected from diverse locations in New South Wales. *Australian Journal of Experimental Agriculture* **32**, 59-63.
Young, R. R. and Lagudah, E. S. (1994). Ribosomal DNA, peroxidase and extensin variation in genomic DNA of *Medicago laciniata* (L.) Miller from Australia and Africa. *Australian Journal of Agricultural Research* **45**, 1013-23.
Zhang, X. and Mosjidis, J. A. (1995). Breeding systems of several *Vicia* species. *Crop Science* **35**, 1200-2.
Zohary, M. and Heller, D. (1984). The Genus *Trifolium*. Jerusalem: Israel Academy of Sciences and Humanities.

CHAPTER 3

Reasons for collecting wild plants

Robert Reid[1] and Sarita Jane Bennett[2]
[1] *Deptment of Agriculture, Stanley, Falkland Islands.*
[2] *Co-operative Research Centre for Legumes in Mediterranean Agriculture (CLIMA), University of Western Australia, Nedlands, WA 6907. Australia*

Abstract

Degradation of the environment is occurring at an alarming rate. At the same time the number of species with agricultural potential is increasing, both as a result of a desire to diversify and the use of biotechnology which has widened the gene pool from which genes can be taken for a particular trait. The risk of global climate change poses further challenges to the future availability of plant germplasm. There are thus a number of principal reasons why collecting wild plants is still important; They are in danger of genetic erosion or extinction, they have been identified as potentially useful to agriculture, the diversity they possess is currently missing or under-represented in existing *ex situ* collections, or more needs to be known about the species in question.

There are a number of different types of collecting mission depending on the end result. Large international agencies tend to be involved in area-driven collecting missions where a particular region is targeted and a wide range of species are collected. Species-specific collections are usually driven by the end-user themselves. These missions are often easier to plan as more ecogeographic detail is known, but can result in a later mission being necessary in the same area, if other species are required. For both types of collecting mission an appropriate sampling strategy will greatly enhance the knowledge gained of the genetic diversity of the species collected.

The paper finishes with a discussion on the amount of passport data to be recorded, the use of climate and morphological data recorded at the collection sites and the responsibilities of germplasm curators in ensuring the availability and ease of access of the material held in their genebank. Current collections of pasture and forage legumes represent only a narrow range of the diversity that is available. Future collecting missions are therefore still important, but should be conducted in a systematic manner to ensure the maximum potential use of the material collected.

Keywords Wild plants, germplasm, *ex situ* collections, genetic erosion, legumes

1. INTRODUCTION

The collection of endemic land races and cultivars of crops is a fairly simple task in that most of the species involved are easy to identify, the area where the crop grows is known and seeds can be obtained from cultivated fields or rural markets. Also information about genetic diversity, agricultural practices and general use of the crop can normally be obtained from local farmers. The collection of wild plants poses a different set of problems. What species should be collected and why collect them? In what ecological niches and in associations with which plants does it grow? How can the collection process be made more efficient and most importantly, how can the benefits be maximised to serve wide ranging interest groups.

In the current state of justification for plant collecting and increasing difficulties in visiting and removing seed from developing countries, why do we need to collect wild species. The aim of this paper is to explain the state of germplasm collections to date and the reason's why the future collection of wild species is so important.

S. J. Bennett and P. S. Cocks (eds.), Genetic Resources of Mediterranean Pasture and Forage Legumes, 32–40.

2. REASONS FOR COLLECTING GERMPLASM

The opportunity to utilise wild plants either directly or as potential breeding material is increasingly limited by the degradation of ecosystems. Habitats are disappearing beneath agriculture, urban sprawl, industrial developments and dams, and are being irreversibly damaged by pollution, overuse and erosion. The principal reasons that are offered for collecting wild plants from a given gene pool or specific environment are; they are in danger of genetic erosion or even extinction, they have been identified as potentially useful to agriculture, the diversity they represent is missing from, or insufficiently represented in, existing *ex situ* collections, or that more needs to be known about them. None of these are mutually exclusive as germplasm can be both threatened and useful, and there may be gaps both in collections of a gene pool and what is known about it (Engels *et al* 1995).

2.1 Degradation of ecosystems

Degradation of ecosystems is the principal reason that many species are suffering from genetic erosion. This can be in terms of the whole species disappearing as a result of urbanisation or over-grazing, or that fragmentation of a species distribution can lead to genes of potential use in breeding being lost before that potential had been realised. There is a single chapter devoted to the threat of genetic erosion in this publication (see chapter 6) and so no further detail will be provided here. If genetic diversity, or a particular species is imminently threatened in an area and *in situ* conservation methods are not feasible or insufficient, collecting is clearly warranted. In the French Mediterranean region, for example, the most serious threats are urbanisation, tourist development, hydro-electric schemes, rice culture, wine growing extensions, use of herbicides, pollution, airfields, conifer plantations, over-grazing, horticultural vandalism, and degradation by collectors (Polunin and Smythies 1988). The most vulnerable areas in the Mediterranean region are undoubtedly the Iberian Peninsula, the Mediterranean Islands, the Appenines and the Balkan Peninsula. The first and last have the richest floras in Europe and many species in these areas are in imminent danger of extinction. (Polunin and Smythies 1988). Other examples include deep ploughing which is depleting the annual forages on arable soils in Algeria (Lesins and Lesins 1979). In Romania, irrigation of former Mediterranean-type dryland in the Dobrogea region has virtually restricted wild annual *Medicago* species to adjacent nature reserves. (Mathison 1983). In the Caucasus region, genetic erosion of crop landraces and wild species is a major problem, due to the extensive use of herbicides and the wide-spread use of high yielding varieties of the principal pasture legumes. Also, the current development of several major hydro-electric and irrigation projects in the region will lead to the flooding of thousands of hectares of montane pasture and will have an undoubtable ecological impact on surrounding areas (Maxted *et al.* 1989).

Small populations with a restricted distribution will obviously be at greatest risk and will, in most cases, be given the highest priority for rescue collection. In the central valley of California the indigenous *Trifolium* species have been reduced to areas of protected remnant vegetation. Most of the area is now under cultivation, with the associated irrigation and farming infrastructures, and some species have become quite rare. (Crampton 1985). Examples of emergency programs that have been undertaken in areas which are being affected are; the Ataturk Dam in south-east Turkey and forage collecting (particularly *Trifolium* spp.) on Mt Kilimanjaro at the high altitudes possibly in danger from global warming (Engels *et al.* 1995)

2.2 Potential for agriculture

Plant collections in the past concentrated on collecting specific species that were known to be important in agriculture, and to collect seed from as many sites as possible in the time allowed. Little time was wasted recording ecogeographical data as each site, and even the site description is often very poor, so that any attempt to re-visit a site is severely restricted. It has now been realised that there are many more species with agricultural potential, both in terms of new species, and wild relatives of current crops that may contain useful genes for a range of characters, such as resistance to diseases, cold tolerance, early flowering etc. In southern Australia on the acid soils, subterranean clover (*T. subterraneum*) was the recommended pasture species for many years, even though it did not persist well at the limits of its range. In recent years, the potential of a number of other species has become apparent, particularly serradella (*Ornithopus* species) for the deep acid sands of Western Australia (Rossiter *et al.* 1985), Biserrula (*Biserrula pelecinus*) for the low rainfall, poor sandy soils (Howieson *et al.* 1995), and Balansa clover (*Trifolium michelianum*) for areas which suffer water-logging (Evans *et al.* 1990; Evans and Snowball 1993). There are many other species that are also currently being evaluated, and some of these are listed in Bennett, Francis and Reid (1998). Ewing (1989) stated that there is sufficient genetic diversity in mediterranean pasture legume species to fill most of the technical needs in Australia. The problem still remains however, which species and where can they be collected.

Modern biotechnological tools have widened the gene pool from which genes can be taken for particular traits. It is now possible to transfer genes from one species to another which are not so closely related taxonomically or genetically. For example, genes can now be taken from the whole section *Falcatae* for breeding into Lucerne (*M. sativa*) rather than just from the immediate relatives of *M. sativa* and *M. falcata*. With this in mind resistance to alfalfa mosaic virus has been achieved in Lucerne.

Within an important species, material may be required from areas not previously collected. Justification for further collection is gained because the populations from the specified area are thought to contain a particular trait that is required for breeding, e.g. cold tolerance from high altitude sites. Prior to the collecting mission, however, it is important to confirm that the material has not already been collected and is held in another genebank.

2.3 Research

Comprehensive knowledge about a target genepool is an important reason for collecting. Germplasm may be required to increase the genetic diversity of a collection so that particular problems can be resolved. These include problems such as the taxonomic boundaries of a particular species or complex of species, the mating system dynamics of a group of species, the ecogeography and genetic diversity of species, particularly those which are endemic to restricted areas or rare. In many cases an alternative sampling strategy will have to be developed (Marshall and Brown 1998), but the information gained will considerably enhance any future breeding programmes of those species.

2.4 Climate change

The possibility of global climate change presents an important challenge to the future availability of plant germplasm. Not only may temperatures rise as a result of the "greenhouse" effect, but rainfall will almost certainly be affected, and in different ways in different regions. Regions and species that are particularly vulnerable to the complex alterations of climatic patterns, which many global warming models predict, will need to be awarded a high priority by germplasm collecting programs. A review of climate change and its effect on plant genetic resources has been undertaken by Jackson, *et al* (1990).

There can be little doubt that for some species it may be too late. The environment overall is changing rapidly, but some regions, such as the Eastern Mediterranean, were radically altered by human activity long ago and many species that were once probably common, have now all but disappeared. This is particularly true of palatable perennial legumes. Lesins and Lesins (1979) postulated that such species as *Medicago glomerata*, from montane, calcareous soils in France and Italy, and *M. glutinosa* which is endemic to sub-alpine, moist areas in the Caucasus region, are particularly under threat. Both species are part of the primary gene pool of *M. sativa* and may provide valuable genes for future breeding programs. Some species are little known and impossible to identify from fragmentary herbarium specimens, an example being *M. x tuneata* from the mountains of Algeria and Tunisia. A detailed search in the latter country in 1994 was unable to locate the species (Reid. unpublished data). Germplasm of this putative hybrid is urgently required as once again it is seemingly in the primary gene pool of *M. sativa* and growing as it does in cold, dry mountains it could have valuable adaptive characteristics.

3. PLANNING

The first step in planning any collecting program for wild species will be a background study of the taxonomy and distribution of the target species/group, often called an ecogeographic survey. Much time and effort can be wasted if collectors do not know enough about the geographic distribution, ecology, phenology and diversity of the plants they are looking for, before setting out in the field. The concept of an ecogeographic survey has been explored in detail by Maxted *et al* (1995) and there are many examples that provide models for future surveys (Bennett and Maxted 1997; Ehrman and Cocks 1990; Maxted 1995; Rihan 1988; von Bothmer 1991). The results of these surveys can be used in clarifying and predicting the collection target. Within the designated area, zones of particular interest can be identified e.g. a high concentration of diverse taxa, areas of low rainfall, particular soil types, etc.

Ecogeographic surveys may proceed in a variety of ways. An individual collector may put together normal background data prior to undertaking a mission, or an international organisation may commission a study of a particular gene pool, e.g. Maxted (1993), or compile ecological information for the exploration of specific species, e.g. IBPGR's studies in forage plants in Central Chile. (Haase pers. comm.). Whichever the case, they are integral to the success of the mission and yet it is still somewhat of a surprise to find that many missions are undertaken with only a minimum of background data. It is often said that the required date is non-existent or difficult to obtain but as pointed out by Auricht *et. al.* (1995) many countries are increasingly making geographic, climatic and socio-economic data available. Furthermore, the international organisations such as FAO, WMO, and UNEP are also linking users requiring information (WRI *et al* 1994).

4. TYPES OF COLLECTING MISSIONS

Many collecting missions organised by the large international agencies often target an extensive range of species, this being somewhat different to these organised by an end-user where there is a focus on particular gene pools, species or even genotypes. In a multi-species collecting mission, a region is targeted and an attempt made to sample as much as possible of the diversity of as many species as possible. These missions are described as being area-driven and are often conducted when no systematic collecting in the area has occurred before. Usually these take place when an area is difficult to reach and future visits unlikely.

Species specific missions tend to be driven by the eventual users of the germplasm. The targets of such missions may be relatively broad (e.g. annual *Trifolium* species in Tunisia) or very specific (e.g. rhizomatous *Medicago sativa* in Central Spain). Generally species-specific missions are less complicated to plan than multi-species missions, as more detail is often known about the ecogeographic distribution of the target material, including data such as the approximate date of seed. The main problem with these specific types of collecting missions is that a return visit may be necessary in the future if germplasm from another species is later required. This has occurred in Sardinia, Italy where subterranean clover was extensively and almost exclusively collected (Francis and Gillespie 1977). It has now been realised that due to its acid soils, which are quite rare in the Mediterranean region, Sardinia may possess many other species of pasture legumes which have potential in the acid sands of Western Australia, and further collections have been undertaken.

Germplasm is sometimes collected on an opportunistic basis during a mission originally targeted for quite different species, characters or ecological conditions. Striking phenotypic features, occurrence in unusual situations and novel or interesting local uses are reasons that can lead to such collecting. These collections are often very important as they allow a more detailed picture of the ecogeography of an area to be determined, and allow a comparison between the originally targeted species and the alternative species to be made, in relation to its adaptation and survival in that area.

There are great differences between collecting wild species and collecting crops. For wild species the collecting opportunity will be narrower because ripe seeds are usually quickly shed and often not available. In contrast, crops seeds usually stay in the plant and in most cases collecting can be done at other times by visiting farmers stores. Populations of wild species are often more difficult to find. More time is spent in looking for collecting sites of wild species than in crop collecting. It may be possible to use herbarium and other records to locate potential collecting sites with precision during the planning stage, but the collector should, where possible, identify the habitat preferences of the target species in order to start developing a "search image" for likely collecting sites, as well as for the species itself (Maxted pers. comm.). A search image simply being a set of indicators that appear to be correlated with the presence of the target species. The indicators are usually associated with the underlying geology or soil association, but also often with a specific geomorphological formations or dominant species.

Until recently, most pasture collectors were also involved in the evaluation of the material they collected. If not engaged in the hands-on initial characterisation stage, then at least they were fully aware of where, and for what purpose, the germplasm would be evaluated. As more organisations, particularly those whose primarily interest is in conservation, become involved in plant germplasm collecting, there is a danger of collecting becoming somewhat divorced from use. Species of little value are collected simply because they are there and valuable samples languish in obscurity in gene banks.

Developing a comprehensive knowledge base of the target gene pool is an important motivation for collecting. Pasture germplasm is still most often needed so that particular research problems may be resolved. Taxonomically unique or isolated, rare and narrowly endemic species often deserve a high priority in research collecting. The development of an appropriate sampling strategy will be considerably facilitated with a knowledge of the distribution of the genetic diversity among and within populations, such as may emerge from preliminary research collecting. Collecting for research purposes will require a somewhat different approach from conservation collecting (Engels *et al* 1995).

A knowledge of ecological preferences can be used to plan the collection of specimens in unexplored areas. An example is provided by *Lens montbretii*, which was collected near Kara-Tash, in the Gaziantep province of Turkey. This species was thought to be rare, and only a few specimens have found their way to European herbaria, most of them from the beginning of the century. During an evolutionary study of the genus *Lens*, it became necessary to obtain living material of *L. montbretii*. The first problem was to locate Kara-Tash, as it did not appear on any of the available maps. Inquiries in the Gaziantep area revealed that Kara-Tash is an abandoned railway station: Kara-Tash means black stone, and indeed basalt is common there. An intensive search for *L. montbretii* in that location met with no success. It was then decided to try another region with similar bedrock, soil type and altitude, near Siverek, about 150km from Kara-Tash. Three populations of *L. montbretii* were detected there, and seeds were collected. Interestingly, following a cytogenetic study of this species, it was decided to transfer it from the genus *Lens* to *Vicia*. (Ladizinsky pers. comm.).

5. VALUE TO END-USERS

For valuable germplasm to reach the end user as quickly as possible, data on the collecting sites are required to facilitate the selection process. Some evaluators argue that only the most detailed site data should be recorded and that approximations may ultimately be misleading. This school of thought advocates the use by the collector of very detailed environmental descriptions, which are deemed to be indispensable for the accurate selection of ecotypes of forage plants adapted to particular conditions. Another school, while recognising the responsibility for recording site information, claims that time and logistical constraints will preclude the noting of all but the most basic site descriptions. The first school is slowly becoming the norm and is certainly the most useful after the collection. Without detailed information about the ecogeography of the collection sites and status of the species collected, only a limited picture of the species ecogeographical requirements and preferences can be determined and this reduces the specificity of a future collecting mission for that species. Also the more information that is obtained at the time of collection then the more useful are the species collected. A plant breeder looking for a new trait in a crop plant can substantially reduce the number of lines of wild relatives to be screened if he has access to the passport data obtained at the time of collection. Examples of plant collecting sheets and the sort of data that should be collected are given in Querol (1993) and Bennett and Maxted (1997).

A description of the climate at the collection site is an important part of the environmental data required for evaluation. Data from the nearest meteorological stations need to be interpolated or extrapolated to get an estimate of climatic variables at the site of collection. This will prove difficult in areas such as steep mountainous terrain, where conditions can vary enormously over short distances. However, in extensive areas with little surface relief (e.g. the *meseta* in Spain or much of the Libyan coastal plain) the annual rainfall and temperature regime at any given site can be readily approximated. In such cases estimates with an error of ±25mm in the annual rainfall total are quite acceptable and have proved invaluable to evaluators. This information is important to breeders as it will again reduce the number of accessions of a particular species to be evaluated. The accessions to be evaluated can be matched to the climate - rainfall, minimum and maximum temperatures - of the area for which they are to be developed. If the length of the growing season at the collecting site is also known, the time of flowering and seed set can be estimated. These are two very important characters in plant breeding, as adaptation of a species to a new area will always be restricted by the length of the required growing season.

At the time of collection the morphological details of the plant should, where possible, be noted as this may be of major consideration in the selection of the accession for inclusion in an evaluation program. This can be illustrated by the genus *Ornithopus* where plant height can determine the end-use of the accession. *O. compressus* typically exhibits a prostrate growth habit, which in turn gives it improved tolerance to grazing, and also makes it a useful plant for controlling soil erosion. However, some accessions from hay fields in Spain and Portugal are very erect, no doubt being 'improved' through farmer selection, and are naturally suitable for hay production (Fu *et al.* 1984). Knowledge of this would facilitate the evaluation, lower the cost of the process and speed up the development of specific cultivars.

6. GERMPLASM ISSUES FOR COLLECTORS

The germplasm collectors, and indeed the curators of genetic resource collections bear heavy responsibilities because their decisions result in what is actually available for all conceivable end-users.

Undoubtedly a major obstacle in getting valuable germplasm to the potential end-user is the failure of a particular species to thrive at its first "bulking-up" site. There are numerous examples of base collections e.g. IPGRI designated collections, which are, because of various environmental constraints, unable to grow some species. Of particular note is the failure of the perennial *Cicer* species at ICRISAT, the cool temperate *Vicia* species at ICARDA and the tropical highland *Trifolium* species at CIAT. Where this situation is likely to occur it is the responsibility of the collector, knowing the site of collection, to suggest alternative sites for first screening. Some success has already been achieved with the successful adaptation of the perennial *Cicer* species to nursery conditions in eastern Washington, U.S.A. (Johnson, pers. comm.).

Another major issue which will have an impact on future collecting strategies is the extent to which a satisfactory sample of the existing variation has been collected already. More sophisticated ecogeographical surveys combined with multivariate analysis may well provide the additional data on variation as well as species distributions required to answer this question. Such an approach, combined with data from characterisation studies of material in genebanks would provide further data on the relationship of diversity to ecogeographical factors and lead to improvements in the design of collecting programs.

It is essential to obtain the correct scientific name of a specimen if it is to be connected to the wealth of information that may be known about the taxon to which it belongs. Germplasm that is not accurately identified in the field may remain uncollected, because its value is not recognised when encountered or if collected may remain in the genebank unused. Misidentification of any material that is collected will lead to spurious results when the germplasm is studied and used. (Maxted and Crust 1995).

In the past, collections have focused on satisfying existing breeding objectives, however, in the future regions may also be chosen for collection because of the novel genes that may be acquired. The efficient identification of regions with novel genes will require extensive research and further use of modern technology. The development of computerised Geographical Informational Systems (GIS) has, in conjunction with the germplasm database, facilitated the identification process. Already precise targeting is possible and some collections are being characterised in greater detail. The days are perhaps not far off when collectors will be able to go into the field with a laptop computer containing a full range of digitised topographic, soil, climate, vegetation maps of the target area in a geographical information system (GIS), perhaps including satellite imagery, with a Global Positioning System (GPS) receiver feeding directly into it. Whilst assays of isozyme variations have allowed the identification of regions containing highly variable germplasm, the application of modern molecular methods of detecting DNA polymorphism's, particularly restriction fragment length polymorphism (RFLP) and random amplified polymorphic DNA (RAPD) will facilitate ecogeographical analysis and the pinpointing of sources of novel genes.

However, the presence or absence of an allele can only rarely be supposed when plants are looked at in the field. Also morphological variations seen by the eye may be environmentally determined. Hence "looking for useful genes" is not tenable (William 1991). Thus, the widely accepted practice of collecting populations of genotypes will continue because forage breeders require certain alleles in plants being used as parents with general adaptation to the environments to which the progeny are aimed. The best method of collection, to ensure that valuable germplasm is available to all end users, is to sample the whole population from the ecogeographic range of the individual species without bias as to the presence or absence of rare alleles.

7. FUTURE COLLECTING STRATEGIES

Plans for future germplasm collections should be characterised into two strategies; to continue broadening the genetic base of those species which on the basis of multi-locational evaluation are recognised as key species for a given ecosystem, with the aim to eventually select superior ecotypes for direct use in plant breeding; and to continue conducting exploratory collecting missions by sampling in as yet overlooked or neglected regions. This latter strategy will both preserve germplasm from genetic erosion, and extend the available species and their ecotypic ranges in the collection for future identification of key species. For southern Australia both types of collecting missions are required throughout the Mediterranean region and adjacent areas (Central Asia) as well as those regions with homologous climates, such as Central Chile, California and the Cape Province of South Africa.

In the future, it can be predicted that *in situ* collection of endangered ecosystems will become increasingly common, as they allow the conservation of the whole ecosystem and all the species within that ecosystem. It must be agreed that for the conservation of the genetic diversity of some species, such as trees for example, *in situ* conservation is a more practical method. However, *in situ* conservation sites are still going to be affected by the same climatic variations as the degraded landscape around them and therefore it is important that the diversity of each of the species is still maintained in genebanks and botanic gardens around the world.

8. CONCLUSION

Despite the substantial progress that has been made in the development of mediterranean pasture and forage legume genetic resources, there is still a narrow collection of diversity (few genera and species) in existing germplasm collections compared with the potential range available in the wild. More legume germplasm must be conserved, but future collection missions should be conducted in a more systematic method than has been used in the past. Collections should be conducted following detailed ecogeographic surveys of the areas to be visited and accessions of each species should be sampled to maximise the amount of genetic diversity in the population. In all cases the future requirements of the end users must be considered. It is evident that there are still large gaps in our knowledge of the genetic diversity that is available, and there are indications from current information that there is still considerable resources which are yet to be utilised.

REFERENCES

Auricht, G.C., Reid, R., and Guarino, L. (1995). Published information on the natural and human environment. In 'Collecting Plant Genetic Diversity'. (Eds. L. Guarino, V Ramanathu Rao and R. Reid) pp. 131 - 152. (CAB International: Wallingford).

Bennett, S.J. and Maxted, N. (1997). An ecogeographical analysis of the *Vicia narbonensis* complex. *Genetic Resources and Crop Evolution*. **44**,411-428.

Bennett, S.J., Francis, F.M. and Reid, R. (1998). Minor and under-utilised legumes. In: 'The plant genetic resources of legumes in the Mediterranean.' (Eds. N. Maxted, S.J. Bennett, M. Bounejmate and L. Robertson) (Kluwer: The Hague) In Press.

Crampton, B. (1985). Native Range Clovers. In 'Clover Science and Technology'. (Ed. N.LK. Taylor). pp. 579-590. (Agronomy Monograph No.25: American Society of Agronomy).

Ehrman, T. and Cocks, P.S. (1990). Ecogeography of annual legumes in Syria: distribution patterns. *Journal of Applied Ecology* **27**, 578-591.

Engels, J.M.M., Arora, R.K. and Guarino, L. (1995). An introduction to plant germplasm exploration and collecting: planning, methods and procedures, follow-up. In 'Collecting Plant Genetic Diversity'. (Eds. L. Guavino, V Ramanathu Rao and R. Reid) pp. 31 - 66. (CAB International: Wallingford).

Evans, PM. and Snowball, R. (1993). Balansa and Persian clover lines outproduce registered cultivars, and subterranean clover and medics, in a 400mm annual rainfall zone in Western Australia. Proceedings of the 7th Australian Agronomy Conference, Perth. pp. 53-56.

Evans, J., Dear, B. and O'Connor, G.E. (1990). Influence of an acid soil on the herbage yield and nodulation of five annual pasture legumes. *Aust. J. Expt. Agric.* **30**, 55-60.

Ewing, M.A. (1989) The development of pasture legumes for the low rainfall cereal-livestock zone of southern Australia. Proceedings of the 5th Australian Agronomy Conference, Perth. pp. 197-211.

Francis, C.M. and Gillespie, D.J. 1977. Ecology and distribution of subterranean clover and *Medicago* species in Sardinia: A report based on a seed collection tour. Western Australia Department of Agriculture Report.

Fu, S.M., Hampton, J.G. and Williams, W.M. (1994). Description and evaluation of serradella (*Ornithopus* L.) accessions. *New Zealand J. Agric. Res.* **37**, 471-479.

Hodges, T.K., Rathore, K.S., and Peng, J. (1993). Advances in genetic transformation of plants. pp. 1013-1023. Proc. XVII International Grassland Congress:, New Zealand.

Howieson, J.G., Loi, A. and Carr, S.J. (1995). *Biserrula pelecinus* - a legumes pasture species with potential for acid duplex soils which is nodulated by unique root-nodule bacteria. *Aust. J. Agric. Res.* **46**, 997-1009.
Jackson, M.T., Ford-Lloyd, B.V and Perry, M.L. (eds). (1990) Climatic Change and Plant Genetic Resources. (Belhaven Press: London).
Lesins, K.A. and Lesins, I. (1979). Genus Medicago (Leguminosae): A Taxogenetic Study. (Dr. W. Junk: The Hague).
Marshall, D.R. and Brown, A.D.H. (1998). Sampling wild legume populations. In 'Proceedings of a CLIMA workshop on the genetic resources of Mediterranean pasture and forage legumes, 14th-18th October 1996'. (Eds. S.J. Bennett and P.S. Cocks). (Kluwer: The Hague)
Mathison, M.J. (1983). Mediterranean and temperate forage legumes. In 'Genetic Resources of Forage Plants'. (Eds. J.G. McIvor and R.A. Bray). (CSIRO: Melbourne).
Maxted, N., Bisby, F., Forde, M., and Munyenyombe, P. (1989). Wild forage legume collection in the Caucasus region of the Soviet Union. IBPGR Report 3/11. IBPGR, Rome.
Maxted, N. (1993). An ecogeographic study of the *Vicia* subgenus *Vicia*. Internal Report IBPGR, Rome.
Maxted, N. (1995). 'An Ecogeographical Study of *Vicia* subgenus *Vicia*'. Systematic and Ecogeographic Studies on Crop Genepools 8. IBPGRI, Rome, Italy.
Maxted, N., and Crust, R. (1995). Aids to taxonomic identification. In 'Collecting Plant Genetic Diversity'. (Eds. L. Guavino, V Ramanathu Rao and R. Reid) pp. 181 - 194. (CAB International: Wallingford).
Maxted, N., van Slageren, M.W. and Rilan, J.R. (1995). Ecogeographic surveys. In 'Collecting Plant Genetic Diversity'. (Eds. L. Guavino, V Ramanathu Rao and R. Reid) pp. 255 - 286. (CAB International: Wallingford).
Polunin, O., and Smythies (1988). Flowers of South West Europe, a field guide. (Oxford University Press: Oxford).
Querol, D. (1993). 'Genetic resources. A practical guide to their conservation'. (Zed Book Ltd.: London)
Rihan, J.R. (1988). An herbarium based ecogeographic survey of forage *Medicago* species in the Mediterranean and adjacent arid/semi-arid areas. Internal Report, IBPGR, Rome, Italy.
Robertson, L.D. and Bounejmate, M. (1998). Genetic erosion of pasture and forage legumes in the Mediterranean basin. In 'Proceedings of a CLIMA workshop on the genetic resources of Mediterranean pasture and forage legumes, 14th-18th October 1996'. (Eds. S.J. Bennett and P.S. Cocks). (Kluwer: The Hague)
Rossiter, R.C., Collins, W.J. and Klein, L. (1985). Winter growth and nutritive quality of Serradella, *Ornithopus* species. *Aust. J. Expt. Agric.* **25**, 362-366.
von Bothmer, R., Jacobsen, N., Baden, C., Jorgensen, R.B and Linde-Laursen, I. (1991). 'An Ecogeographical Study of the Genus *Hordeum*'. Systematic and Ecogeographic Studies on Crop Genepools 7. IBPGR, Rome, Italy.
Williams, J.T. (1991). Plant Genetic Resources: Erme New Directions. *Advances in Agronomy* **45**, 61 - 91.
WRI, UNEP and UNDP (1994). 'World Resources 1994 - 95. A Guide to the Global Environment'. (Oxford University Press: Oxford).

CHAPTER 4

Using collections to describe ecological relationships

Sarita Jane Bennett
Co-operative Research Centre for legumes in Mediterranean Agriculture (CLIMA), University of Western Australia, Nedlands, WA 6907. Australia

Abstract Both herbarium and genebank passport data provide a wealth of information on species distributions and their ecogeographic requirements that is rarely used to its full potential. Within this paper, the history of legume pastures in southern Australia is discussed, emphasising the unsystematic method of selection of many cultivars to date. Another common problem is that cultivars are expected to be productive over large areas of southern Australia and over a wide range of edaphic conditions. This is unrealistic and the amount of natural variation that is present in a colonising species, *Trifolium glomeratum*, is shown. Using previous collection data to plan future plant collections and plant breeding programmes provides a more systematic method of decision making that will increase the productivity and success of cultivar development.

Examples of ecogeographic relationships drawn from previous collections exist in both herbarium collections and genebank passport data. Herbarium collections of the perennial *Medicag* and the *Vicia narbonensis* complex highlight the importance of using this data to establish the ecogeographic relationships of the species within each group. Passport data from plant collections also enable the edaphic requirements of the species to be established and examples are shown from a collection in Crete in June 1995 and from the South Australian Research and Development Institute *Medicago* genebank in Adelaide. Finally, the importance of using Geographical Information Systems in establishing ecogeographic relationships is highlighted and the many important uses of establishing such a system are explained.

Keywords Herbarium collections, genebanks, Geographical Information Systems, ecogeography, legumes

1. INTRODUCTION

The early European settlers to Australia knew that legumes increase the nitrogen content of soil, and brought with them *Trifolium repens* (white clover) and *M. sativa* (lucerne). Of these *T. repens* was killed in many years by drought, and lucerne would only grow in the fertile alluvial plains where it was occasionally killed by flooding (Davidson and Davidson 1993). However, many other species were accidentally introduced either in hay, with cereals or with seeds of *Lolium perenne* (perennial ryegrass) or *T. repens*. By 1890 the floristic composition of southern woodland pastures had changed as a result of increased grazing pressure to include mediterranean annuals as well as native species. On acid soils the new flora included the pioneer legumes, *T. glomeratum* and *T.* (subterranean clover); on neutral soils it included the annual burr medics (*M. minima* and *M. polymorpha*) and *T. glomeratum*; and on alkaline soils it was almost exclusively annual herbs and grasses of Mediterranean origin (Michael 1981). However these annuals were regarded as inferior to the native perennials because they dried off in summer and were thought to produce less fodder in periods of drought. The fact that the seeds were also eaten by livestock and that this provided a high quality diet was not appreciated until the 1920s when subterranean clover was introduced to commerce.

It was about this time that farmers became convinced that adding super-phosphate would alleviate the problem of low phosphate in the soil and increase the yields of subterranean clover. During the 1920s much work was carried out in Western and South Australia on the naturalised subterranean clovers as it was realised that there may be differences in their

S. J. Bennett and P. S. Cocks (eds.), Genetic Resources of Mediterranean Pasture and Forage Legumes, 41–52.

growing season. As all varieties germinated with the first rains it was possible to estimate the length of the growing season by noting their time to flowering. By 1924 three different varieties had been established (Adams 1924), and by 1932 another 4 varieties had been added to the list (Adams 1934). All had been accidentally introduced to southern Australia.

Subterranean clover is only adapted to acid soils where the annual rainfall is greater than 485 mm. It fails to survive on alkaline soils, which include the mallee soils of Western Australia, South Australia and Victoria (Trumble and Donald 1938). However it was noticed that medics, including barrel medic (*M. truncatula*), lucerne (*M. sativa*), black medic (*M. lupulina*) and burr medic (*M. polymorpha*) thrive on these soils. Barrel medic was the most interesting species because it grows in much drier areas than the Mt Barker variety of subterranean clover, has a higher proportion of hard seed and therefore re-establishes after two to three years of cropping. Its burrs do not stick to wool like the burr medics. The first commercial variety was released in 1939 (Davidson and Davidson 1993).

Another problem in Western Australia is that approximately 600,000 ha of the high rainfall areas in the south-west and 1 million ha of the eastern wheatbelt have naturally acid soils (Coventry 1985). Adding lime tends to reduce the availability of some trace elements such as zinc and copper (Porter and Yates 1984). Resolution of this problem has focused on methods of using these soils rather than trying to modify them using lime. As acid soils also tend to result in a reduced colonisation of rhizobia, work has concentrated on the selection of acid tolerant legumes and their associated acid-tolerant rhizobia. Examples of these are lupins (*Lupinu* sp.) and more recently serradella (*Ornithopus* sp.). Although some of the medics are well adapted to the drier areas of the eastern wheatbelt, their associated rhizobia are not. Selection for more tolerant medic rhizobia from the acid soils of Sardinia has shown that *M. polymorpha* and *M. murex* from these areas nodulate better in acid soils than other medics (Howieson and Ewing 1984)

Subterranean clover is very successful down to 450 mm annual rainfall. Therefore the selection for alternatives to subterranean clover should concentrate on the drier areas of the wheatbelt. Other important considerations are for legumes or new varieties capable of growing on acid or saline soils, or help to prevent the increase of salinity. One possible solution is to use deep-rooted legumes, although it is likely that high salt areas will preclude the development of rhizobia. In order to select these alternative legumes it is important to understand the ecology and climate of their natural habitat.

The aim of this paper is to show that plant collections can be used to describe the ecological relationships of legumes to their native habitats and to assist in the selection of legumes for the farming systems of mediterranean Australia. Germplasm collection is fundamental to improving the species diversity in Australia and also provides an insight into the relationship between a species distribution, its frequency over that distribution, and the prevailing edaphic and climatic variables. Examples of the above relationships will be given in a range of pasture and forage legumes. Finally the value of GIS is briefly discussed.

2. GERMPLASM COLLECTION IS FUNDAMENTAL TO IMPROVING SPECIES DIVERSITY IN AUSTRALIAN FARMING SYSTEMS

Legumes are important components of Australian farming systems, but their productivity is limited by an inadequate genetic base within species and to a small number of species in use. For example, one cultivar of *Medicago littoralis* was the only recommended cultivar in use over thousands of square kilometres of the Eyre Peninsula of South Australia for many years (Cocks 1994). This narrow genetic base restricts the ability of legumes to colonise the range of microhabitats that are present and increases their susceptibility to pests and diseases as well as to natural variations in the climate.

Southern Australia has a similar climate to that found in the Mediterranean basin, characterised by warm to hot summers and cold wet winters with a more or less well defined summer drought. However the Mediterranean is a centre of diversity for legumes containing all the species that are present in Australia plus many more that are potentially important. For example in Turkey alone there are 900 legumes (Davis 1970) compared with 167 in the Perth region of Western Australia (Marchant *et al.* 1987). The legumes of Turkey include nearly 100 species of *Trifolium*, 30 species of *Medicago* (including lucerne) and all the annual species currently in use, 59 species of *Vicia* and 58 species of *Lathyrus*. This compares with the flora of the Perth region in which half of the recorded legumes are originally from the Mediterranean area and consist of 24 species of *Trifolium*, 9 species of *Medicago*, 3 species of *Vicia* and 1 species of *Lathyrus*. Only a limited number of these are used in cultivation. Thus the Mediterranean is and will continue to be an important source of diversity for Australian farming systems (see also chapter 1 of this volume).

Another important factor is the soil types of the two regions. The Mediterranean has much newer soils with a high proportion of limestone. In Australia the soils are generally very old, weathered, often highly acid and nutrient poor. One area of acid soils in the Mediterranean is Sardinia and thus the flora of Sardinia and other smaller pockets of acid soils will contain important genetic diversity for the agriculture of southern Australia, and Western Australia in particular.

Ehrman and Maxted (1989) suggested that in the genus *Vicia* some species are generalists and show a high level of diversity combined with a wide distribution, while others are specialists with a specific ecological niche and highly localised patterns of distribution. Examples of generalists are *V. sativa* and *V. narbonensis*, which are widely distributed in relation to climate and soil, whereas specialists such as *V. hyaeniscyamus* and *V. kalakhensis* are endemic species found only in the foothills of the Jebel al-Ansariyeh in Syria on neutral black soils over basalt. Both groups may have important characteristics for agriculture.

An insight into the importance of genetic diversity in Australian farming systems is gained by looking at the distribution and importance of some of the naturalised species in Australia and their adaptations to different habitats. Cocks (1994) in a paddock in South Australia found that *T. clusii* is adapted to the wet conditions at the lowest point of the paddock and that *T. glomeratum* is found only in the drier sections. He suggests that other species within the same sections of *Trifolium* may exhibit similar ecological adaptations, e.g. *T. resupinatum* and *T. fragiferum* form part of the section *Vesicaria* to which *T. clusii* belongs and that these may also be adapted to wet sites. A second study by Fortune *et al.* (1995) on the distribution and abundance of annual legume seeds in the wheatbelt of Western Australia found that *T. glomeratum* was the most frequently occurring species, followed by *T. subterraneum* and *T. tomentosum*. Of the other species frequently found, *T. arvense*, *T. suffocatum*, *T. hirtum*, *T. dubium* and *T. cernuum* were found at high rainfall sites and *T. cherleri* and the medics, in particular *M. minima*, at drier sites. Perhaps of most importance is the result that wherever the naturalised species occurred they contributed more seed to the seed bank than the sown species, and that on average they flowered 20 days later than the sown species. In no case was one species found alone.

This emphasises the importance of mixtures and the subsequent increase in genetic diversity. No site is homogenous so why should one species be expected to colonise the whole area successfully? With mixtures the various species can colonise the micro-habitats in which they are most successful. Other benefits of mixtures are increased survival during environmental disasters, such as drought or water-logging, greater tolerance to epidemic diseases and stabilisation of production levels.

3. ECOTYPIC DIFFERENTIATION OF NATURALISED LEGUMES IN WESTERN AUSTRALIA

Within Western Australia there is a wide range of rainfall, temperatures and soil types. For example annual rainfall at Pemberton in the south-west is 1200 mm while at Merredin in the eastern wheatbelt and Mullewa in the northern wheatbelt it is 310 and 337 mm respectively (Hobbs 1992 and Cramb *et al.* 1991). The length of the growing season varies accordingly from 10 months at Pemberton to 3-4 months at Merredin and Mullewa (Rossiter and Ozanne 1970).

The naturalised legumes within these areas have undergone natural selection to cope and survive with the range of climatic and edaphic conditions such that different ecotypes have evolved. This is illustrated in a study of *T. glomeratum,* which clearly reveals that at least two ecotypes are present (Bennett 1997).

Seed was collected from two sites, one in the south at Mount Barker (747 mm annual rainfall) and one in the wheatbelt near Nungarin (315 mm annual rainfall). The seed was collected in grids so that within site genetic variation could be established as well as that between sites. Plants were grown at Perth, and 17 morphological traits were scored during the spring. The means and variances of the quantitative characters are shown in Table 1. Plants from Mount Barker flowered later and produced more dry matter than plants from Nungarin. Also the variance of the majority of characters is greater for the Mount Barker plants. At Nungarin plants put more energy into seed production than into vegetative growth, a sensible strategy for plants coming from a risky, low rainfall site (see figure 1).

Table 1. Quantitative characters measures for *T. glomeratum*

	Mount Barker		Nungarin	
	Mean	Variance	Mean	Variance
Days to flowering after sowing	126.6	48.4	100.3	17.1
Dry weight of plant (g)	51.2	583.8	16.3	92.4
Node of 1st inflorescence	3.29	0.35	3.35	0.27
Leaf size (mm)	2.78	0.33	3.20	0.26
Petiole length (mm)	2.38	0.91	1.88	0.40
Plant height (mm)	3.04	0.57	3.81	1.79
Plant width (mm)	31.3	42.3	21.3	20.6
Florets per head	30.9	8.1	37.2	1.8
Seeds per head	53.1	32.9	68.8	14.5
Seed weight (mg/50 seeds)	14	0	21	0

Clearly, the two sites have produced different ecotypes of *T. glomeratum.* The results suggest that the harsher the environment, the less room there is for genetic variation, and the greater the importance of successful reproduction. Environmental stress is strong in the semi-arid areas of the wheatbelt and only those species or ecotypes that flower and set seed when conditions are favourable will survive. Competition between plants is less important. In Mount Barker where the climate is less severe, competition between plants becomes more important, and therefore a range of flowering times and an increase in vegetative growth improves the chances of success.

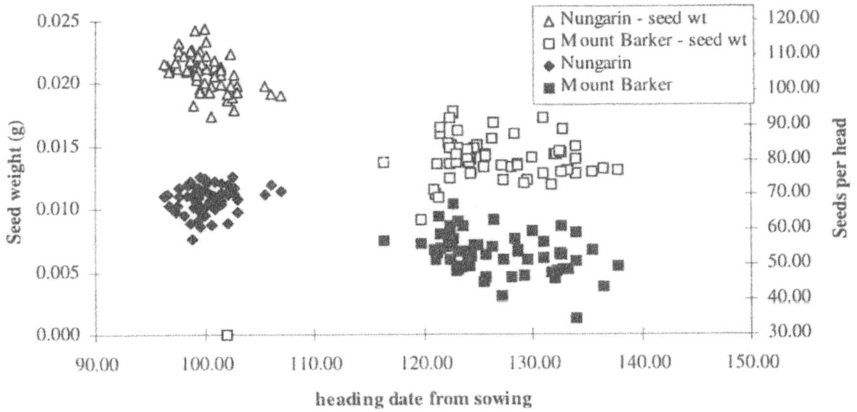

Figure 1. Number of seeds per head and seed weight as a response to flowering time in *T. glomeratum*

It is important to realise the significance of these and other results showing ecotypic variation in introduced annual legumes (eg Woodward and Morley 1974, Cocks *et al.* 1982, Rossiter and Collins 1988, Smith *et al.* 1995). They illustrate the importance of understanding Australian farming systems and how genetic variation can be used to accommodate climatic and edaphic variation. Only species that show a wide range in environmental distribution will grow over a large area. It is more realistic to develop a number of cultivars that are more area or climate specific. Referring back to the paper by Ehrman and Maxted (1989) on generalist and specialist species, it is important when collecting taxa to understand their ecological and climatic distribution as this will help to determine their genetic variation and possible adaptation to new environments. A number of questions need to be answered; how variable is its natural habitat, and in what climatic and edaphic ranges will it be used in Australia? There is little purpose spending time and money developing a specialist species for a variable habitat, as there will be a high failure rate wherever the habitat differs from that where the species originated.

4. USING COLLECTION AND HERBARIUM DATA TO DETERMINE THE IMPACT OF CLIMATE AND SOIL ON SPECIES DISTRIBUTION AND FREQUENCY

In order to gain an insight into distribution and frequency, it is important that a minimum of information is gathered by plant collectors. Using geographical information systems (GIS) the data can then be mapped to determine ecogeographical preferences. The important data to collect along with each specimen are listed below;

Latitude and longitude	Area sampled
Altitude	Parent rock (coded)
Rainfall	Soil type/ texture (coded)
Site details	Soil depth
Slope angle and aspect	Water relation
Site habitat (coded)	Salinity and soil pH
Grazing pressure	Plant density

With this information a detailed picture can be drawn about each individual species such that future collecting trips can be targeted for specific species or for a specific area that is known to be relatively similar to the target area in Australia. However the majority of collections tend to be lax about the information they collect, which makes it difficult to determine detailed ecological relationships and renders the collection of less use than it could otherwise have been.

Some of the information that can be gained from both herbarium specimens and seed collections will be detailed below with examples for both forage and pasture legumes.

In June 1995 forage and pasture legumes were collected in Crete with the aim of finding species suitable for the heavy alkaline soils usually inhabited by medics. Crete was deemed suitable because limestone is a predominate feature of the geology and Crete has the high rainfalls often associated with such soils (Francis *et al.* 1995). For Australian conditions *T. angustifolium* and *T. lappaceum* are thought to show the most promise and in Figure 2 it can be seen that *T. lappaceum* was only collected from one site with a pH of less than 7.5. *T. angustifolium* was collected from a total of 10 sites of which 6 were above pH 7 and all were above pH 6. *T. resupinatum* and to a lesser extent *T. pallidum* and *T. nigrescens* also showed some tolerance of high pH. However, it must be pointed out that since only high pH areas were targeted, it cannot be concluded that these species are adapted specifically to alkaline conditions. They may occur over a wide pH range. Despite this it can be predicted that seed from high alkaline areas will contain some of the diversity needed for adaptation to these conditions in Australia.

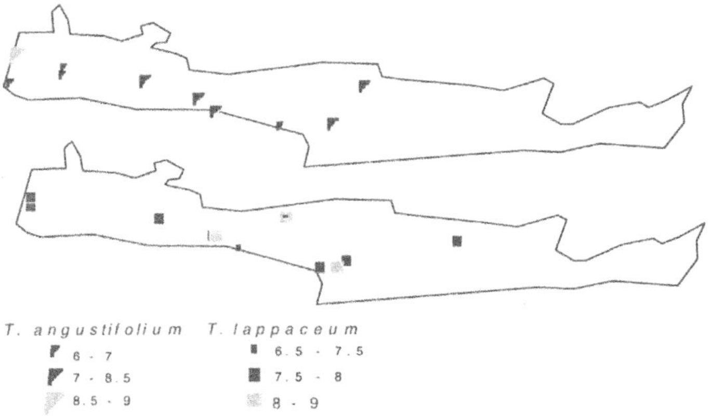

Figure 2. Soil pH of sites where *Trifolium* species were collected in Crete

Another important consideration when collecting pasture or forage species is grazing pressure. In Crete it was found that only *T. nigrescens* and *T. resupinatum* were collected in heavily grazed areas, although most other species were adapted to moderate grazing (Francis *et al.* 1995). Again care must be taken in interpreting this data as few of the sites visited were in fact subject to heavy grazing.

Ecological relationships can also be explored using herbarium specimens. In particular, herbarium data should be examined during the planning phase to assist in the selection of sites during the subsequent collection trip. Unfortunately the quality of data accompanying herbarium specimens tends to be less than for seed collections and in some cases there can be a tendency to collect only rare or unusual specimens. Despite this a general picture of a species distribution can be deduced and at least altitude plotted.

An ecogeographical analysis of the *Vicia narbonensis* complex (Bennett and Maxted 1997) was conducted using a mixture of herbarium specimens and seed collections. The complex consists of 19 taxa. The analysis also included *V. bithynica,* which has recently been removed from the complex (Maxted 1995). *V. narbonensis, V. serratifolia* and *V. bithynica* have the widest distribution, and the greatest diversity of species is found in Turkey. Those species with the widest distribution latitudinally often also have a wide altitudinal range. This occurred in the above study, with those taxa that were rarely collected, such as *V. kalakhensis*, *V. hyaeniscyamus* and *V. eristalioides*, revealing specific altitudinal requirements (Figure 3).

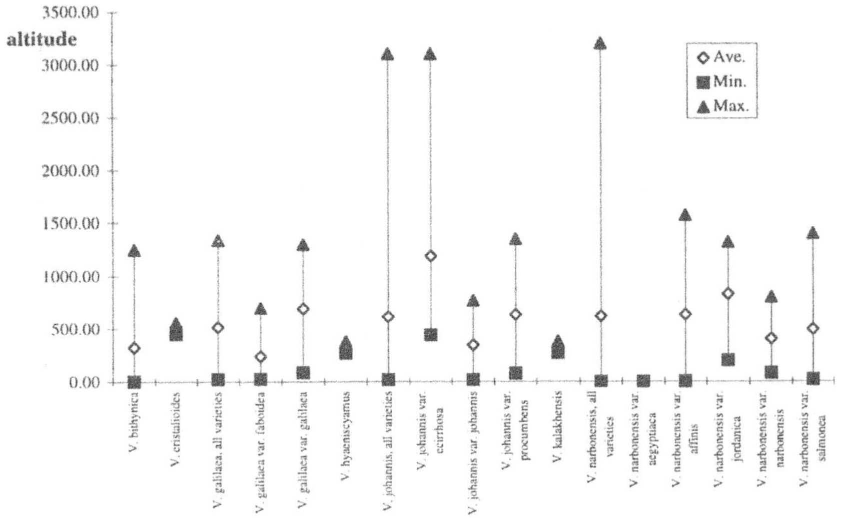

Figure 3. Altitudinal variation of the *Vicia narbonensis* complex specimens by taxa

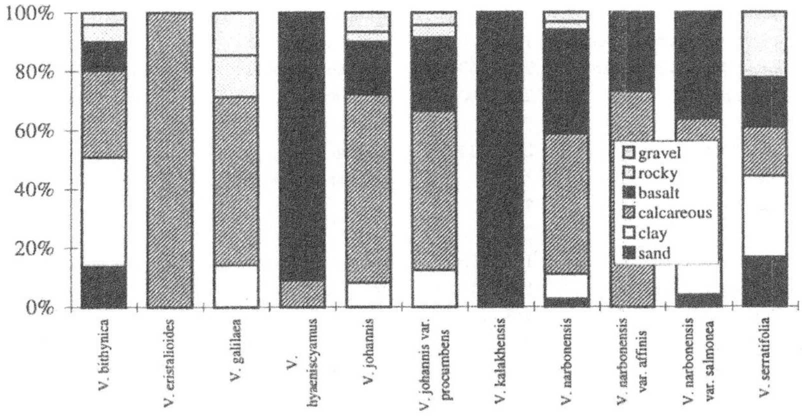

Figure 4. Soil types of the *Vicia narbonensis* complex

Unfortunately the tendency of many collectors of herbarium specimens not to collect detailed habitat and soil data with the associated specimens prevailed here and, as a consequence, it is hard to draw firm conclusions of the preferred ecological requirements of each taxa. Figure 4 shows that the majority of taxa occur mostly on calcareous soil and that *V. eristalioides* and *V. narbonensis* var. *narbonensis* occur solely on calcareous soils. *V.*

kalakhensis and *V. narbonensis* var. *jordanica* occur exclusively on basaltic soils and a number of other species are frequently found on these soils. A few of the species occur rarely on sandy soils. From this it can be concluded that the *V. narbonensis* complex is unlikely to be successful on the sandy soils of Western Australia.

The perennial *Medicago* have been the subject of a detailed herbarium-based ecogeographic study (Bennett unpublished data; summary in Bennett 1996). *Medicago sativa* (lucerne or alfalfa) is the only species cultivated in Australia. However there are 21 other perennial species in the genus, some of which are currently used as fodder in their natural habitats. As they all occur in the mediterranean areas of Europe, north Africa and west Asia there is potential that some of the other species may be suitable for agricultural use, either in their own right or as part of a lucerne breeding programme. Following collection of all the passport data it was possible to establish the habitat and soil type preferences of each species, along with their altitudinal range and their natural distribution. Table 2 shows that there is a wide range in altitude, but less difference in habitat and soils. Most species show a preference for grasslands and for alkaline soils, often calcareous.

Table 2. Perennial *Medicago* species and their geographic requirements

Perennial *Medicago* species	Altitude range	Habitat preference	Soil requirement
M. arborea	1 - 780m	seashore and gardens	calcareous
M. cancellata	?	?	?
M. carstiensis	47 - 650m	mountains and woodland	calcareous
M. cretacea	?	mountains	rocky, calcareous
M. daghestanica	?	?	?
M. falcata	3 - 1800m	variety, especially grassland	sand and calcareous
M. glomerata	350 - 650m	variety	calcareous
M. glutinosa	?	?	?
M. hybrida	?	?	?
M. lupulina	1 - 3571m	variety, especially grassland	calcareous and gravel
M. marina	1 - 149m	seashore	sand
M. papillosa	200 - 2530m	grassland	rocky
M. papillosa ssp. *macrocarpa*	1786 - 2530m	mountain grassland	?
M. papillosa ssp. *papillosa*	200 - 2530m	mountain grassland	rocky
M. pironae	250 - 600m	woodland	calcareous
M. platycarpa	?	?	?
M. prostrata	230 - 1350m	grassland	rocky
M. rhodopea	?	grassland hills	?
M. rupestris	?	?	?
M. ruthenica	?	?	rocky
M. sativa	1 - 3869m	variety, much cultivated	calcareous
M. sativa ssp. *caerulea*	1000 - 1520m	grassland	calcareous
M. sativa ssp. *sativa*	417 - 2150m	cultivated	?
M. saxatilis	?	?	?
M. suffruticosa	175 - 2750m	mountains	rocky, calcareous
M. suffruticosa ssp. *leiocarpa*	225 - 4500m	mountain grassland	calcareous
M. suffruticosa ssp. *suffruticosa*	208 - 1800m	mountain grassland	rocky, calcareous
M. x *varia*	18 - 2083m	grassland	rocky, sand, calcareous

M. lupulina is a species that shows wide variation in both habitat and soil types (see Figure 5). It is currently cultivated for fodder and for green manure, although to date its yield is low (Lesins and Lesins 1979; Townsend 1974). However, it persists even under severe grazing and is known to reseed readily (Small and Jomphe 1989). Another species that shows a wide range of habitats and soil preferences is *M. arborea*. This is a shrubby species growing to 4m and is the only medic with a woody growth habit. It is used as a fodder for goats in its natural habitat, which tends to be dry, calcareous soils. There is thought to be potential for harvesting the young shoots before they become too woody. Figure 5 shows that its most common habitats are gardens, hill slopes and the seashore. It is rarely found in pasture or open grazing land as it is highly palatable and cannot survive heavy grazing.

M. marina is a specialist species that occurs almost exclusively by the sea and in loose sand. It is the only medic with such habitat requirements, although sand of more solid constituency is inhabited by some annuals eg *M. littoralis* (Lesins and Lesins 1979). It has been reported but not confirmed that the species is unpalatable to sheep and cattle because of the hairiness of the leaves and pods (Lesins and Lesins 1979). However hairiness varies widely and no doubt less hairy genotypes could be selected. Another use for this species is to regenerate salt pans. However, its tolerance to low pH is not known, and for use on saline sands in Western Australia, it may not be sufficiently well adapted.

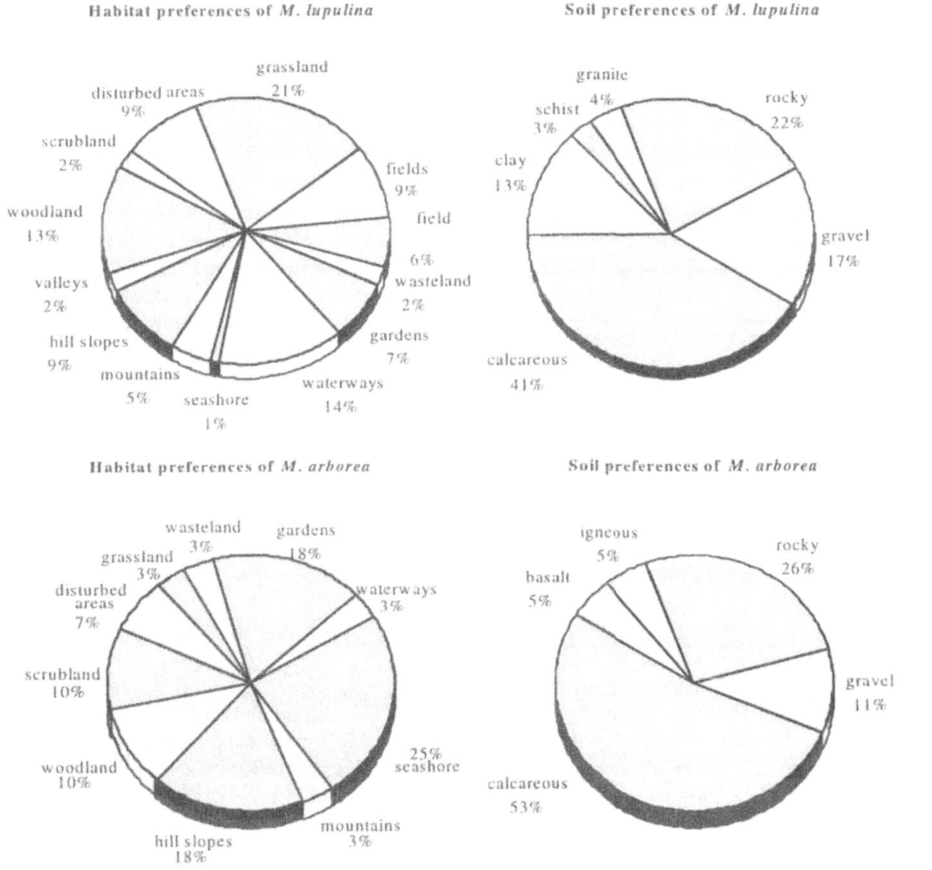

Figure 5. Habitat and soil preferences of two perennial species of *Medicago*

The final example of using collections to describe ecological relationships uses data from the annual *Medicago* genebank in the South Australia Research and Development Institute (SARDI) in Adelaide (Geoff Auricht and Steve Hughes, *pers comm.*). As this database has been developed from seed collections the amount of ecological data accompanying each collection is much greater and possibly more accurate than data from herbaria. As a result a number of different ecological factors can be mapped and a much greater understanding of ecological relationships can be generated. The examples presented below are the geographic and edaphic requirements of *M. orbicularis* and *M. minima* var. *minima*.

M. orbicularis has been collected from Portugal to Iran and on all coasts of the Mediterranean Sea. Most of the seed was collected at sites below 900 m, apart from some mountainous sites in Algeria, Turkey, Syria and Iran where it was found up to 1850 m (at one site in Iran). The majority of sites are in low rainfall areas (to 560mm/year), but there are sites with up to 1200 mm in the mountains. The soils of all but a few of the sites in south-west Turkey were assessed to be freely drained. This is in contrast to remarks made by Lesins and Lesins (1979) and Heyn (1963) who observed that *M. orbicularis* occurs in heavy waterlogged soils, possibly a reflection of the time of year at which the sites were visited. *M. orbicularis* also appears to prefer highly alkaline soils, with a large proportion of the sites in Turkey, Algeria and Syria being in areas of pH 9 or above. While the majority of medics prefer alkaline soils relatively few are found in areas of very low rainfall. *M. orbicularis* has the added advantage that its pods are spineless. It is reported to have forage value in the grasslands of Iraq but that it is of low nutritional value (Townsend 1974). Be that as it may, it clearly has potential for parts of south-eastern Australia.

M. minima var. *minima* also inhabits the drier areas of the Mediterranean basin. It has colonised large areas of the Goldfields in Western Australia, and is widespread in low rainfall areas of south-eastern Australia, including the high rainfall margins of the pastoral zone. The spiny pods get caught in wool causing some economic loss. No doubt this characteristic and its adaptation to the limestone soils facilitated its capacity to colonise such large areas of Australia. In the Mediterranean basin it has been collected from altitudes of 10 to 1500 m, with annual rainfalls ranging from 175 mm to 1250 mm, and where drainage varies from freely drained (preferred) to a swamp. The pH is generally alkaline, ranging from pH 6 to pH 9. The ecological range exhibited by this species suggests that it has large genetic variation and, if the spines could be removed from the pods, could be of considerable value to agriculture.

The information obtained for *M. minima* var. *minima* can also be used when considering other species. *M. minima* belongs to the subgenus *Spirocarpos*, section *Leptospirae*. There are another nine species in this section. Do they also have wide ecological ranges? What are their pod and growth characteristics if they do? Some groups of species have similar morphological and ecological requirements and this information can be used to direct research for new species.

5. THE VALUE OF GIS

Geographical Information Systems (GIS) are "....computer hardware and software packages designed to store, analyse and display spatially referenced data. They deal with information that can be related to some form of map." (Haslett 1990)

They are becoming increasingly important in ecogeographic surveys and in the description of ecological and climatic relationships of species and genera. They describe objects in terms of their position with respect to a known co-ordinate system, i.e. latitude and longitude. They can contain attributes that are unrelated to position, e.g. colour, soil pH, date of flowering etc. Spatial interrelationships (topological relations) can be set up that describe how attributes are linked together or how one can travel between points.

All of the maps used in this paper have been produced using a GIS system. These are useful as they allow the interpretation of distribution patterns of target taxa and so enable accurate predictions of sites where taxa are likely to be found, or more specifically, where taxa with particular attributes are likely to occur.

Guarino (1995) has listed a number of uses of GIS that are particular to plant germplasm studies. These are:

- delimitation of areas where a particular set of conditions or combinations are found. This is important when material or sites with particular characteristics are being targeted, e.g. calcareous soils and an annual rainfall above 1000 mm;
- derivation of combined data sets to predict or enhance descriptive value, thus generating mathematical models, e.g. to predict the risk of genetic erosion among landraces;
- calculation of climatic 'envelopes' for species, landraces or genotypes. This enables the identification of homoclimes, e.g. between mediterranean Europe and Australia;
- passport data can be updated by overlaying collection site positions onto other digitised base maps;
- verification of data, e.g. a map of collection sites or species distributions can spot outliers occurring as a result of data handling errors;
- production of sample lists to satisfy different criteria, e.g. collections with a particular set of attributes. This is similar to querying in a database system, except that the end result is both a map and a list;
- generation of high quality analyses with manipulation available to decide which layers or attributes to show on each map.

At present GIS systems are not sufficiently used and their full potential in the management of genetic resources has not been realised. Many people are put off by the 'unfriendly' nature of the GIS programmes available, but this is being slowly rectified with each new release. Also as PC computers become more powerful and with larger memories, greater use can be made of PC GIS systems, which are often Windows-based and therefore more user friendly. Another advantage of GIS is that the high quality maps produced are easier to understand by those not up to date with other frequently-used analytical methods. Finally, more information can be depicted in a single map than can be illustrated using tables and graphs. Therefore they may have potential for describing to farmers, for example, why a new cultivar or species should be successful in terms of its natural climatic and ecological requirements.

6. CONCLUSION

This paper gives a brief history of farming in Australia and the importance of pasture legumes. The lack of diversity found in the majority of pasture legumes is discussed and its impact on sustainability and productivity. Germplasm collection plays an important part here and using existing collections to describe ecological relationships of legume species should help to ensure that future collections will yield useful germplasm and new species. A number of legume genera are used as examples to show how ecological relationships can be established using both herbarium specimens and their associated passport data and genebank collections. Particular attention is paid to the annual and perennial medics. Finally the value of GIS in ecogeographical research is discussed.

REFERENCES

Adams, A. B. (1924). Subterranean clover. *Journal of the Department of Agriculture W.A., Series 2* **1**, 514-518.

Adams, A. B. (1934). Subterranean clover: Response of different strains to seasonal conditions. *Journal of the Department of Agriculture W.A., Series 2* **11**, -596.

Bennett, S. J. (1997). Genetic variation between and within two populations of *Trifolium glomeratum* (cluster clover) in Western Australia. *Australian Journal of Agricultural Research* **48**, 969-976.

Bennett, S. J., and Maxted, N. (1997). An ecogeographic analysis of the *Vicia narbonensis* complex. *Genetic Resources and Crop Evolution* **44**, 411-428.

Cocks, P. S. (1994). Colonization of a south Australian grassland by invading Mediterranean annual and perennial pasture species. *Australian Journal of Agricultural Research* **45**, 1063-1076.

Cocks, P. S., Craig, A. D. and Kenyon, R. V. (1982). Evolution of subterranean clover in South Australia. 2. Changes in genetic composition of a mixed population after 19 years grazing on a commercial farm. *Australian Journal of Agricultural Research* **33**, 679-695.

Coventry, D. R. (1985). Changes in agricultural systems on acid soils in southern Australia. Proceedings of the 3rd Australian Agronomy Conference, Hobart , 126-145.

Cramb, J., Perry, M. W., McKeague, S., Perry, M. W., and Mulcahy, M. J. (1991). The environment of Western Australia. In 'The wheat book. A technical manual for wheat producers'. (Eds M. Perry and B. Hillman.) pp. 1-51. (Department of Agriculture, WA: Perth, Australia.)

Davidson, B. R., and Davidson, H. F. (1993). 'Legumes: The Australian experience. The botany, ecology and agriculture of indigenous and immigrant legumes.' (Research Studies Press Ltd.: Taunton, England.)

Davis, P. H. (1970). 'Flora of Turkey and the East Aegean Islands. Vol. 3.' (Edinburgh University Press: Edinburgh.)

Ehrman, T., and Maxted, N. (1989). Ecogeographic survey and collection of Syrian Vicieae and Cicereae (Leguminosae). *Plant Genetic Resources Newsletter* **77**, 1-8.

Fortune, J. A., Cocks, P. S., Macfarlane, C. K., and Smith, F. P. (1995). Distribution and abundance of annual legume seeds in the wheatbelt of Western Australia. *Austalian Journal of Experimental Agriculture* **35**, 189-197.

Francis, C., Maxted, N., Kyriakakis, S., and Bennett, S. (1995). Crete: Observations on the ecogeography of grain, forage and pasture legumes. CLIMA Occasional Publication No. 11.

Guarino, L. (1995). Geographic information systems and remote sensing for plant germplasm collectors. In 'Collecting plant genetic diversity. Technical guidelines'. (Eds L. Guarino, V. Ramanatha Rao and R. Reid.) pp. 315-328. (CAB International: Wallingford, UK.)

Haslett, J. R. (1990). Geographic Information Systems: A new approach to habitat definition and the study of distributions. *TREE* **5**, 214-218.

Heyn, C. C. (1963). 'Annual species of *Medicago*. Scripta Hierosolymitana. Volume XII.' (Publications of the Hebrew University: Jerusalem.)

Hobbs, R. J. (1992). Function of biodiversity in mediterranean ecosystems in Australia: Definitions and background. In 'Biodiversity in mediterranean ecosystems in Australia'. (Eds R. J. Hobbs.) pp. 1-25. (Surrey Beatty and Sons Pty Limited: England.)

Howieson, J. G., and Ewing, M. A. (1984). Soil acidity and legume nodulation. *Journal of the Depertment of Agriculture W.A., Series 4* **25**, 125-127.

Lesins, K. A., and Lesins, I. (1979). 'Genus *Medicago* (Leguminosae). A taxogenetic study.' (Dr. W. Junk Publishers: The Hague.)

Marchant, N. G., Wheeler, J. R., Rye, B. L., Bennett, E. M., Lander, N. S., and Macfarlane, T. D. (1987). 'Flora of the Perth region. Part One.' (Western Australian Herbarium: Perth, Australia.)

Mathison, M. J. (1983). Mediterranean and temperate forage legumes. In 'Genetic resources of forage plants'. (Eds J. G. McIvor and B. A. Bray.) pp. 63-81. (CSIRO: East Melbourne.)

Maxted, N. (1995). 'An ecogeographical study of *Vicia* subgenus *Vicia*. Systematic and ecogeographic studies on crop genepools. 8.' (IPGRI: Rome, Italy.)

Michael, P. W. (1981). Alien plants. In 'Australian vegetation'. (Eds R. H. Groves.) pp. 44-64. (Cambridge University Press: Cambridge.)

Porter, W. M., and Yates, J. S. (1984). The effects (of acidity) on plant growth. *Journal of the Department of Agriculture W.A., Series 4* **25**, 123-124.

Rossiter, RC. and Collins, WJ. (1988). Gentic diversity in old subterranean clover (*Trifolium subterraneum* L.) populations in Western Australia. 2. Pastures sown initially to the Mount Barker strain. *Australian Journal of Agricultural Research* **39**, 1063-1074.

Rossiter, R. C., and Ozanne, P. G. (1970). South-western temperate forests, woodlands and heaths. In 'Australian grasslands'. (Eds R. M. Moore.) pp. 199-218. (Australian National University Press: Canberra.)

Small, E., and Jomphe, M. (1989). A synopsis of the genus *Medicago* (Leguminosae). *Canadian Journal of Botany* **67**, 3260-3294.

Smith, F. P., Cocks, P. S. and Ewing, M. A. (1995). Variation in the morphology and flowering time of cluster clover (*Trifolium glomeratum* L.) and its relationship to distribution in southern Australia. *Australian Journal of Agricultural Research* **46**, 1027-1038.

Townsend, C. C. (1974). Leguminales. In 'Flora of Iraq. Vol. 3'. (Eds C. C. Townsend and E. Guest.) pp. 112-573. (Ministry of Agriculture: Baghdad.)

Trumble, H. C., and Donald, C. M. (1938). Soil factors in relation to the distribution of subterranean clover and some alternative legumes. *Journal of Australian Institute of Agricultural Science* **4**, 206-208.

Woodward, R. G. and Morley, F. H. W. (1974). Variation in Australian and European collections of *Trifolium glomeratum* L. and the provisional distribution of the species in southern Australia. *Australian Journal of Agricultural Research* **25**, 73-78.

CHAPTER 5

Ecogeography and genetic conservation

Nigel Maxted
School of Biological Sciences, The University of Birmingham, Edgbaston, Birmingham, B15 2TT, U.K.

Abstract The current threat to plant genetic diversity from global genetic erosion, coupled with the use by contemporary biotechnological techniques for the utilisation of genetic material from diverse species, has lead to increased professional and public demand for more efficient and effective species conservation. This demand was most eloquently expressed at the 'Earth Summit' held in Brazil in 1992 and is enshrined in the articles of the Biodiversity Convention. One of the important prerequisites of any efficient genetic conservation programme, whether *ex situ* or *in situ* is the identification of centres of plant genetic diversity for the target taxon being conserved.

A simple methodology for identification of these centres has recently been established; this involves the collation and analysis of ecogeographic data. These studies or surveys involve the collection of large and complex ecological, geographic and taxonomic data sets from literature sources and the passport data for herbarium specimens and germplasm accessions. These data are synthesised to produce three basic products: the database which contains the raw data for each taxon, the conspectus which summarises the data for each taxon and the report which discusses the contents of the database and conspectus, as well as proposing future collection and conservation strategies. Ecogeographic studies are an essential component of any conservation programme, not only because they aid the location of genetic diversity, but also because they provide critical assistance in the formulation of appropriate and effective conservation programmes for target plant species.

1. INTRODUCTION

The ecogeographic passport data associated with herbarium specimens and germplasm accessions can be used predictively. If the passport data of a particular species indicates that the species has been previously found on limestone scree slopes, above 2000 metres in South-west Asia, then material of this species is likely to be currently found within these constraints. It is this combination of ecological and geographical passport data from historical collections that can be used to predict where species may be currently located. The efficiency of conservation (*ex situ* and *in situ*) can be enhanced by the accurate location of centres of genetic diversity.

This need for increased efficiency of conservation effort is underlined in the Convention on Biological Diversity (CBD), where the CBD calls on nations to:

> *"Develop, where necessary, guidelines for the selection, establishment* and *management of protected areas or areas where special measures need to be taken to conserve biological diversity."*
>
> Article 8 - CBD (UNCED 1992)

Ecogeographic investigations are essential for the selection of areas that contain taxa that require active conservation. This is particularly pertinent for the conservation of wild species, where much time can be wasted if the habitat preferences or geographical distribution of the targeted species are not known prior to formulation of the conservation strategy.

What are ecogeographic studies? Maxted *et al.* (1995) provide the following definition of an ecogeographic study:

> "An ecogeographic study is an ecological, geographical and taxonomic information gathering and synthesis process. The results are predictive and can be used to assist in the formulation of collection and conservation

S. J. Bennett and P. S. Cocks (eds.), Genetic Resources of Mediterranean Pasture and Forage Legumes, 53–66.

priorities."

In the literature the terms ecogeographic study and survey are both used, the difference between the two is one of degree, a study involves a more detailed data analysis and interpretation phase than a survey. A study will involve detailed collation of fresh environmental data and multivariate analysis of the patterns of distribution and may take several years to complete (e.g. Ehrman and Cocks 1990; Bothmer *et al.* 1991; Maxted 1995). Surveys will focus on collating data recorded by other plant collectors, rather than collecting fresh data and the survey may be restricted to a media search and collating passport data from herbarium specimens or germplasm accessions (e.g. Rihan 1988; Edmonds 1990). For this reason, as a routine perquisite to plant conservation, ecogeographic surveys are likely to prove more applicable.

2. AN ECOGEOGRAPHIC MODEL

One procedure for undertaking an ecogeographic survey is outlined in the simple model proposed by Maxted *et al.* (1995), shown in Figure 1. The model is divided into three phases: project design; data collation; and production.

- Phase I - project design:
  - Project commissioning
  - Identification of taxon expertise
  - Selection of target taxon taxonomy
  - Delimitation of the target area
  - Identification of taxon collections
  - Designing and building the ecogeographic database structure
  - ---
- Phase II - Data collection and analysis:
  - Listing of germplasm conserved
  - Media survey of geographical, ecological and taxonomic data
  - Collection of ecogeographic data
  - Selection of representative specimens
  - Data verification
  - Analysis of geographic, ecological and taxonomic data
  - ---
- Phase III - Product generation:
  - Data synthesis
  - Ecogeographic, database, conspectus, report
  - Identification of conservation priorities

Figure 1. An Ecogeographic Paradigm (taken from Maxted *et al.* 1995)

3. PROJECT DESIGN

3.1 Project commissioning

The project commission may vary from an individual collector gathering ecogeographic data prior to an individual collecting trip to an international conservation agencies (BGCS, CLIMA, CSIRO, ICARDA, IPGRI, IUCN, WWF, etc.) commissioning a full ecogeographic study of a particular target taxon. The target taxon and target area may also vary from a single species in a restricted area (e.g. *Trifolium subterraneum* in Turkey) to a whole genus worldwide (e.g. *Medicag* species throughout the world). In either case a group of taxa from a defined geographical area must be considered sufficiently of interest to warrant an ecogeographic study or survey and subsequent conservation, the first step of which is to produce a project commission statement.

The commission statement will outline the objectives, breadth of the target taxon and target area to be investigated, whether a survey or study is to be undertaken, as well as outlining any specific requirements, such as whether the products should focus on suggestions for *in situ* or *ex situ* strategies to be recommended.

An example of a project commission follows:

> An ecogeographic survey is commissioned for the genus *Trigonella* L. in Syria by the Centre for Legumes In Mediterranean Areas. The survey has the objective of identifying areas that contain novel genetic diversity not already conserved that could be utilised in selection or breeding programmes for the benefit of Australian agriculture. The report should contain a detailed conservation strategy for the genus, including: collecting routes, timing, suitable local counterparts and details of any permits required. It should also attempt to identify those *Trigonella* species of immediate and medium-term potential value to Australian agriculture.

Along with the commission statement, the commissioning agency must also consider the selection of the researcher to undertake the ecogeographic study. Due to the difficulty in the identification of wild species, the poor quality of many identification aids (the various forms of keys and illustrations available) and the extra problems associated with identifying older herbarium specimens, emphasis should be placed on finding appropriate specialists to either undertake or supervise ecogeographic studies. The ecogeographer should have some background knowledge of the group and be experienced in the use of identification aids. Misidentification of material will diminish the predictive value of the data as a whole. The ecogeographer should also, if possible, have a good understanding of the geography of the area to be studied, especially in the case of regional or local studies. Local geographical knowledge may also prove invaluable when trying to decipher location details from a specimen label written by hand several decades previously.

3.2 Identification of taxon expertise

The acquisition of ecogeographic data can be enhanced by discussion with appropriate specialists. They will be able to suggest relevant literature, recommend local floras and monographs, suggest which herbaria and/or genebanks be visited, put the ecogeographer in contact with other specialists and provide general assistance based on their own experiences. The appropriate taxonomic and phytogeographic specialists to approach can be identified from the authorship of taxonomic treatments of target taxa, floras and phytogeographic studies of the target region. Index Herbariorum (Holmgren *et al.* 1990) may be of use in finding which researchers are associated with which herbaria. It often lists their specialist group and region. On the Internet, a World Wide Web site maintained by the University of Helsinki lists more than 1,500 links to botanical resources available over the Internet. The web site is updated weekly and can be found at:

http://www.helsinki.fi/kmus/botmenu.html

Contact name: Dr. R. Lampinen, email: Raino.Lampinen@helsinki.fi

3.3 Selection of target taxon taxonomy

The importance of having a good taxonomic understanding of the group prior to undertaking an ecogeographic study is underlined by IBPGR (1985) and Maxted *et al.* (1995). This can be obtained from various sources: target taxon specialists, recent classifications of the group, various taxonomic studies (using both morphological and other sources of data), revisions and monographs. These will help the ecogeographer to distinguish the generally accepted classification of the group. The classification will provide leads to other taxonomic literature: accepted taxa, taxon descriptions, synonymised lists, distributions, identification aids (keys and illustrations), ecological preferences, bibliographies and critical taxonomic notes. More obscure groups may lack a recent revision or monograph, but the researcher must collate the published taxonomic data that is available to provide the taxonomic backbone to the study.

The majority of wild species have been described and classified using a combination of morphological characteristics: the degree of genetic differentiation between the species is more difficult to assess and has less often been considered by traditional taxonomists. If the aim is to conserve maximum genetic variation in the target taxon, then a classification using the biological species concept, where genetic differentiation is given greater importance are more appropriate. However, there are few biologically based classifications available. They tend to be restricted to a few major crop plants and their allies, where the genetic relationships among the taxa have been extensively studied and the make-up of the gene pool is relatively well understood.

3.4 Delimitation of the target area

The target area of the taxon being studied may be restricted by the terms of reference of the project commission, but if unspecified the taxon should be studied throughout its range. The commissioning agent may restrict the survey to a specific region to save resources in the short term, but this may ultimately prove to be a false economy. Multiple studies of the same taxon, possibly with separate authors, are likely to form a less coherent whole and may unnecessarily limit the predictive value of the ecogeographic study. Having established which region is to be included in the target area, extra ecogeographic information on the target taxon can be obtained from local floras. Guides to which floras cover which parts of the world are provided by Frodin (1984) and Davis *et al.* (1986).

3.5 Identification of taxon collections

The researcher undertaking the study will need to visit the major herbarium and germplasm collections of target taxon from the target area. The broader the sampling of ecogeographic data associated with herbarium specimens or germplasm, the more likely the data will prove ecologically and geographically predictive. *Index Herbariorum* (Holmgren *et al.* 1990) records where important dried plant collections are held. The important collections to be seen during the study fall into two categories: major international herbaria and local herbaria in the target area. The relative advantages and disadvantages of two categories of herbarium for the ecogeographer are as follows:

Some of the ecogeographic data included in the database compiled by Maxted (1995) were taken direct from the author's own germplasm collection database, which illustrates that ecogeographic data can equally well be obtained from the passport data recorded by previous germplasm collectors. Details of what material is currently being conserved can be obtained from botanical garden, genebank and *in situ* conservation area catalogues and databases. Also, directories and databases of conserved material have been and are being established by a variety of national and international agencies. The International Plant Genetic Resources Institute (IPGRI) produces international directories of germplasm collections on a crop basis and also maintains a database, which may be queried on demand.

	Advantages	Disadvantages
Major international herbaria	1. broad taxonomic coverage, possibly material used in the production of revisions and monographs 2. broad international geographical coverage, possibly material used in the production of local Floras 3. skilled researchers available to provide general advice 4. appropriate taxonomic and geographical specialists 5. type material of target taxa 6. good botanical library	1. predominance of old collections, making extraction of passport data more difficult and likely predictive value lower 2. geographical names associated with older collections sites may have changed more recently
Regional herbaria	1. good local regional coverage of target area 2. better documented material, as the herbarium is likely to have been more recently established 3. regional specialists present, who can assist in deciphering local geographical names	1. limited resources for herbarium maintenance 2. lack of target taxon specialists 3. limited botanical library

3.6 Designing and building the ecogeographic database structure

The ecogeographic data for the target taxon can be collated on paper, but is more efficiently collated directly into a computer database. A portable personal computer combined with database and word-processing software can be used for data entry and editing. Time and effort would clearly be wasted by first copying the data onto paper then transferring it to the database at a latter stage. The process of data transfer is as time consuming as the original data collection and may introduce errors into the database. Generally, the fewer steps between data gathering and record entry into the database, the more efficient the process.

Before starting the process of data collation the ecogeographer will need to design an appropriate structure for the database which will hold the data. In general database structures should be kept as simple as possible. A database is composed of records, equivalent to the horizontal lines in a report, and fields, each equivalent to a single column in a report. Each record could describe a separate herbarium sheet or accession, for example. Each field would then provide specific details about some aspect of the sample.

It is important to standardise the data held in the database wherever possible. This will save storage space, speed up retrieval times and help avoid errors in content and interpretation of the contents. The codes used should be accepted TDWG (International Working Group on Taxonomic Databases for Plant Sciences) standards and used consistently.

4. DATA COLLECTION

4.1 Listing of germplasm conserved

Before embarking on the detailed data collation phase of the project, current conservation activities should be reviewed. If sufficient genetic material is already safely conserved either *in situ* or *ex situ* further activity may not be required. Details of what material is currently being conserved can be obtained from botanical garden, genebank and *in situ* conservation area catalogues and databases. Searching these sources to decide if the material is being actively conserved may prove time consuming, but consulting taxon experts may guide the ecogeographer to the appropriate catalogues or databases. To alleviate this problem national and international directories and databases of conserved material have been and are being established by a variety of national and international agencies, for example the European Co-operative Programme on Genetic Resources (ECP/GR), the International Plant Genetic Resources Institute (IPGRI), the Genetic Resources Information Network (GRIN) of the United States Department of Agriculture (USDA) and the Commission on Plant Genetic Resources of the Food and Agriculture Organisation of the United Nations (FAO). As these and other databases become more widely available so the ecogeographer will be able to more easily assess the amount of genetic variation of any target taxon that is already adequately conserved.

Care must be taken when interpreting information on current genebank or botanical garden holdings. The material held may be misidentified, though it should be possible to check the identification by consulting voucher material or identifying living material. The actual quantities of germplasm available could also be misleading: genebanks and botanical gardens are encouraged to duplicate their collection in other genebanks or botanical gardens and so duplicated accessions can give a false impression of the genetic diversity conserved. The ecogeographer should consider that, although accessions may be held in a genebank, the material may for various reasons be unavailable to potential users and so create a false impression of a taxon's conservation status.

4.2 Media survey of geographical, ecological and taxonomic data

The collation of these data could be undertaken while visiting the major herbaria which often have good botanical libraries attached. The appropriate literature will include: monographs, revisions, floras, gazetteers, articles, papers, soil, vegetation and climatic maps, atlases, etc. However, increasingly, as new technologies develops, literature sources will include other media than printed books, such as: microfiche (abstracts of papers); on-line databases (e.g. CAB abstracts, AGRIS, AGRICOLA, Science Citation Index); CD-ROM discs (e.g. PlantGene CD, TreeCD, CAB abstracts); videos; and other contemporary data storage media. The data that might be obtained from different media will include:

a. accepted taxon name*
b. locally used taxon name*
c. where in the target area the species is reported to grow*
d. timing of local flowering and fruiting*
e. habitat preference*
f. topographic preference*
g. soil preference*
h. geological preference*
i. climate and micro-climatic preference*
j. breeding system employed*
k. genotypic and phenotypic variation (are local variants found, is this variation genetically or environmentally based?)
l. biotic interactions
m. archaeological evidence
n. ethnobotanical evidence
o. conservation status* (e.g. Red Data Book status)

* indicates data that could be coded

The development of Geographical Information Systems (GIS) will also prove invaluable to future ecogeographic projects. GIS packages generally include a sophisticated graphics programme for handling digitised cartographic data, interfaced with a database management system for storing and manipulating data associated with particular map features. The whole is a highly adaptable mapping system, which can easily cope with a wide range of geographical, ecological and biological data sets. Among other uses, GIS will make it easier for the ecogeographer to interpret the distribution patterns of the target taxa and so accurately predict sites where the taxon is likely to be found. In the same manner GIS may be used to predict or assess levels of genetic erosion.

4.3 Collection of ecogeographic data

The kinds of information that the ecogeographer can obtain from the passport data associated with herbarium specimens and germplasm accessions are listed below:

a. herbarium, genebank or botanical garden where specimen is deposited*
b. collectors name and number
c. collection date* (to derive flower and fruiting timing)
d. phenological data* (does specimen have flower or fruit)
e. particular area of provenance*, latitude and longitude or even greater detail if possible
f. altitude*
g. habitat type*
h. soil type*
i. vegetation type*
j. site slope and aspect*
k. land use and/or agricultural practice*
l. phenotypic variation
m. biotic interactions
n. competitive ability*
o. palatability*
p. ability to stand grazing*
q. vernacular names
r. plant uses

* indicates data that could be coded

The listing is extensive and it is unlikely that all of these data categories will be recorded during a single study. Within this list, however, there are certain data items that should be recorded for the study to yield predictive results, these are: taxon identification, geographical provenance (the more detailed the better), collection date, habitat preference and altitude.

Early plant collectors could not have predicted the detailed analysis that would subsequently be based on the provenance information they recorded. This is therefore often very basic: site location data may be ambiguous and ecological details missing. Older specimen labels are almost invariably hand written, which adds the problem of the need to decipher the script, that may also be written in a foreign language. Botanists from local herbaria may provide invaluable help, not only in herbarium label translation, but also in the precise location of specimen provenance if a gazetteer of local geographical names and locations is unavailable. There is no comprehensive world gazetteer yet available, but the *Atlas of the World* (Times Books 1988) contains an extensive gazetteer and the *Official Standard Names Gazetteer*, is being constructed and is available in country volumes from the U.S. Board of Geographical Names.

Care must be taken in accepting scientific names written on herbarium sheets; the identification should always be checked. Specimens are more likely to be selected if they have detailed ecogeographic passport data or if they show features of particular taxonomic, ecological or geographic interest, *i.e.*, they are odd or rare forms, come from unusual environments or are found on the edges of their natural distributional range. Maxted (1995) found that data from about a third of the specimens seen during the study were finally included in the ecogeographic database. Recently collected specimens often have higher-quality passport data which is easier to read, being often type-written. These data are also more likely to have remained current. The database will inevitably contain many gaps. In general, it is much easier to record curatorial or geographic than ecological data from herbarium specimens.

4.4 Selection of representative specimens

The fact that all ecogeographic surveys will be limited by time and that the herbaria of the world contain millions of specimens means that the researcher is forced to select representative specimens of the target taxon. It is important that the ecogeographer places emphasis on obtaining reliable specimen provenance data for those specimens to be included in the database. Ideally, only specimens that either have latitude and longitude data available, or for which these data can be established, should be selected for inclusion in the database. Specimens should be positively selected to represent the breadth of geographic and ecological conditions under which the target taxon is found. It is desirable to collect detailed passport data from a broad range of representative specimens, rather than duplicate data already included in the database from previous specimens. Entering data from more than one specimen sampled from the same population is likely to add little to the predictive value of the data set as a whole.

The ecogeographer may, however, be able to infer various features (latitude and longitude, geology, soil, altitude, etc.) of collection sites from location data by reference to Geographical Information Systems or various kinds of maps. Whether this is possible will depend on the precision of the location data available, the topography of the collection area and the precision of the environmental data required. For example, if the collection site is situated on a gently undulating plain, then a crude estimate of altitude may be gained from the location data, as the altitude is unlikely to vary significantly in the vicinity of the collecting site. However, if the collection site is situated in a mountainous area, then the altitude is likely to vary markedly within relatively short distances and so estimates of site altitude based on location will be misleading unless this is extremely precisely specified. This kind of secondarily-derived data should be flagged in the database to distinguish it from data derived directly from herbarium labels. Specimens should also be flagged, if possible, to denote the accuracy of locational data, whether available in the original record or derived from other locality information; in practice, this means giving a code for the scale of map used to pin-point a locality and read off co-ordinates (Rhoades and Thompson 1992).

Ecogeographers are faced with the question, how many specimens should be entered into the database before the amount of extra information gained from each specimen does not increase the predictive value of the data set? There is no specific answer to this question. However, the compiler should be on the look-out for the point when novel ecogeographic combinations no longer occur in the specimens being examined and the latitudinal/longitudinal extent of the distribution of the species has ceased to expand: the full range of geographical and ecological niches that the taxon inhabits will then probably have been recorded in the database.

4.5 Data verification

Before the database can be deemed complete, the ecogeographer must search for and correct errors. Indexing the database (i.e. re-arranging the records in alphabetical or numerical order) on each field in turn may highlight typing errors or invalid entries. Mapping latitude and longitude data may reveal errors if particular localities are shown up as obvious outliers in impossible places. Herbarium specimen and germplasm collectors often send duplicate sets of their material to different international collections. The ecogeographer should search the database for these duplicates and be aware of their possible effect on data analysis.

4.6 Geographic, ecological and taxonomic data analysis

The raw ecological and geographical data included in the database can be analysed to help identify the particular geographical locations and habitats favoured by the target taxa. Interpretation can be aided by the use of bar charts, which simply indicate the number of specimens seen from a particular geographical / ecological unit, or distribution maps, which pin-point collection sites for specimens representing taxa. These maps or charts can be used in conjunction with geographical, rainfall, geological, soil maps, etc. to predict current locations of the target taxon.

Stace (1989) stresses the importance of the means of visually displaying plant distributions. This can take two forms, shading or enclosing an area with a single line or using various kinds of dot distribution maps. The use of an enclosing line is ambiguous, as it provides no indication of the concentration of the taxon within the region. A single outlying specimen might erroneously suggest that the taxon is present throughout an entire region. The occurrence of a species is often sparse at the periphery of its range and there is rarely a distinct cut-off line. Indicating presence in this manner also means that any variation due to local ecological and geomorphological factors within the individual provinces or countries cannot be indicated. Enclosed line maps may be of more use in indicating general distributional patterns. Practically, the procedure for displaying the distributions of taxa can be enhanced by the use of one of the mapping programs currently available. Large scale maps of the world are readily available, but if more detailed maps are required they may need to be customised or even digitised from scratch. Mapping programs will allow the import of latitude and longitude co-ordinates from the ecogeographic database and their plotting onto customised maps.

Morphological or ecological information can also be superimposed onto a dot distribution map, as Strid (1970) has done for the uppermost internode length of various populations of *Nigella arvensis* in the Aegean region. The position of a rectangle indicates the population location, while its height shows the relative length of the internode. Pie-charts placed on maps can be used to display the relative frequency of a character in different places. They are commonly used to show the distribution of allelic variation from different geographical locations.

Enclosing line maps can also be used to indicate concentration of species. These maps, known as Isoflor maps, do not show actual species distributions, but each line is a contour delimiting a greater or lesser concentration of species. Various species distributions are superimposed onto a single map. Then contour lines are then drawn around areas of the map to indicate the number of species found at any one point. An example of an isoflor map for *Vicia* section *Hypechusa* is shown in Figure 2 (taken from Maxted 1995).

One of the simplest means of ecogeographic data analysis is to use frequency distributions: that is the number of specimens (which can be expressed as percentages) collected from sites characterised by different biotic and abiotic features e.g. climate, soil types, aspect, shading characteristics, habitat types, etc. Data arranged in this fashion will identify the particular niche that the target taxon prefers and so can be used to relocate previous collection areas and indicate other areas where the taxon is likely to be found. Frequency distributions can often be compiled directly by querying the database. Correlation of the occurrence of specimens along environmental gradients, such as altitude, latitude, and soil pH will give error terms and can therefore be used predictively. Correlation of morphological characters with environmental gradients will help indicate possible ecotypic adaptation, both in wild and cultivated material.

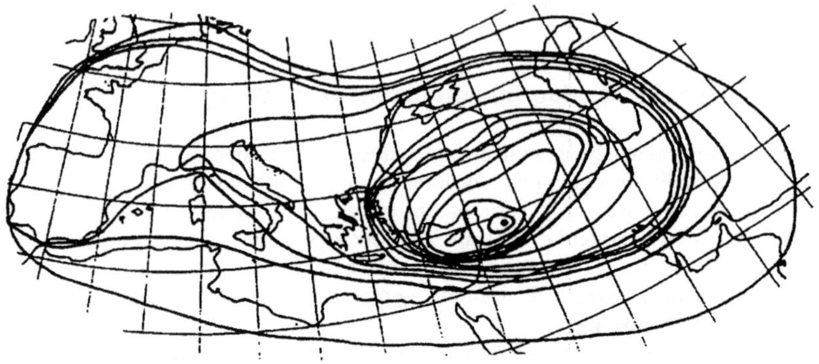

Figure 6. Isoflor map for *Vicia* section *Hypechusa* (from Maxted and Douglas 1996).

All of the above methods of analysis are based on consideration of a few environmental factors at a time, or a single morphological variable. Ecogeographic data, however, is multivariate, in that many items of data are available for each record (*e.g.* each collecting site, germplasm accession or herbarium specimen). Where an ecogeographic study is undertaken, the data collated during the study is likely to of much higher quality than that collected solely from herbarium specimen passport data during a survey. One of the most thoroughly statistically tested ecogeographic data sets is that reported by Cocks and Ehrman (1987), Ehrman and Cocks (1990) and Ehrman and Maxted (1990) for the annual legumes in Syria. These authors undertook comprehensive field work over several years, during which time they gathered extensive ecogeographic data and were able to use this data to predict potential areas of conservation. Ehrman and Cocks (1990), for example, used various methods of cluster analysis on their environmental data to classify the collecting sites into groups or classes (clusters) the members of which had climates which were more similar overall (rather than as regards any one single variable) to one another than they were to members of any other class.

It should be noted, however, that the resources required to undertake a study are relatively expensive. Therefore, if an ecogeographic investigation is to be used as a routine part of collection and conservation activities then the quicker, less expensive, option of undertaking a survey is likely to be favoured. As surveys focus on collating passport data from herbarium specimens or germplasm accessions, the data is unlikely to be sufficiently robust to permit detailed multivariate analysis.

5. PRODUCTION

5.1 Data synthesis

The final production phase of the project commences with the synthesis of all the data collected during the study. The researcher should also be aware of the degree of completeness of the database, or the collections on which it was based, in terms of how fully the target area has been effectively covered. If a particular habitat is under-represented in the database, is it because the taxon is absent from that habitat or because that type of habitat has not been sampled, or even because the target taxon has not been recognised in such a habitat? This problem must be considered if the results of the analysis and the inferences drawn from them are not to be misleading.

5.2 The ecogeographic database, conspectus and report

The ecogeographic database, conspectus and report should be seen as the three essential products of an ecogeographic study. The ecogeographic database contains the raw data of the project. The conspectus summarises the available ecological, geographical and taxonomic information for the target taxon through part or the whole of its range. While the report interprets the data held in the other products and will aid the ecogeographer select conservation priorities.

The conspectus is arranged by plant names and they can be listed either alphabetically or systematically. When possible the following information should be included:

a. accepted taxon name, author(s), date of publication, where published,
b. reference to published descriptions and iconography,
c. short morphological descriptions, or keys, for important taxa or those that may be difficult to identify,
d. phenology, flowering season,
e. vernacular names,
f. geographical distribution, countries from which the taxon is recorded, which includes reliable records from the literature. Presence indicated by listing the country ISO-codes units or other standard codes,
g. distribution maps, preferably dot a distribution map produced directly from the latitude and longitude data held in the database,
h. ecological notes, including altitude (minimum and maximum elevation), habitat, topographic, soil, geological, climate and micro-climatic preference, biotic interactions and other habitat details,
i. geographical notes, containing general data and interpretation of the taxon's geographical distribution,
j. taxonomic notes, containing general data and interpretation of the taxonomic data, noting any distinct genotypic and phenotypic variation within the taxon,
k. conservation notes, containing an assessment of the potential genetic erosion faced by and the conservation status of the taxon.

An example of the level of detail that might be included in an ecogeographic conspectus is given in Appendix 1.

The report discusses the contents of the database and conspectus and must draw general conclusions concerning the group's ecogeography and present a concise list of conservation priorities. If possible, the following points should be covered:

a. the delimitation of the target taxon,
b. the classification of the target taxon that has been used,
c. the mode of selection of representative specimens,
d. the choice of hardware and software,
e. the ecogeographic database file structures and inter-relationships,
f. discussion of database content,
g. discussion of the target taxon ecology,
h. discussion of the target taxon phytogeography, discussion of the distribution patterns and a summary of the distribution in tabular form,
i. discussion of any interesting taxonomic variants encountered during the study,
j. discussion of the current and potential uses of the target taxon,
k. discussion of the relationship between the cultivated species and their wild relatives,
l. discussion of any particular identification problems associated with the group, presentation of identification aids to vegetative, floral and fruiting specimens,
m. discussion of *in situ* and *ex situ* conservation activities associated with the target taxon, including the extent of diversity already conserved,
n. discussion of the genetic erosion threat facing the group,

o. discussion of priorities and suggested strategy for future conservation of the target taxon,
p. the ecogeographic conspectus may be included within the report as an appendix or as a separate entity.

5.3 Identification of conservation priorities

The various products of the ecogeographic survey provide a basis for the ecogeographer to formulate future conservation priorities and strategies for the target taxon. Within the target area, zones of particular interest may be identified, e.g. areas with high concentrations of diverse taxa or low rainfall or high frequency of saline soils or extremes of exposure, etc. If a taxon is found throughout a particular region then the researcher can use the ecogeographic data to actively select a series of diverse habitats to sample, in this way hoping to sample distinct sectors of the gene pool. If a taxon has been found at one location, but not at another with similar ecogeographic conditions, then the ecogeographer may suggest that these similar locations should be searched. Within the target taxon, specific variants can be identified which warrant conservation priority, e.g. populations that may have crop potential not previously noted, taxa that are particularly endangered from genetic erosion, etc.

The ecogeographic survey or study must conclude with a clear, concise statement of the proposed conservation strategy for the target taxon and proposed conservation priorities. This will answer questions such as: what part can local workers play in conservation activities, should population levels be closely monitored to assess the threat of genetic erosion, should a national or international collecting team be directed to collect the priority target taxa, is it possible to conserve the taxa *in situ*, is *in vitro* conservation required or is a more detailed study required before these questions can be answered? Once specific areas of genetic variation have been highlighted, a route that covers the maximum number of sites in the minimum time or location for an *in situ* reserve can be suggested.

6. CONCLUSION

Herbaria, genebanks and botanical gardens, as well as containing plants and plant propagules are storehouses of botanical data that can be used to facilitate plant conservation. The analysis of a taxon's geography, ecology and taxonomy allows its conservation status to be assessed and permits the prediction of areas and habitats that are likely to contain genetic diversity of the target taxon. Once located, populations of the taxon can be monitored, sampled if necessary and effectively conserved.

REFERENCES

Bothmer von, R., Jacobsen, N., Baden, C., Jφrgensen and Linde-Laursen, I. (1991). 'An ecogeographic study of the genus *Hordeum*'. Systematic and Ecogeographic Studies on Crop Genepools 7. (IBPGR: Rome, Italy).

Cocks, P.S. and Ehrman, T.A.M. (1987). The geographic origin of frost tolerance in Syrian pasture legumes. *Journal of Applied Ecology* **24**, 678-683.

Davis, S.D., Droop, S.J.M., Gregerson, P., Henson, L., Leon, C.J., Lamlein Villa-Lobos, J., Synge, H. and Zantovska, J. (1986). Plants in danger: what do we know? International Union for Conservation of Nature and Natural Resources, Gland, Switzerland.

Edmonds, J.M. (1990). 'Herbarium Survey of African *Corchorus* L. Species'. Systematic and Ecogeographic Studies on Crop Genepools 4. (IBPGR: Rome, Italy).

Ehrman, T. and Cocks, P.S. (1990). Ecogeography of annual legumes in Syria: distribution patterns. *J. App. Ecol.* **27**, 578-591.

Ehrman, T.A.M. and Maxted, N. (1990). Ecogeographic survey and collection of Syrian Vicieae and Cicereae (Leguminosae). *Plant Genetic Resources Newsletter* **77**, 1-8.

Frodin, D.G., (1984). 'Guide to the standard Floras of the world'. (Cambridge University Press: Cambridge).

Holmgren, P.K., Holmgren, N.H. and Barnett, L.C. (1990). 'Index Herbariorium, part 1: the herbaria of the world'. (New York Botanical Garden: New York).
IBPGR, (1985). Ecogeographic surveying and *in situ* conservation of crop relatives. Report of an IBPGR task force meeting held at Washington D.C., U.S.A. International Board for Plant Genetic Resources, Rome, Italy.
Maxted, N. (1995). 'An ecogeographic study of *Vicia* subgenus *Vicia*'. Systematic and Ecogeographic Studies in Crop Genepools 8. (IBPGR: Rome, Italy).
Maxted, N. and Douglas, C. (1996). A phenetic investigation of *Vicia* section *Hypechusa* (Alef.) Ashers. and Graebner (Leguminosae, Papilionoideae, Viceae). *Lagascalia* **18**, 283-307.
Maxted, N., van Slageren, M.W. and Rihan, J. (1995). Ecogeographic surveys. In 'Collecting plant genetic diversity: technical guidelines.' (Eds. Guarino, L., Ramanatha Rao, V. and Reid, R.) pp. 255-286. (CAB International: Wallingford).
Rihan, J.R. (1988) An herbarium based ecogeographic survey of forage *Medicago* species in the Mediterranean and adjacent arid/semi-arid areas. Unpublished report. IBPGR, Rome, Italy.
Rhoades, A.F. and Thompson, L. (1992). Integrating herbarium data into a geographic information system; requirements for spatial analysis. *Taxon* **41**, 43-49.
Stace, C.A. (1989). 'Plant taxonomy and biosystematics'. (Edward Arnold: London).
Strid, A. (1970). Studies in the Aegean flora, XVI. Biosystematics of the *Nigella arvensis* complex with special reference to the problem of non-adaptive radiation. *Op. bot. Soc. bot. Lund.*, **28**, 1-169.
Times Books. (1988). 'Atlas of the world'. (ed. 7). (Times Books: London).
United Nations Conference on the Environment and Development (UNCED). (1992). Convention on Biological Diversity. UNCED, Geneva.

APPENDICES

Appendix 1.Example of ecogeographic conspectus construction. Compiled by Bennett (pers. comm.).

Medicago marina L.

Ref. Pub. Description: Sp. pl. 2:779.(1753).

Morphological description: Prostrate to decumbent perennial, 20-60cm long, often branched at base; Vegetative parts densely white-tomentose giving the plant a greyish appearance. Stipules nearly entire. Leaflets 8-16mm x 7-11mm, obovate-cuneate. Peduncle, 7-12 flowered, head-shaped, longer than petiole. Inflorescence capitulate, florets fragrant, 6-10mm long. Calyx 2-4mm long, teeth about equal to tube. Corolla lemon-yellow, 6-8mm, at least 2x calyx; standard obovate. Young fruit protruding sideways from calyx. Mature pods yellowish-grey to dark grey, densely white-tomentose, cylindrical, spiny to almost spineless. Coils 2-4, 4-6mm in diameter, loose with small hole in centre. Surface of coils with a small number of radial veins branching near lateral vein. Seeds brownish, 1-2 in each coil, 3-3.8mm x 1.5-2mm, separated by thin parenchymous wall. Radicle less than half seed length.

Phenology: February to July.

Chromosome No.: 2n=16.

Vernacular name: None recorded.

Geographical distribution: ALB; BUL; CYP; DZA; EGY(N); ESP (inc. Balearic Ils. and Canary Ils.); FRA (inc. Corsica); GRC (inc. Crete); ISR; ITA (inc. Sardinia and Sicily); LBN; LIB(N); MAR; PRT; ROM; SUN; SYR(W); TUN; TUR; YUG.

Geographical notes: Found on the shores of the Mediterranean and Black Seas, and the Atlantic coast of Iberia and France.

Ecological notes: Almost exclusively on seashores; usually in loose sand. Very occasionally on river banks or fields near sea. Furthest distance from sea recorded is 2km. Soil type exclusively sand. The only *Medicago* of such habitat requirements, though seashore soils of more solid consistency may be inhabited by other species, e.g. *M. littoralis*.
Alt: 1-149m.

Taxonomic notes: Easily distinguishable from all other *Medicago* species by pubescent leaves, stems and pods giving plants a greyish appearance.
Agricultural notes: May contain important characters for salt tolerance.
Conservation notes: Not threatened.

CHAPTER 6

Genetic erosion of pasture and forage legumes in the Mediterranean basin

Larry D. Robertson[1] and Mustapha Bounejmate[2]

[1]*Genetic Resources Unit, International Center for Agricultural Research in the Dry Areas (ICARDA), P.O. Box 5466, Aleppo, Syria.*

[2]*Pasture, Forages and Livestock Program, ICARDA, P.O. Box 5466, Aleppo, Syria.*

Abstract Genetic erosion occurs as a result of disruption of the natural habitat by humankind or through the replacement of traditional landraces or ecotypes by modern cultivars. It may occur because of a decline in a species' diversity through the reduction of its geographical distribution or through the complete loss of a species. Within a species a decline in genetic diversity is caused by the loss of genetic information through the reduction of its distribution and population size. Comparison of the current geographical distribution of a species with its past distribution can be used to assess the degree of genetic erosion that has occurred. Quantified assessment of genetic erosion is possible using species diversity indices and measures of genetic diversity.

Although there is some evidence of genetic erosion in pasture and forage legumes of the Mediterranean, it is only circumstantial. Much of this evidence is from observations of loss of habitats to development and intensification of agriculture, and in particular, the increase of grazing pressure in many areas where it has been equated with a reduction in the distribution of species. There is an urgent need to quantify the extent of genetic erosion in the Mediterranean more accurately by determining a baseline of the original species distributions in the region and precisely defining the current distribution of species. Genetic diversity of older original collections of species should be quantified and compared with that of current natural populations to determine if there has been a reduction in the genetic diversity of natural populations.

Pasture and forage species within a genus have different ecological niches, and as such are not interchangeable. Increasing genetic erosion may result in the loss of species or genes that may have potential in Mediterranean agricultural systems, or could be useful in extending the geographical range of a species currently used in agriculture.

Key Words Genetic erosion, genetic diversity, species diversity, Mediterranean, pasture and forage legumes, *in situ* conservation

1. INTRODUCTION

The Mediterranean region has a rich flora, with a large number of endemics (Table 1). The steppe region is reported to be particularly rich (Houerou 1991, Houerou and Boulos 1991). The Mediterranean is the centre of origin and primary diversity of many temperate pasture and forage legumes and thus is an important genetic resource of these species (Mathison 1983). Vetch (*Vicia* species), chicklings *(Lathyrus* species), annual medics *(Medicago* species) and clovers *(Trifolium* species), along with other lesser-known genera are thought to have originated in the Near East and Mediterranean centres of origin, with major diversity present in both West Asia and North Africa (WANA), for the most economically useful annual species (Mathison 1983, Zohary and Hopf 1988). Species of vetch which have immediate agricultural potential are V. *ervilia, V. hybrida, V. narbonensis, V villosa,* and *V sativa subsp. sativa* and subsp. *amphicarpa,;* species of *Lathyrus* of potential use are *L. sativus, L. ochrus,* and L. *cicera;* medic species currently used or of potential use include, *M. intertexta, M. laciniata, M. littoralis, M. polymorpha, M. rigidula, M. rotata, M. scutellata, M. orbicularis,* and M. *truncatula,* and those of *Trifolium* include *T. alexandrinum, T.*

S. J. Bennett and P. S. Cocks (eds.), Genetic Resources of Mediterranean Pasture and Forage Legumes, 67–77.

cherleri, T. subterraneum, T. campestre, and *T. angustifolium* among the annuals and *T. fragiferum* and *T. repens* among the perennials, to name but a few. Examples of other genera include serradella *(Ornithopus compressus)* and sulla *(Hedysarum spinosissimum).*

Table 1. The rich flora of the western Mediterranean (After Heywood 1991)

Country	No. species	No. endemics
Algeria	3150	250
France	4500	73
Libya	1800	134
Morocco	3700	650
Portugal	2600	127
Spain	4900	720
Tunisia	2200	14

Each species has its own edaphic requirements and therefore fills different niches in an agricultural system. For example, *Medicag tornata* and *M. murex* are adapted to low fertility and sandy soils, *M. aculeata* is adapted to low temperatures, *M. intertexta* persists better than other medic species on poorly-drained saline soils and *M. laciniata* is adapted to areas with warm winters (annual rainfall between 100 and 400 mm). *Vicia* species are also adaptated to different niches; *Vicia sativa* subsp. *nigra* predominates in cold high regions, *V. villosa* subsp. *dasycarpa* is also adapted to cold high altitude environments, but it is adapted to more acid soils. *Vicia sativa* subsp. *amphicarpa* is common on heavily grazed marginal lands in arid zones. In production systems one species cannot easily be substituted for another, and therefore, it is important to preserve as much of the diversity as possible.

Since its inception, the International Center for Agricultural Research in the Dry Areas (ICARDA) has conducted over 40 collection missions, in collaboration with national programs of WANA, for the conservation of genetic resources of pasture and forage legumes (Bounejmate *et al.* 1998). Though much germplasm has been collected, many researchers express a growing concern that the rate of genetic erosion of these resources has been accelerated due to the intensification of grazing, adoption of improved varieties, industrialisation and urbanisation. The International Board of Plant Genetic Resources Working Group on Forages in the Mediterranean Region and Adjacent Arid/Semi-arid Areas (Anon. 1985) reviewed the present situation of genetic diversity of pasture and forage species in the Mediterranean and concluded that there were high levels of genetic erosion in many species and that there was great potential for many of these species to continue to suffer high rates of genetic erosion. Thus, there is a clear need to collate and quantify the extent of that erosion.

2. CONCEPTS OF GENETIC EROSION

2.1 Diversity

Diversity is the variability or the existence of alternative forms (IBPGR 1991). In genetic resources there are two generally used conventions to quantify diversity: genetic diversity and species diversity. Genetic diversity and species diversity both concern variation in genetic information; species diversity additionally considers ecological diversity.

Species diversity is a concept further developed from the concept of distribution of species in different geographical regions (ecogeography), and is commonly used by forestry scientists in dealing with issues of clear-cutting and meeting governmental regulations concerning maintaining diversity in forests. Species diversity is defined as the average rarity of a species at a given location (community), that is, proportional abundance (Patil and

Taillie 1982). In the most diverse community most species will be rare, the highest diversity would be with many species, each at a low frequency.

Genetic diversity is a further measure of variability of genetic information. For genetic resources the term genetic diversity is defined as "the range of a gene pool; the amount of genetic variation present in a population or species as a consequence of its evolutionary pathways" (IBPGR 1991). This concept can be applied to various levels of taxonomic unit as well as to a geographical context.

2.2 Definition of Genetic Erosion

There are few definitions of genetic erosion to be found, though the tern is often used in many papers. The Elsevier Dictionary of Plant Genetic Resources (IBPGR 1991) defines genetic erosion as "The loss of genetic material (genes, genotypes) from individuals or populations". This definition does not adequately describe what genetic erosion is, nor does it describe its dynamics. Genetic erosion is a process of wearing away the genetic resources of the biosphere. It is an ongoing process where the dynamics change, and not only refers to the loss of genes and genotypes, but also refers to the loss of habitat of a specific taxa or the loss of a taxa completely.

One major factor in defining genetic erosion is to determine the baseline of the measurement under discussion. To be accurate, any statement concerning genetic erosion should describe the baseline, as it enables genetic erosion to be quantified over a specified period of time. Equally important in quantifying genetic erosion is the need to describe the context that is being used to measure genetic erosion from a particular baseline. Usually this is a combination of taxonomy and geography, such as stating that the species diversity of a particular genus has been reduced in a country or region, or that the genetic diversity of a particular species has been reduced in a particular area of a country.

2.3 Measurement of Genetic Erosion

There are various methods of measuring genetic erosion. These range from subjective observations to the newest techniques of molecular characterisation.

The most common form of information on genetic erosion found in the literature is of the subjective type, and often refers to species abundance observed during collection missions, where the species are found to be rare. Often this type of measurement refers to loss of habitat and makes the connection to loss of species. Many statements can be found that grazing pressure was high, or that the land has been changed into intensive agricultural use, such as orchards, where pastures existed previously. Other factors noted include the development of man-made lakes, urban sprawl, and tourist developments. From these assessments it can be concluded that the loss of habitat can lead to either a complete loss of a species, or a reduction in its distribution. A reduction in a species distribution leads to further degradation of the genetic diversity of a species.

A more factual assessment is the use of floras to determine the distribution of a species which is compared to its current range following an ecogeographical survey. This ecogeographical approach can give a quantitative assessment of the loss of a species or the reduction of its distribution. Again, the conclusion is made that a reduction of the range of a species is correlated to a reduction in its genetic diversity. A quantitative ecological approach for ecological diversity is to calculate parameters of species diversity. The simplest application of this approach is the use of species counts at various locations. A more analytical approach is obtained by calculating parameters of diversity from species distributions. The most common of these are; the species count index (Lewis *et al.* 1988), the Simpson index (Simpson 1949) and the Shannon-Weaver index (Bowman *et al.* 1971) of species diversity. These are calculated as follows:

Species Count: $\Delta(C) = s - 1$

Shannon Index $\Delta(C) = -\Sigma_{pi} \ln (p_i)$

Simpson Index $\Delta(C) = 1 - \Sigma p_i^2$

where $\Delta(C)$ is the estimated species diversity for a specific location (community), *s* is the number of species at a specific location, and p_i represents the proportional abundance (frequency) of a particular species at a specific location which is summed over all species at a location (Σ). All three parameters assign a species diversity of zero to a single-species community. An additional refinement of assessing species diversity has been the development of the concept of 'diversity profiles' (Patil and Taillie 1982, Swindel *et al.* 1984, Lewis *et al.* 1988). Diversity profiles compare the diversities in communities using plots of left tails sums of ranked proportional abundance of two communities relative to each other. These profiles are useful for illustrating intrinsic diversity orderings, the relative contribution of abundant species and the relative contributions of rarer species to diversity. They can also show trends over time as affected by abundant and rare species.

To measure the variation of genetic diversity within a taxonomic unit, or a combination of a taxonomic unit and a geographical location, several parameters of genetic diversity have been developed. The purpose of these is to measure genetic diversity as close as possible to the DNA level. These parameters have been applied to categorical morphological data, categorised quantitative data, isozyme allelic frequencies, and more recently, to the products of molecular markers such as Random Fragment Length Polymorphism's (RFLP's) and Random Amplified Polymorphic DNA (RAPD's). The two most commonly used parameters for this are Nei's genetic diversity index (Nei 1973) and the Shannon-Weaver information index (Brown and Weir 1983). These parameters are defined as:

Shannon-Weaver information index: $GD_s = - \Sigma_{pi} \ln(p_i)$

Nei's genetic diversity index: $GD_n = 1 - \Sigma p_i^2$

where, GD is the estimated genetic diversity for a population or group of populations based on taxonomic ordering, geographical region, or a combination of these, p_i is the frequency of a particular allele or class (which could be categorical data, normalised quantitative data or presence or absence of a particular band in a RAPD profile, etc.) which is summed (Σ) over all populations or individuals for the grouping in question (1).

Both parameters assign a genetic diversity of zero to a group with only one category. The above formulae for Nei's genetic diversity index and the Simpson index are genetic and ecological versions of the same calculations. In the former the frequency is of alleles, RAPD bands or morphological categories and in the latter the frequencies are proportions of species. For both species diversity and genetic diversity the Shannon-Weaver formulae are equivalent. In both ecological and genetic diversity measures these formulae have the same characteristics. Nei's genetic diversity and Simpson's index are relatively insensitive to sample size; however, they both measure richness of forms and are largely dependent on the frequencies of the two most common forms (whether genetic information or species) and mainly emphasise number of forms (richness). The Shannon-Weaver Index is sensitive to both the relative proportions of forms, evenness, in addition to the number of forms, richness, and will be largest when the largest number of forms are equally frequent. Hutcheson (1970) has devised a test for comparing Shannon-Weaver diversity measures.

3. ASSESSMENT OF PASTURE AND FORAGE GENETIC EROSION IN THE MEDITERRANEAN

3.1 Contributing Factors

Most factors contributing to genetic erosion can be attributed to disturbance or change of the natural environment (IBPGR 1986, Goodrich 1987, Guarino 1995). In the Mediterranean, the main factors have been a loss of habitat, intensification of agriculture and in particular, an increase in grazing pressure (Bounejmate *et al.* 1992, Parkin 1966, Carter 1981, Francis 1987, Harlan 1983). Seed size is also important when quantifying genetic erosion, as larger seeded species are considered to be more vulnerable (Smith and Robertson 1991). In Morocco, Bounejmate *et al.* (1992) found a correlation in *Medicago* species between seed size and frequency of occurrence. Small-seeded medics such as *M. polymorpha* were more frequent than large-seeded species such as *M. scutellata* and *M. intertexta.*

3.2 Identifying Germplasm at Risk

The risk of genetic erosion can be 'quantified'. In general, cultivated landraces of feed legumes are more at risk than wild or weedy populations of pasture species, as they are entirely dependent on humankind for their continued existence. Higher-input crops, such as vetch and alfalfa, grown in favourable conditions are generally more affected than low-input crops grown in marginal conditions. Of the cultivated landraces, those species of higher economic value and grown on larger areas are usually at a greater risk than minor crops, since they will attract more attention from breeders and extension agents and offer the farmer more incentive to risk higher inputs. However, the reduction in cultivation of a minor crop increases its risk considerably. The risk of genetic erosion of wild and weedy species is related to the usefulness of its habitat. Factors such as proximity to large human populations, use of herbicides by farmers, and grazing pressure affect the risk of genetic erosion. Methods for assessing the danger of erosion for a particular species have been developed (IBPGR 1986, Goodrich 1987, Guarino 1995). These models allow greater weight for factors considered more important. When data are filled for all factors an overall score of risk of genetic erosion can be determined. These models require reliable data for the different factors and are subject to the various weights given to different factors.

3.3 Difficulties in assessment

As stated earlier, genetic erosion has to be defined relative to a baseline and in a particular context. Quantifying this baseline with pasture and forage species in the Mediterranean is difficult. Even where there are data on the distribution of species, such as in floras, they often have incomplete coverage and do not specify areas where species are found but usually generalise with statements such as the species is found in the northern part of the country or it is found in low-rainfall environments, for example. The difficulty in using floras to develop baselines for assessing genetic erosion in a quantitative manner is illustrated by Figures 1 and 2. Figure 1 shows the species diversity of *Medicago* using the 'Flora of Tunisia' (Pottier-Alapetite 1979). When compared to species diversity derived from recent ecogeographic surveys in Tunisia (Figure 2) it appears that the species diversity has increased over the last 40 years, which is not the case. The problem is the lack of detailed distribution data in floras. Often there are variations in the naming of taxa, with different sources using different names for the same taxa, and there is the problem of incorrect identification of collected samples. It is rare to find data on the relative abundance of species at different locations, usually the best that can be obtained is the presence and absence of species.

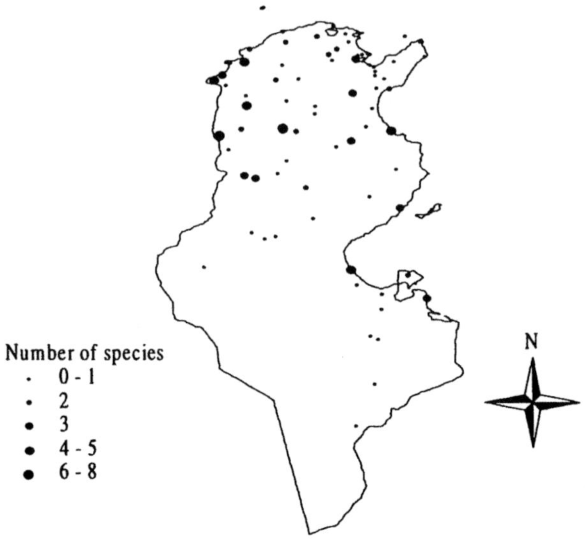

Figure 1. Species diversity of *Medicago* in Tunisia (developed from Pottier-Alapetite 1979); size of circles represent number of species at a location, 1 to 8

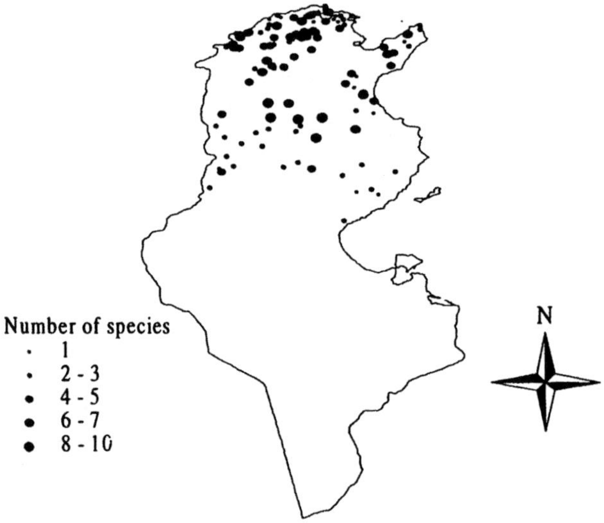

Figure 2. Species diversity of *Medicago* in Tunisia (derived from germplasm collection missions since 1991); size of circles represent number of species at a location, 1 to 10

Even in ecogeographic surveys data on the relative proportion of species is rarely recorded. Species count is therefore the only measure of species diversity that can be used and this often results in entirely different results in different years, Table 2 summarises the distribution of medic species in Morocco obtained from floras and from 10 different collecting missions (Bounejmate *et al.* 1992). If any one mission was taken to represent present-day distribution, the assumption could be made that genetic erosion has occurred. But when aggregated, these missions support the same number of species of medic in Morocco.

Table 2. ***Medicago*** **species collected in Morocco at 10 collecting missions, and according to Negre (1959) (from Bounejmate *et al.* 1992)**

Species	Negre (1959)*	1	2	3	4	5	6	7	8	9	10	TM*
01-M. aculeata Willd.	+	+	+	+	+	+	+	+	+	+	+	+
02-M. arabica (L.) Huds.	+					+			+			+
03-M. constricta Dur.				+							+	
04-M. intertexta (L.) Miller	+				+	+		+			+	+
05-M. laciniata (L.) Miler	+				+	+	+	+	+	+	+	+
06-M. littoralis Rhode	+			+	+	+	+	+	+	+	+	+
07-M. minima (L.) Bart.	+				+	+	+	+	+	+	+	+
08-M. murex Willd.	+					+			+		+	+
09-M. orbicularis (L.) Bart.	+			+	+	+	+	+	+	+	+	+
10-M. polymorpha L.	+				+	+	+	+	+	+	+	+
11-M. praecox DC.								+				+
12-M. rigidula (L.) All.	+				+	+		+				+
13-M. rotata Boiss.	+							+				+
14-M. sauvagei Nègre	+				+							+
15-M. scutellata (L.) Miller	+			+							+	+
16-M. secundiflora Durieu	+											+
17-M. tornata (L.) Miller	+			+		+		+	+		+	+
18-M. truncatula Gaertn.	+	+	+	+	+	+	+	+	+	+	+	+
Total	16	2	5	9	14	11	7	12	9	7	11	18

N* Indicate the species listed by Nègre (1956) and Nègre (1959) for Morocco.
TM represent the total observed within Morocco from all sources.
The numbers 1 to 10 correspond to various collections.

3.4 Extent of Genetic Erosion

Most examples of genetic erosion that can be substantiated are where the extinction of a species has been documented. These have usually been of species which have a limited distribution and have been relatively rare. One such case is reported by Lesins and Lesins (1979) who discovered that an endemic of Albania, *Medicago pseudorupestris,* can no longer be found. They report the loss of the species and the possible consequences of this. Through the use of herbarium specimens they found that the species would have been of interest as a possible source of resistance of insects due to its thick hair cover.

Most reports of genetic erosion are speculative and very general. A few examples are given below;

- "genetic erosion in general and especially for wheat was found to be higher than was originally thought" (Perrino *et al.* 1986) in Morocco;
- "results suggest that genetic erosion is occurring but occasional sites can be found where medics are abundant" (Bounejmate *et al.* 1992) in Morocco;
- "twenty species of *Vicia* have previously been described from Morocco. In our study, only 11 were positively identified. It is apparent, nevertheless, that species such as *Vicia narbonensis, V. erratifolia, V. cedrotorum, V. vicioides, V. durandii* and V. *glauca* are at least rare in Morocco, some of these species may well be victims of genetic erosion" (Francis *et al.* 1994) in Morocco;
- In Italy, Hammer and Perrino (1995) found *Vicia articulata* had disappeared from south Italy and Sicily;
- Annicchiarico (1990) found that the production of seed of traditional Ladino clover landraces by farmers in the Po Valley in Italy had drastically decreased because of the expansion of maize production for silage. He found that landraces of Ladino clover from

the province of Milan had been lost. Both of these losses are caused by the disappearance of old permanent meadows;

- Cowling (1986) found severe genetic erosion of *Lupinus* species in north-eastern Greece. *L. angustifolius* had disappeared from Thrace. The reason for the genetic erosion was suggested to be due to an increase in wheat production and from over-grazing;
- Gustave Gintzburger (*pers comm.*) states that *Medicago rigidula* can be found in the Artemisia Herba-Alba range of the northern foothills of the Jebel Sinjar in Iraq but not in the same environment of Jebel Abdal Aziz in Syria only 80 km west (though it is reported in floras).

There are reports (Ehrman and Cocks 1990), however, which question the extent of genetic erosion in the case of some genera or species of pastures and forages in the eastern Mediterranean.

Hammer *et al.* (1996) have estimated genetic erosion of some pasture and forage species in Albania and southern Italy based on the number of landraces found compared to their former number. In Albania they found 100% genetic erosion of *Vicia sativa*. In southern Italy they found genetic erosion of *Hedysarum coronarium* L. and *Medicago sativa* L. of 100%, *Pisum sativum* L. of 67% but no genetic erosion of *Lupinus albus* L.

4. CONSEQUENCES OF IGNORING GENETIC EROSION

The consequences of ignoring genetic erosion for agriculture can be difficult to see in the short term. Often, the consequences can be hidden. The loss of a potential pasture species from genetic erosion may never be recognised if the species is lost before there is any evaluation of its potential for use. The same can be said for reduction of the genetic base for supplying important characteristics, such as tolerance or resistance to biotic and abiotic stresses, hardseededness, winter growth and summer dormancy. Examples could be cited where a source of resistance was found in only one accession after screening many thousand of accessions. The loss of species diversity could be catastrophic to ecosystems, and a more direct consequence of genetic erosion in the form of habitat clearance can easily be seen with resulting soil erosion, sedimentation of dams, floods and other natural disasters.

5. FUTURE NEEDS

There is a need to develop a baseline of present diversity of species and of the genetic diversity of important species. Species diversity needs to be quantified for the important pasture and forage species by determining their precise distributions in natural ecosystems and agro-ecosystems. This should be done for present-day ecosystems to give a baseline for measuring present rates of genetic erosion. Where older data exist, they should be used to try to develop baselines to estimate the extent of genetic erosion that has occurred up to the present time.

An example of species diversity measures are given in Table 3 using the *Medicago* species recorded in Morocco. This table shows differences in species diversity within Morocco, where according to the Shannon-Weaver index the region of Oujda shows the highest species diversity and the Tanger region shows the lowest diversity. In Figures 3 and 4, species diversity, measured as species count is shown using recent collection missions for *Vicia* and *Medicago* species, respectively. The larger the circle at a given site, the greater the number of species. Regions with greater species diversity can be identified from these maps. For example, in *Vicia* there is greater species diversity in the northern part of Morocco and in the central region of Fez and Meknes.

Table 3. Species diversity in various regions of Morocco (developed from Bounejmate *et al.* 1992)

Region	Species count	Simpson's index	Shannon-Weaver
Oujda	6	0.8040	1.7539
Marrakech	6	0.7502	1.6227
Beni-Mellal	7	0.8166	1.7011
Settat	7	0.8228	1.5246
Tanger	8	0.7673	1.3518
Middle-Atlas	5	0.7910	1.5364

A precise method of determining the genetic diversity of species held in gene bank collections would be using molecular markers, as these measure variation at the nucleotide level. Using collections from the same region the levels of genetic diversity could be compared to determine if there have been any changes or shifts over time. Correlations can be made with land use, and any areas where special care is required can be earmarked for *in-situ* conservation projects. The effectiveness of in-situ conservation sites can also be monitored using this method.

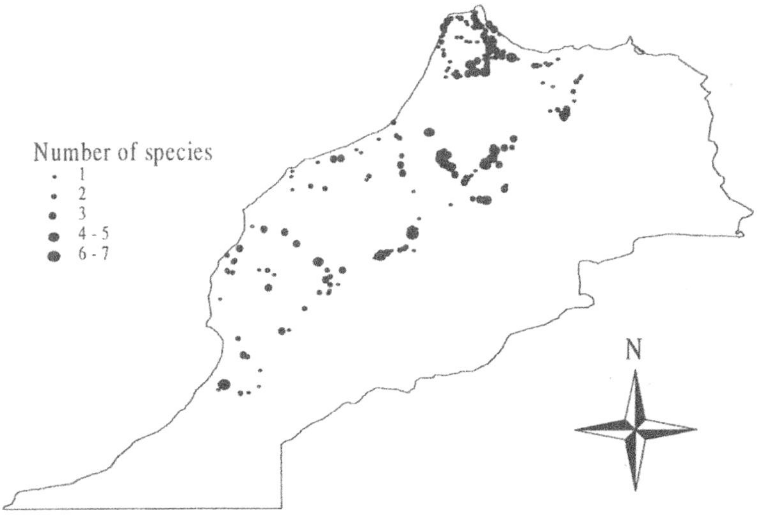

Figure 3. Species diversity of *Vicia* in Morocco (developed from data from recent germplasm collection missions in 1992 1993 and 1994); size of circles represent number of species at a location,1 to 6

In all studies of genetic diversity using gene bank collections, the results must be interpreted with care. The whole genetic diversity of a population may not have been sampled at the time of collection or the diversity of the population may have been changed during regeneration. Also, in order to determine the levels of genetic diversity using molecular markers, polymorphism's have to be found, and in some species this can be difficult.

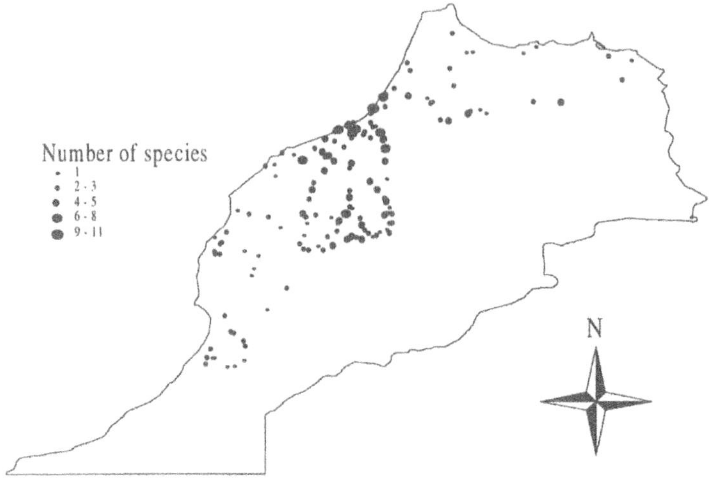

Figure 4. Species diversity of *Medicag* in Morocco (developed from data from recent germplasm collection missions in 1974 to 1994); size of circles represent number of species at a location, 1 to 11

6. CONCLUSION

Genetic erosion has been observed for pasture and forage legumes in the Mediterranean region. There has been complete loss of species from specific regions and also a reduction in the genetic variability for specific species as their area has been reduced or the replacement of traditional landraces by improved varieties. The major factors contributing to genetic erosion have been a loss of habitat, intensification of agriculture and an increase in the grazing pressure. Species which are sown for feed have lost traditional landraces as they have been replaced by improved varieties. Seed size is also important when quantifying genetic erosion, as larger seeded species are considered to be more vulnerable. However, much of the data on genetic erosion is spotty and often circumstantial based on personal observations rather than quantifiable indices of diversity which are available. There is an urgent need for the establishment of baselines to measure future genetic erosion of pasture and forage legumes in the Mediterranean region using surveys of species abundance in each country of the region.

REFERENCES

Annicchiarico, P. (1990). Collecting Ladino white clover ecotypes in the Po Valley. *Plant Genet. Res. Newslett.* **85,** 25-27.

Anon, (1985). Forages for Mediterranean and adjacent arid/semi-arid areas. Report of a Working Group, held at Limassol, Cyprus, 24-26 April 1985. IBPGR Secretariat, Rome.

Bounejmate, M., Robson, A.D. and Beale, P.E. (1992). Annual *Medicago* species in Morocco I. Species and their Abundance. *Aust J Agric. Res* **43,** 739-749.

Bounejmate, M., Gintzburger, G., Christiansen, S., Robertson, L.D., Abd El Moneim A. and Konopka, J. (1998). ICARDA invests heavily in forage and pasture legume genetic resources conservation and utilisation for the future of semi-arid zones in West Asia and North Africa. In 'Proceedings of a CLIMA Workshop on Genetic Resources of Mediterranean Pasture and Forage Legumes', 14-18 October 1996. (Eds. S.J. Bennett and P.S. Cocks) (Kluwer: The Hague).

Bowman, K.O., Hutcheson, K., Odum, E.P. and Shenton, L.R. (1971). Comments on the distribution of indices of diversity. *Stat Ecol* **3**, 315-366.
Brown, A.D.H. and Weir, B.S. (1983). Measuring genetic variability in plant populations. In 'Isozymes in Plant Genetics and Breeding, Part A'. (eds. S.D. Tanksley and T.J. Orton) pp. 219-239.(Elsevier Science: Amsterdam).
Carter, E.D. (1981). Seed and seedling dynamics of annual medic pastures in South Australia. Proc. XIV Int. Grassland Congress, pp.447-450.
Cowling, W.A. (1986). Collection of wild *Lupinus* in Greece. *Plant Genet Res. Newslett.* **65**, 20-22.
Ehrman, T. and Cocks, P.S. (1990). Ecogeography of annual legumes in Syria: distribution patterns. *J Appl Ecol* **27**, 578-579.
Francis, C.M. (1987). Morocco, a plant collection tour. A report compiled for the Grain Research Development Corporation, Agriculture Western Australia, Perth.
Francis, C.M., Bounejmate, M. and Robertson, L.D. (1994). Observations on the distribution and ecology of *Vicia* and *Lathyrus* species in northern Morocco. Al *Awamia* **84**, 17-27.
Goodrich, W.J. (1987). Monitoring genetic erosion: detection and assessment. IBPGR/87/33. IBPGR, Rome.
Guardino, L. (1995). Assessing the threat of genetic erosion. In 'Collecting Plant Genetic Diversity: Technical Guidelines' (Eds. L. Guarino, V.R. Rao, and R. Reid). pp. 67-74 (CAB International; Wallingford, UK).
Hammer, K. and Perrino, P. (1995). Plant genetic resources in south Italy and Sicily – studies towards *in situ* and on-farm conservation. *Plant Genet. Res. Newlett.* **103** **19**-23.
Hammer, K., H. Knüpffer, L. Xhuveli and Perrino, P. (1996). Estimating genetic erosion in landraces – two case studies. *Genet. Res. Crop Evol.* **43**, 329-336.
Harlan, J.R. (1983). The scope for collection and improvement of forage plants. In 'Genetic Resources of Forage Plants' (Eds. J.G. Mclvor and R.A. Bray). pp. 63-84 (CSIRO: East Melbourne, Australia).
Houérou, le N.N. (1991). Forage species diversity in Africa: an overview of the genetic resources. In 'Crop Genetic Resources of Africa, Vol. 1'. (Eds. F. Attéré, H. Zedan, N.Q. Ng and P. Perrino). pp. 99-117. (IBPGR: Rome).
Houérou, le H.N. and Boulos, L. (1991). Bioclimatic and phytogeographic characteristics of the and rangelands of northern African and the Near East. In 'IVth International Rangeland Congress Vol 1, Montpellier 22-26 April 1991.' (Eds. A.N. Gaston, M. Kernick and H.N. le Houérou.) pp. 535-542. (CIRAD: Montpellier, France).
Hutcheson, K. (1970). A test for comparing diversities based on the Shannon formula. *J. Theor. Biol.* **29**, 151-154.
IBPGR. (1986). Genetic Erosion: monitoring and assessment. AGPG:IBPGR/86/89. IBPGR, Rome.
IBPGR (1991). 'Elsevier's Dictionary of Plant Genetic Resources'. Elsevier Science Publishers, B.V., Amsterdam.
Lesins, K.A. and Lesins, I. (1979). Genus *Medicago* (Leguminosae): A Taxonomic Study. (Dr W. Junk bv Publishers: The Hague).
Lewis, C.E., Swindel, B.F. and Tanner, G.W. (1988). Species diversity and diversity profiles: Concept, measurement, and application to timber and range management. *J. Range Man.* **41**, 466-469.
Negre, R. 1956. Les luzernes du Maroc. *Travaux de l'Institut Scientifique Cherifien. Maroc, Serie Botanique* **5:xxii**, 1-120.
Negre, R. 1959. Revision del *Medicago* d'Afrique du Nord. *Bulletin de la Societe d'Historie Naturelle d'Afrique du Nord* **50**, 267-314.
Mathison, M.J, (1983). Mediterranean and temperate forage legumes. In 'Genetic Resources of Forage Plants' (Eds. J.G. Mclvor and R.A. Bray). pp. 63-84 (CSIRO: East Melboume, Australia).
Nei, M., (1973). Analysis of gene diversity in subdivided populations. *Proc. Nat. Acad. Sci.* **70**, 147-157.
Parkin, R.J. (1966). Stocking rates on Cyprus barrel medicin the northern wheatbelt. *J AgricWA* **7**, 276-279.
Patil, G.P. and Taillie, C. (1982). Diversity as a concept and its measurement. *J Am. Stat. Assoc.* **77**, 548-561.
Perrino, P., Polignano, G.B. Sui-Kwong, J. and Khouya-Ali, M. (1986). Collecting Germplasm in Southern Morocco. FAO/IBPGR *Plant Gen. Res. News.* **65**, 26-28.
Pottier-Alapetite, G. (1979). 'Flore de la Tunisie, Vol 1: Angiosperms - Dicotyledons: Apteatles – Dialypetales'. Publications Scientifiques Tunisiennes No. 644. Imp. Officielle de la Republique Tunisie.
Simpson, E.H. (1949). Measurement of diversity. *Nature* **163**, 688.
Smith, A. and Robertson, L.D.(Eds.) (1991). 'Legume Genetic Resources for Semi-Arid Temperate Environments'. Proceedings of an International Workshop on Genetic Resources of Cool-season Pasture, Forage, and Food Legumes for Semi-arid Temperate Environments, Cairo, Egypt 19-24 June 1987. (ICARDA: Aleppo, Syria).
Swindel, B.F., Conde, L.F. and Smith, J.E. (1984). Species diversity: Concept, measurement, and response to clear-cutting and site-preparation. *For Ecol. Man.* **8**,11-22.
Zohary, D. and Hopf, M. (1988). 'Domestication of plants in the old world'. (Clarendon Press: Oxford, UK).

CHAPTER 7

Sampling wild legume populations

D.R. Marshall[1] and A.H.D. Brown[2]
[1]*Plant Breeding Institute, The University of Sydney, Sydney N.S.W. 2006*
[2]*CSIRO, Division of Plant Industry, GPO Box 1600, Canberra ACT 2601*

Abstract. The sampling of wild legume populations poses special challenges for plant collectors. Such populations often vary markedly in their density, their age structure, and their flowering and ripening times, making them more difficult to sample than crop plants. In addition, many species either shatter or bury their seeds to aid dispersal or avoid predation. It is important, therefore, that the limited time and resources at the disposal of collectors of these species be as effective as possible in collecting the maximum amount of useful variability. We argue that locally common alleles - alleles with a frequency of >5% in a set of adjacent populations - are the prime target for collectors interested in sampling genes for use directly or in breeding programs.

A basic sampling strategy, that maximises the number of locally common alleles collected is (i) to collect 50-100 individuals per site; (ii) to sample individuals at random at each site or microsite visited; (iii) to sample sufficient seeds or other propagates per plant to assure representation of each original plant in all duplicates; (iv) to sample as many sites as possible with 50 per region a notional aim; and (v) to cluster sampling sites in obviously different edaphic and climatic zones so as to sample both the geographic and micro-geographic components of variation. When there is insufficient easily procurable material to collect 50 plants, more sites should be sampled.

This basic strategy is recommended when information on the mating system and population genetic structure of the target species is unknown. Where this information is available it is possible to refine the basic strategy to take this information into account. We disagree with recent arguments in the literature that downplay population divergence, and limit total sample size to fewer than 200 per species.

1. INTRODUCTION

The sampling of natural populations of wild plants, particularly legumes, poses significant challenges for plant collectors. As noted by Frankel (1970) and Marshall and Brown (1983) natural plant populations differ from cultivated populations in a number of ways which can make the task of the plant collector more difficult :-

(i) Natural populations generally show greater differentiation on both a geographic and a micro-geographic scale than crop plants, since they are not subjected to the homogenising influence of an annual re-sowing and bulk harvest.
(ii) Wild plants are often genetically more variable for traits such as heading, flowering and reliability of seed set than crop plants where uniformity in these characters is of compelling economic importance.
(iii) The population density of wild species can vary markedly in comparison with crop plants where this factor is under the control of farmers. Wild species are generally members of mixed communities rather than monocultures.
(iv) Wild plants display a greater range of breeding systems and, more particularly, contain a greater proportion of outbreeding species than crop plants which have been subjected to intense selection for uniform and reliable seed set.
(v) Most crop plants are annuals, whereas many wild legume species are perennials and their populations possess a complex age structure.

S. J. Bennett and P. S. Cocks (eds.), Genetic Resources of Mediterranean Pasture and Forage Legumes, 78–89.

(vi) Wild species generally disperse, or in the case of legumes occasionally bury, their seed as they mature. This may make the collection of even modest sized samples very difficult.

In recent years, many authors have considered the question of the most appropriate sampling strategies for plant collectors (Oka 1975; Qualset 1975; Marshall and Brown 1975 1981 1983; Damania and Porceddu 1983; Yonezawa 1985; Namkoong 1988; Chapman 1989; Damania 1989; Porceddu and Damania 1995b; Brown and Marshall 1995). However, few of these studies have been targeted at wild plants and for those that were, the focus has been on the conservation of threatened species, rather than on economic use. In addition, the conclusions arising from these studies have often varied markedly.

One example of the disparity in conclusions that can occur is the difference in the recommended number of plants to collect per site visited, which has ranged from 1 to 500. A second, even more startling example, comes from two recent papers. In the first Van Rheenan *et al.* (1993) extended the one locus model of Marshall and Brown (1975) to several (two to eight) independent loci and concluded that to be 95% sure of conserving all alleles with frequency >0.05 required increasing sample sizes. For highly inbreeding or apomictic species, they calculated that sample sizes of 59 for 1 locus, 1195 for 2 loci and 23510 for 3 loci were required to collect all combinations of variants, where the rarest allele at each locus occurred at a frequency of 0.05. In sharp contrast, Lawrence *et al.* (1995a and b) from an analysis of a similar model concluded "A sample of about 172 plants, drawn at random from a population of a target species, is of sufficient size to conserve at a very high probability, all or very nearly all of the polymorphic genes that are segregating in the population, provided that their frequency is not less than 0.05, irrespective of whether the individuals of the species set all of their seed by self- or by cross-fertilisation or a mixture of both. When samples are taken from a number of populations the size of the sample drawn from each need be no larger than 172 divided by the number of populations visited." It is worth noting the authors assumed 20,000 loci were polymorphic in the target species.

The aim of this chapter is to critically review the current literature on germplasm sampling and to use this to develop guidelines on how to respond, in terms of sampling strategy, to the practical problems faced by field collectors of wild legume germplasm. We would emphasise that the final decision as to the sampling strategy to use in any particular case must rest with the collector. No theory, or analyses of the results of previous missions, can provide ironclad rules that apply in all situations. They can, however, provide the sorts of information collectors need to make rational choices when faced with difficult decisions in the field.

2. THEORETICAL BACKGROUND

We assume here;

(i) that the collector is sampling populations to obtain material for use directly or in breeding programs. The collector is therefore interested in collecting useful genes - genes that have the potential to improve the agronomic performance of a species or its range of adaptation.

(ii) the target species is not rare and endangered. Plant explorers sampling such species face special problems (see Brown and Briggs 1991; Brown 1992). We assume there is sufficient material in the target region so that the collector has a choice with respect to the numbers and composition of the samples collected.

(iii) the species is not genetically depauperate and does not consist of one or a few fixed genotypes as is occasionally the case with highly inbreeding or apomictic exotics. We assume the target species , in common with most wild species in their native habitats, contains substantial levels of genetic variation.

(iv) the collector has limited time and resources and there are strict limits to the amount of material which can be collected and used.

Under these assumptions the problem facing the plant explorer is to collect the maximum amount of useful variation within a specified and limited number of samples (Allard 1970; Marshall and Brown 1975 1983). However, in order to determine how to collect, we must first define what to collect - that is, what is meant by 'useful genetic variability' and how it is to be measured.

3. MEASURING GENETIC VARIATION

Genetic variation can be described by a large array of parameters e.g. gene and genotype frequencies, heterozygosity levels, diversity indices, and disequilibrium coefficients (see Weir 1990). We have argued earlier (Marshall and Brown 1975 1983) that the actual number of alleles is the most appropriate measure of genetic diversity and is the parameter to maximise in the collected samples. This parameter provides a direct measure of the variation at each locus within and between populations unlike parameters based on quantitative characters that measure only the proportion of genetic variation expressed phenotypically in a particular environment or set of environments and confound the effects of individual loci. Further, as an ideal, we are interested in preserving at least one copy of each allele regardless of its frequency, because later users can adjust the frequencies of specific alleles. Thus the number of alleles in a sample or its allelic richness is a direct measure of its value (Marshall and Brown 1995).

4. USEFUL GENETIC VARIATION

We have also argued previously that since the aim of sampling is to ensure the current and future availability of adequate germplasm, and since these needs are diverse and unpredictable, no particular group of genes can be regarded as generally more useful and deserving of a higher priority in sampling. Rather the aim should be to preserve as many variants as possible at each locus. The issue of defining "useful" variation that deserves priority in collecting reduces, under this argument, to discriminating among the alleles that occur at each locus. This led us (Marshall and Brown 1975) to make two arbitrary divisions of alleles at each locus based on their frequency of occurrence within and between populations, viz. those that are common (frequency >5%) and those that are rare (frequency <5%) and those that are widespread occurring in many populations, or local, and occurring in one or a few adjacent populations. Of the four classes of alleles generated by these subdivisions we suggested that it was the locally common alleles that should receive priority in sampling. In support of this viewpoint we argued that widespread common alleles are inevitably collected regardless of sampling strategy. Further, the probability of including alleles that are rare in any given population but occur in many populations in a sample, depends largely on the total number of plants collected and is, therefore, also largely independent of sampling strategy. Finally, alleles that are rare in any given population and also localised in their occurrence will be difficult to collect, and presumably represent recent mutants or deleterious genes maintained at low frequency and, therefore, of less interest to breeders. In contrast, locally common alleles will also be more difficult to collect than their widespread counterparts. However they are presumably adaptive and selectively maintained in populations. They also presumably represent adaptations to significant local variations in the biotic or abiotic environment, and are therefore likely to be of special interest to agronomists and plant breeders.

5. HOW TO COLLECT

The key elements in any sampling strategy are :
(i) the number of plants to sample per site;
(ii) the method of sampling;
(iii) the number and type of propagule per plant to collect;
(iv) the number and distribution of sampling sites within the target area.

Each of these will be considered in turn below.

However, before we do this, it should be stressed that information on the kinds and amounts of genetic variation in target species populations and its distribution in the target region is critically important in developing fully optimal sampling strategies. Knowledge of the genetic structure of plant populations has increased markedly in recent years (e.g. Doebley 1989; Hamrick and Godt 1989; Frankel, Brown and Burdon 1995). This knowledge provides a more secure platform on which to develop widely applicable sampling technologies. However, detailed information on the population structure of particular target species is usually lacking. Indeed, for many wild legumes we often lack adequate information on such basic traits as the mating system. We will, therefore, first define sampling procedures for use when the collectors lack information on the genetic structure and mating system of the target species. We will then consider how these procedures vary when these assumptions are relaxed.

5.1 Number of plants per site

Since excessive sampling at any one site limits an explorer's opportunities to collect locally common variants at other sites, the sample at each site should be kept as small as practicable. We (Marshall and Brown 1975) defined the optimum sample size as the number of plants required to collect, with 95% certainty, at least one copy of all the common alleles - those with frequency >0.05.

Analysis of a wide range of diallelic and multiallelic models suggests that a sample of about 60 to 80 unrelated gametes from a population is sufficient to allow this objective. This would be assured by collecting and bulking seeds or vegetative material from 30 to 40 randomly chosen individuals in a fully outbreeding sexual species, or from 30 to 40 random genotypes in an apomictic species, or from 60 to 80 random individuals in a self-fertilising species. We therefore set a sample of 50-100 individuals from each population as a realistic and pragmatic benchmark.

It should be emphasised, however, that this number and the criteria on which it is based are arbitrary and can, and have been, questioned on both biological and philosophical grounds (see Marshall and Brown 1981 1983; Marshall 1989; Marshall and Brown 1995). It has frequently been suggested the number of plants per population collected should be substantially increased either to increase the chance of collecting rare alleles or allelic combinations (e.g. Qualset 1975; Krusche and Geburek 1991; Van Rheenan *et al.* 1993) or to increase the probability above 95% of capturing at least one copy of each allele (Lawrence *et al.* 1995a). However, we would stress that increasing the certainty level higher than 95% or dropping the critical allele frequency below 0.05 quickly and dramatically increases sample size with only marginal gains in sampling efficiency.

5.2 Method of sampling

Having settled on the number of plants to sample the next issue is to choose the individuals in the sample.

A much discussed question in this regard is whether individuals should be taken strictly at random or biased as to phenotype or microsite (e.g. Marshall and Brown 1975 1981; Porceddu and Damania 1992a). Random sampling is generally the most reliable and desirable method, particularly in species which show no or limited sub-population structure. In contrast, wild species often show local sub-population structure and hence stratified random sampling is appropriate for them (e.g. separate random sampling of individuals in different microsites). Biased sampling of rare phenotypic variants in a population is to be avoided, except when such plants clearly merit separate and distinct recognition (e.g. a rare disease-free individual in a heavily diseased field). Such samples may be taken in addition to, but not instead of, population samples, and should receive separate collecting numbers.

Strict randomness of sampling requires that every plant at the site has an independent and equally likely chance of inclusion in the sample. However, this is often impossible or impractical to achieve (but see Ward 1974). In practice, collectors usually sample at systematic or random intervals along a number of transects. Systematic sampling is easiest and spreads the sample over the population but can be biased if variation is periodic in the population (Brown and Briggs 1991). The starting-points and directions of the transects and the position of sampling points along them can be chosen according to a randomisation procedure. It is usually a good idea to keep a minimum distance between sampling points to avoid excessive sampling of closely related individuals and repeated sampling of clones.

5.3 Number and type of propagules per plant

A third decision for the plant collector is the kind(s) and amount(s) of material to be collected from each plant to be sampled.

As noted by Brown and Briggs (1991), collected material ranges from pollen, seeds and vegetative cuttings or propagules (such as tubers, bulbs or corms) to whole individuals. Collecting seeds differs from collecting vegetative material in a number of ways (Brown and Marshall 1995). In particular, it is more restrictive for the timing of the mission. Otherwise, seed as the organ of dispersal and storage, enjoys a number of advantages. For example, it is generally less bulky and easier to handle and store. Pollen is even less bulky, but its storage can be difficult and it is generally not possible to recover plants from pollen. Cuttings are often the most convenient vegetative samples to make, but can require the use of *in vitro* techniques.

The genotypes in a sample of vegetative material are exactly those of the sampled parent plants, whereas those in a seed sample depend on the breeding and pollination system. With uniparental reproduction (self-fertilisation or agamospermy), seeds will closely resemble the parent plant. In contrast, the seeds of an outbreeder will differ from the parent and show within-family diversity (Yonezawa and Ichihashi 1989). In outbreeding species, seeds from several fruits should be gathered from each individual sampled, if possible, rather than from a single fruit, to increase the diversity of genes in the sample. Also, fruits should be collected from all parts of the crown of trees, because these may have been pollinated by different pollen sources. Similar numbers of seeds should be collected from each individual sampled.

The need for sufficient material, rather than genetic principles, largely determines the number of propagules per plant to sample (Brown and Marshall 1995). If available, enough seeds or cuttings should be sampled to provide for the division of the samples among collaborators and other recipients and to avoid immediate multiplication. However, quarantine facilities or the space available in the collecting vehicle may limit the total volume of material that can be collected. When this is the case, it is generally better to maximise the number of plants sampled and reduce the number of propagules per plant.

5.4 Number of sites

Since the aim is to collect as many variants as possible in the target species, and since each new site offers the prospect of sampling a new set of locally common alleles, then the optimum number of sites to sample is the maximum possible. However, this begs the question: Is there a benchmark figure for the total number of sampling sites for each species and region? Brown and Marshall (1995) suggested that analogous statistical and cost/benefit arguments could apply at the mission level to those made at the population level. This would suggest that a set of about 50 sites comprises a justifiable sample. This would ensure that we would capture at least one copy of all locally common alleles that occur in more than 5% of populations with a reasonably high level of probability.

However, this can only be a weak guide. Unlike the members of a single population, the potential sites in a given area could conceivably be completely different from each other and yield no redundancy upon sampling. Moving on and collecting an individual at a new site is usually preferable to collecting an additional individual at the current site. Nevertheless, it is helpful to start with a specific target of 50 sites per species per region, and vary this target up or down when clear reasons exist for doing so.

5.5 Distribution of sites in the target area

Delimiting the population, and thus the sampling site, can be a problem in wild species collecting. This, plus the fact that populations of wild species are likely to be less dense, to vary greatly in heading data and seed ripening, and to shatter their seeds immediately they are ripe, makes them more difficult to sample than crop plants. As a result the distribution of sampling sites within the region of interest is likely to be of greater importance in wild plants than in crop plants.

Before the mission begins therefore, the collector should assemble the available information on the kinds of environments occurring in the target region and on the pattern of distribution of the target species. Based on these data, the region should then be roughly divided into a limited number of areas, clearly distinct because of ecological (i.e. physiographic, edaphic, climatic), botanical, agricultural or cultural differences. The total time available can then be divided among the areas according to travelling convenience, prevalence of the targets and any perceived or known differences in genetic diversity among areas.

It is now well established (e.g. Jain 1975; Bradshaw 1975; Snaydon 1978) that most patterns of genetic differentiation within plant species result from, and are strongly correlated with, environmental heterogeneity. To maximise the chances of collecting locally common alleles, the collector should therefore sample many diverse environments. As we noted previously (Marshall and Brown 1975), there are many ways of achieving this objective. In practice, however, the key decision is the relative emphasis to give to macro-geographic versus micro-geographic variation. If the collector's aim is to capture the maximum amount of genetic variation associated with broad geographic differences in environmental factors, then the sampling sites should be over-dispersed: that is, more or less evenly distributed over the target area. Alternatively, if the collector aims to capture variability associated with both geographic and micro-geographic differences of soil and climatic factors, then the sampling sites should be clustered in small groups (say three to five sites per group), and the groups distributed according to the obvious geographic heterogeneity in the environment.

The clustering of samples has several benefits in wild species apart from maximising the collection of common alleles associated with micro-geographic variation. It reduces the time of travelling between sites and allows more time for sample collection. It forces the explorer to search consciously for difference edaphic or climatic habitats. This avoids the possibility, which exists when only a single sample is taken per region, of unconsciously collecting samples from similar, say all lush or all arid, environments.

6. ADAPTING THE STRATEGY TO SPECIFIC TARGETS

The above analysis indicates that the basic sampling strategy for wild species is (i) to collect 50-100 individuals per site; (ii) to sample individuals at random at each site or microsite visited; (iii) to sample sufficient seeds or vegetative material per plant to assure representation of each original plant in all duplicates; (iv) to sample as many sites as possible, with 50 per region a realistic aim; (v) and to cluster sampling sites in obviously different edaphic and climatic zones so as to sample both the geographic and micro-geographic components of variation. However, this strategy is based on a number of assumptions, which if they are not met, can reduce the efficiency of sampling. Here we examine ways to modify the above procedure when some of the more important of these assumptions are not met.

6.1 Constraints on the number of plants sampled per site

An important, but not necessarily obvious, assumption implicit in the framing of the basic sampling strategy was that the effort required to sample an additional plant at a site is small in comparison to the effort required to sample an entirely new site. In other words, we assumed it would be relatively easy for the collector to sample 50-100 plants at each site. This assumption is usually valid for sown crops where large areas are planted densely to a single species which sets seed readily and uniformly and holds that seed until harvest. However, it is often invalid for wild plants. Indeed, it may be extraordinarily difficult, if not impossible, to collect 50-100 plants per site in some wild species. The reasons may be many and varied. For example :

- The target species may be sporadic and sparsely distributed, and populations may consist of few individuals (e.g. many populations of *Glycine* species in Australia).
- As most wild species readily shatter or disperse their seeds, it may be difficult to find 50-100 plants carrying ripe seed, even in a dense population.
- The collector may simply lack the facilities to carry the seed or propagules of more than 50 plants from each target site, particularly in rough and inaccessible terrain.

If there is a strict upper limit to the total number of plants sample, so that the sampling of >50 plants at each stop would seriously restrict the number of sites visited, we would strongly recommend a reduction in the number of plants collected at each site. For example, if only 1000 plants could be collected, it would be better to take 20 plants from 50 populations than 50 plants from 20 populations. The justification is simply that the collection of fewer plants from more sites allows greater opportunities to collect locally common alleles.

6.2 Mating systems

The general strategy assumes that the collector lacks detailed knowledge of the mating system of the target species and its variation. This led us to a highly conservative recommendation with respect to the number of plants to sample per site. In particular, the finding that a random sample of about 60-80 gametes would include 95% of all common alleles in a population under most circumstances was formulated as a recommendation to collect 50-100 plants. However, this would only be necessary in apomictic species where progeny are genetically identical with their maternal parent, or in inbreeding species where the bulk of the population is homozygous at most loci. In self-incompatible or largely outbreeding species, seed from a smaller sample of plants would be adequate. In fact, if the species was a truly random-mating one, a sample of 100 seeds from any plant would suffice since, in theory, those seeds would represent 100 randomly chosen gametes on the male side. The point is that if an explorer knows that the species he is sampling is outbreeding, he may be able to improve this sampling efficiency by collecting more seed off fewer plants. However, we emphasise two points. First, the number of plants sampled in known outbreeders should be kept at a reasonable level of, say, 20 plants per population. Very few species in reality practise random mating. This applies particularly to wild species, which often show marked population differentiation over short distances. In these circumstances, too drastic a reduction in the number of plants sampled in a population could conceivably lead to a drastic reduction in the number of alleles collected. Second, this procedure is only valid in *known* outbreeding species. In obligate apomictics and heavily inbreeding species the only way to ensure the collection of 100 random gametes is to sample 100 random plants.

6.3 Genetic knowledge of the target species

We have assumed up to this point that the collector has no knowledge of the genetic structure of the target species. While this has commonly been the case in the past it is increasingly less so. Collectors can gain information on population structure from two sources - observations of conspicuous polymorphisms in the field whilst collecting or information from previous collecting missions. We consider each below :

6.3.1 Conspicuous polymorphism

Populations of many legumes show a range of levels of morphological polymorphisms (leaf markings, leaf shape pubescence, flower colour, seed colour etc.). These differences are often evident to the observant collector during sampling. This is particularly so in perennials, less so in annuals which often die quickly after seed maturity. Clearly, when some populations of a species appear to be much more polymorphic than others, it would be sensible to take larger samples from the richer populations. This is especially important in inbreeding species, where the range in the level of polymorphism among populations is likely to be wider than in outbreeders (Schoen and Brown 1991) and the link between genetic variation in marker genes and genetic variation throughout the whole genome is stronger (Schoen and Brown 1993).

6.3.2 Knowledge of the genetic structure of the target species

A knowledge of the level of genetic variation in the target species and region may come from various sources. It may come from studies of material from earlier collecting missions to the same area. Alternatively, collections of the same species from different regions or related or similar species from the same region may be available which can provide useful information on the population structure of the target species. This information can greatly increase the efficiency of sampling because it allows the collector to concentrate his effort on the most diverse populations.

Oka (1975), Marshall and Brown (1975 1983) and Yonezawa (1985) have considered the issue of the optimal allocation of resources in field sampling where the plant collector has information on the genetic structure of the target species. These studies indicate that where a collector has limited total resources, the optimal allocation of resources between sites visited and sample size per site depends on the relative levels of genetic variation within and between populations in the target region. For wild species, these models suggest that it is generally the number of populations visited rather than the sample size per population that determines the overall efficiency of a collecting mission. This point was emphasised by Yonezawa (1985) who concluded that without a special reason, intensive sampling of a few populations was not justified.

7. RADICAL ALTERNATIVE STRATEGIES

The sampling methods outlined here are conservative and broadly in line with current practices. However, several authors have recently proposed quite radically different strategies to those advocated here. These differences derive from greater emphasis either on collecting "rare" alleles (frequency <0.05) in populations, or on collecting common alleles with greater certainty.

7.1 Sampling low-frequency alleles

Many breeders, genebank managers and collectors have problems with the notion that rare or uncommon alleles, arbitrarily defined here as alleles with frequency <0.05, should be disregarded in framing sampling strategies. There are two reasons for their concern. One is the fact that rare alleles are the most numerous class of alleles in most species. For example Doebley *et al.* (1985) surveyed 13 enzyme systems in 94 collections of maize. In a total sample size of 1,128 individuals (2,256 gametes) they found 163 alleles at 23 loci (an average of a little over 7 alleles/locus). Of these, 118 or more than 72 per cent, occurred at a frequency of less than 0.05. The second is that rare alleles, especially at disease resistance loci, have often proved to be of great economic values.

We do not dispute either of these facts. What we would emphasise, however, is the difficulty in terms of sample size of collecting rare alleles, especially if they are localised. Table 1 gives the number of gametes which must be sampled to be certain of including at a range of probability levels at least one copy of each allele present at a range of low frequencies in a population. The point to be emphasised is that the sample size required increases at an exponential rate as the allele frequency decreases. Samples of 50-100 individuals (100-200 gametes) may therefore often fail to capture rare alleles, with a frequency of 0.01 or less, at a single site.

However, many such alleles will be captured if 50 plants are collected from 50 sites, so that the total regional sample is 2,500 plants or 5,000 gametes if they are widespread and occur in most populations. In fact, a sample of this size is 99% sure of capturing at least one copy of all alleles with frequency >0.001 in the regional gene pool. This, however, does not resolve the problem with respect to locally rare alleles which occur at low frequency at one or a few sites. To capture at least one copy of an allele that occurs at a frequency of 0.001 in a population with 95% certainty requires a sample of 2,994 gametes or 1,497 individuals from that population. We concluded earlier that it would be seldom worthwhile searching for localised rare variants in natural populations (Marshall and Brown 1975). We still think this is the case.

Table 1. Numbers of gametes that must be sampled to be certain of including, at a range of probability levels, at least one copy of an allele at various frequencies in a population

	Allele frequency in population			
Probability	0.05	0.02	0.01	0.001
0.90	45	114	229	2301
0.95	59	148	298	2994
0.99	88	228	458	4603

7.2 Near certainty in sampling

A second radical approach to sampling for the conservation of plant genetic resources has been proposed by Lawrence *et al.* (1995a and b). These authors argued that it is axiomatic that the aim of genetic conservation is to conserve all the genes of a species of interest, and therefore, that the sample size should be sufficiently large to be "virtually certain" of capturing all alleles at each polymorphic locus in a population or series of populations. Using a single locus model with two alleles they calculated that a sample of 172 plants, would, regardless of the breeding system, capture at least one copy of both alleles provided they occurred with a frequency ≥0.05 with a probability of at least 99.99 per cent or "near certainty". They then considered several single locus multi-allelic examples and showed that a sample of 172 allowed them to capture all alleles at a multi-allelic locus with frequencies ≥0.05 at very high probabilities.

Lawrence *et al.* (1995 a and b) then extended their model to k loci in a similar manner to Rheenan *et al.* (1993) and considered two quantitative cases, a species polymorphic for 1,000 loci and one polymorphic for 20,000 loci. They concluded that "the calculations in these papers show that a sample of about 172 plants is sufficient to conserve all or very nearly all of the alleles at 20,000 polymorphic loci provided their frequency is not less than 0.05." Finally, they argued that since the aim of genetic conservation is to conserve all of the alleles of a target species, it follows that it is the species, rather than its constituent populations, that is the statistical population from which the sample are to be drawn. On this basis they suggested that 172 plants drawn from species is sufficient to conserve all or nearly all polymorphic genes provided the gene frequency is >0.05 in the species as a whole. This is regardless of the number of populations sampled.

This is a startling conclusion and if valid, is fraught with implications for plant collectors, breeders and conservationists alike. It suggests, for example, that large collections of germplasm that are highly prized by so many, are unnecessary, and are in fact expensive and wasteful mistakes. It also suggests that genebank holdings which now appear to total well over 2 million accessions for our 60 or so most important crops can safely be reduced to about 10,000 samples. Closer to home it would mean no more collecting was required for any species currently of major economic importance in Australia.

However, we believe these papers, especially the first, and the sweeping conclusions they draw are seriously flawed.

First, the authors claim that they are interested in conserving "all" the genes in a species. This is naive. As noted by many authors over the last 25 years (e.g. Allard 1970; Marshall and Brown 1975) most plant species contain very substantial stores of genetic variation and consist of millions, if not hundred of millions of different genotypes. It is clearly impractical to collect all of them in a sample of 50 individuals from 50 sites, let alone one of 172 from one or more sites. What Lawrence *et al.* (1995) have calculated is, in our terminology, the sample size required to collect common alleles from a population, or *widespread common* alleles from a species. Again it has long been recognised that (i) such alleles can be collected with ease, (ii) they will be included in a sample at high frequency regardless of strategy, and (iii) for most species of economic interest, they undoubtedly already exist in collections (Marshall and Brown 1975 1983; National Research Council 1993, Chapter 4).

Second, the authors conclusion that if samples are taken from a number of populations the size of each need be no larger than 172 divided by number of populations visited, implies that all populations are equivalent in the kinds and levels of genetic variation they contain. In effect it says that population differentiation is non-existent. Even if this assumption were to hold, and this is against all evidence in plants, there would be no logical reason to sample or conserve more than one population. However, the fact is that cultivars of cultivated plants and populations of wild plants have diverged from one and another genetically. It is these differences, not their similarities, that are of concern to users. Population divergence is of particular importance in wild plants which are not subject to the homogenising influence of cultivation. The fact that populations of a species are differentially adapted and contain different unique alleles or allele combinations at some fraction of their loci is therefore the fundamental starting point for the sampling for genetic conservation.

Third, the authors claim that a sample of 172 plants is sufficient to conserve all or very nearly all of the alleles at 20,000 loci, provided that their frequency is not less than 0.05, is dependent on a number of simplifying assumptions and can be shown to lack general validity especially for highly self-pollinated plants. It assumes all loci are inherited independently of each other which for a few genes is sustainable but for whole genomes is not. It also assumes that all populations of the target species are in inbreeding equilibrium, that there is no sub-population structure, and the genes of interest have only one allele with frequency close to 0.05. For the 172-plant sample in inbreeding species the authors themselves calculate that the probability of conserving at least one copy of each allele at a frequency of ≥0.05 is 0.0524 or about 5% for 20,000 loci. If there are two alleles with a frequency of 0.05 the probability is 0.0292 or about 3% and if there are three alleles then the probability is 0.0028 or less than 1%. The point is that at a given probability level whether it be 0.95, 0.99 or 0.9999 and a given minimum allele frequency, larger samples are needed for joint assurance at increasingly more loci.

It is also instructive to consider the sensitivity of these models to changes in the lower limit of allelic frequencies to be conserved. As noted above, if this is set at $\geq$0.05 for a self-pollinating plant, then the probability of conserving the least common allele of each of 20,000 loci is 5.24 x 10^{-2}. If this lower limit is increased to $\geq$0.06 the probability becomes 0.62 and if it is reduced to $\geq$0.04 it reduces to 1.7 x 10^{-8} or virtually zero. Thus changing the frequency of the least common allele of interest by a mere 1% radically alters the chance that all will be collected especially at 20,000 loci.

However, the critical problem with the focus on the 172 plants-per-species approach is not that it is likely to mislead scientists working in the field. Their evidence and experience is sufficient to indicate that the central conclusion is wrong. Rather, the approach is likely to be used by administrators and politicians to radically cut funding for plant collection and conservation on the grounds that the models and analysis in this paper proves that, for many species, (i) all that needs to be collected has been collected and (ii) there are numerous opportunities to reduce the size of genebanks and their maintenance costs. It is imperative, therefore, that the sorts of misleading and erroneous statements made in these papers be exposed and corrected.

REFERENCES

Allard, R.W. (1970). Population structure and sampling methods. In 'Genetic Resources in Plants - Their Exploration and Conservation'. (Eds O.H. Frankel and E. Bennett.). pp.97-107. (Blackwell Scientific Publications: Oxford).

Bradshaw, A.D. (1975). Population structure and the effects of isolation and selection. In 'Crop genetic resources for today and tomorrow.' (Eds. O.H. Frankel and J.G. Hawkes). pp. 37-52. (Cambridge University Press: Cambridge, UK).

Brown, A.H.D. (1992). Human impact on plant gene pools and sampling for their conservation. *Oikos* **63**, 109-118.

Brown, A.H.D. and J.D. Briggs (1991). Sampling strategies for genetic variation in *ex situ* collections of endangered plant species. In 'Genetics and Conservation of Rare Plants'. (Eds D.A. Falk and K.E. Holsinger). pp.99-119. (Oxford University Press: New York).

Brown, A.H.D. and D.R. Marshall (1995). A basic sampling strategy: Theory and practice. In 'Collecting Plant Genetic Diversity: Technical Guidelines'. (Eds L. Guarino, V. Ramanatha Rao and R. Reid). pp.75-92. (CAB International: Wallingford, U.K).

Chapman, G.C. (1989). Collection strategies for the wild relatives of field crops. In 'The Use of Plant Genetic Resources'. (Eds A.H.D. Brown, O.H. Frankel, D.R. Marshall and J.T. Williams). pp.263-279. (Cambridge University Press: Cambridge).

Damania, A.B. and E. Porceddu (1983). Variation in landraces of *Turgidum* and bread wheats and sampling strategies for collecting wheat genetic resources. In 'Proceedings of Sixth International Wheat Genetics Symposium'. (Ed S. Sakamoto). pp.123-136. (Plant Germplasm Institute: Kyoto).

Damania, A.B. (1989). 'Crop Genetic Resources Sampling: A Training Handbook for Germplasm Collection'. (ICARDA: Aleppo).

Doebley, J.W. (1989). Isozymic evidence and the evolution of crop plants. In 'Isozymes in Plant Biology'. (Eds D.E. Soltis and P.S. Soltis). pp.165-191. (Dioscorides Press: Portland).

Doebley, J.W., Goodman, M.M. and Stuber, C.W. (1985) Isozyme variation in races of maize from Mexico. *American Journal of Botany* **72**, 629-639.

Frankel, O.H. (1970). Genetic conservation in perspective. In 'Genetic Resources in Plants - Their Exploration and Conservation'. (Eds O.H. Frankel and E. Bennett). pp.469-489. (Blackwell Scientific Publications: Oxford).

Frankel, O.H., Brown, A.H.D. and Burdon, J.J. (1995). 'The Conservation of Plant Biodiversity'. (Cambridge University Press: Cambridge).

Hamrick, J.L. and Godt, M.J. (1989). Allozyme diversity in plant species. In 'Plant Population Genetics. Breeding and Genetic Resources'. (Eds A.H.D. Brown, M.T. Clegg, A.L. Kahler and B.S. Weir). pp.43-63. (Sinauer Associates: Sunderland).

Jain, S.K. (1975). Population structure and the effects of breeding system. In 'Crop Genetic Resources for Today and Tomorrow'. (Eds O.H. Frankel and J.G. Hawkes). pp.15-36. (Cambridge University Press: Cambridge).

Krusche, D. and Geburek, Th. (1991). Conservation of forest gene resources as related to sample size. *Forest Ecology and Management* **40**, 145-150.

Lawrence, M.J., Marshal,l D.F. and Davies, P. (1995a). Genetics of genetic conservation. I. Sample size when collecting germplasm. *Euphytica* **84**, 89-99.

Lawrence, M.J., Marshal,l D.F. and Davies, P. (1995b). Genetics of genetic conservation. II. Sample size when collecting seed of cross-pollinating species and the information that can be obtained from the evaluation of material held in genebanks. *Euphytica* **84**, 101-107.

Marshall, D.R. (1989). Crop genetic resources - current and emerging issues. In 'Plant Population Genetics, Breeding and Genetic Resources'. (Eds A.H.D. Brown, M.T. Clegg, A.L. Kahler and B.S. Weir). pp.370-391. (Sinauer: Sunderland).

Marshall, D.R. and Brown, A.H.D. (1975). Optimum sampling strategies in genetic conservation. In 'Crop Genetic Resources for Today and Tomorrow'. (Eds O.H. Frankel and J.G. Hawkes). pp.15-36. (Cambridge University Press: Cambridge).

Marshall, D.R. and Brown, A.H.D.(1981). Wheat genetic resources. In 'Wheat Science - Today and Tomorrow'. (Eds L.T. Evans and W.J. Peacock). pp.21-40. (Cambridge University Press: Cambridge).

Marshall, D.R. and Brown, A.H.D. (1983). Theory of forage plant collection. In 'Genetic Resources of Forage Plants'. (Eds J.G. McIvor and R.A. Bray). pp.135-148. (CSIRO: Melbourne.

Namkoong, G. (1988). Sampling for germplasm collections. *HortScience* **23**, 79-81.

National Research Council (1993). 'Managing Global Genetic Resources : Agricultural Crop Issues and Policies' pp.131-152. (National Academy Press: Washington D.C.).

Oka, H.I. (1975). Consideration on the population size necessary for conservation of crop germplasm. In 'Gene Conservation - Exploration, Collection, Preservation and Utilization of Genetic Resources'. (Ed T. Matsuo). pp.57-84. JIBP Synthesis Volume 5. (University of Tokyo Press: Tokyo).

Porceddu, E. and Damania, A.B. (1992a). 'Sampling Strategies for Conserving Variability of genetic Resources in Seed Crops'. Technical Manual No.17. (ICARDA: Aleppo).

Porceddu, E. and Damania, A.B. (1992b). Sampling variation in genetic resources of seed crops: a review. *Genetic Resources and Crop Evolution* **39**, 39-49.

Qualset, C.O. (1975). Sampling germplasm in a centre of diversity : an example of disease resistance in Ethiopian barley. In 'Crop Genetic Resources for Today and Tomorrow'. (Eds O.H. Frankel and J.G. Hawkes). pp.15-36. (Cambridge University Press: Cambridge).

Schoen, D.J. and Brown, A.H.D. (1991). Intraspecific variation in population gene diversity and effective population size correlates with the mating system in plants. *Proceedings of the National Academy of Sciences*, U.S.A. **88**, 4494-4497.

Schoen, D.J. and Brown, A.H.D. (1993). Conservation of allelic richness in wild crop relatives aided by assessment of genetic markers. *Proceedings of the National Academy of Sciences U.S.A.* **90**, 10623-10627.

Snaydon, R.W. (1978). Genetic changes in pasture populations. In 'Plant relations in pastures.' (Ed. J.R. Wilson). pp. 253-269. (CSIRO: East Melbourne, Australia).

Van Rheenan, H.A., Bhah, S. and Rao, K.V.S. (1993). The sustainable preservation of biodiversity in self-pollinating plant species: Sample size and collection methodology. *Plant Gen. Res. Newsletter* **96**, 1-4.

Ward, D.B. (1974). The 'ignorant man' technique of sampling plant populations. *Taxon* **23**, 325-330.

Weir, B.S. (1990). 'Genetic Data Analysis'. (Sinauer Associates: Sunderland).

Yonezawa, K. (1985) A definition of the optimal allocation of effort in conservation of plant genetic resources - with application to sample size determination for field collection. *Euphytica* **34**, 345-354.

Yonezawa, K. and Ichihashi H. (1989). 'Sample size for collecting germplasm from natural plant populations in view of the genotypic multiplicity of seed embryos borne on a single plant'. *Euphytica* **41**, 91-97.

CHAPTER 8

The need to collect new pasture and forage species

C M Francis
Co-operative Research Centre for Legumes in Mediterranean Agriculture, University of Western Australia, Nedlands, WA 6907. Australia.

Abstract Most world collections have too few pasture species. Despite the numerous Australian plant collection tours, accessions have been dominated by *Medicago* and *Trifolium subterraneum*, and they are by no means strongly placed to meet the changing demands of new systems and environmental constraints.

There is thus adequate opportunity and need for strategic seed exchange and collection tours which focus on species for specific systems or environments. The improving world databases and the substantial ecological knowledge gained from our own collections allows the definition of likely sites of many target species. Modern technology for climatic comparison and GIS mapping analysis will greatly aid the plant collections of the future.

A priority list is proposed which has an emphasis on free standing, easily harvestible species which will provide alternatives to annually regenerating pasture species and a greater range of environmental adaptation.

1. INTRODUCTION

Jack Harlan (1984) noted that no collection of annual forage species is as adequate as it should be and most are absurdly inadequate. The very small genetic base of useful species like *Trifolium michelianum* (balansa clover) is a good example. In Australia, collections of *Medicago* and *Trifolium subterraneum*(subterranean clover) are more than adequate for the foreseeable future. However, the very size and concentration on these species (with 26,000 *Medicago* and 7000 *Trifolium subterraneum* in the collections), has meant that potentially very valuable alternatives have been neglected. We are just beginning to throw off the yolk of medics and subterranean clovers with quite spectacular results. The question arises however as to how we assemble and prioritise amongst dozens of potentially useful species. We are faced with a dilemma and often a pretty sketchy knowledge base on which to select species for further research and development.

Jack Harlan (1984) also posed two questions. Have we surveyed the potential forages of the world in sufficient detail to be able to point out geographic centres of the most useful plants? Can we, with present knowledge, define some gene centres for forage plants where we might most profitably concentrate our collecting efforts by following the techniques which have been so profitable exploited by the wheat and barley world collections? The answers to both these questions will depend on the number and intensity of our immediate collection experiences and the relative degree of assessment on the germplasm collected.

Some forage legumes appear to have been in use for thousands of years. Species first used as grain have later been exclusively used for forage and vice versa (Ladizinsky 1987). The changing pattern of agriculture makes the maintenance of a wide genetic base essential. The question of setting research priorities amongst the hundreds of potentially useful forage species will always arise. How much progress we have actually made toward this end and how do we go about it? No mechanism is

S. J. Bennett and P. S. Cocks (eds.), Genetic Resources of Mediterranean Pasture and Forage Legumes, 90–95.

in place whereby forage researchers formally review their priorities on the basis of the latest agronomic challenges, research or plant exploration experience. The potential does exist and there are signs, at least for legumes adapted to Mediterranean regions, that the present broad base of national and international experience does allow some progress with priority setting. Priorities may be set according to an existing or emergent agronomic requirement, the size of the present genetic resource base of a defined useful species, the accessibility and cost of retrieval from the wild and whether the species is facing effective extinction. The resource required may be for a restricted ecological situation, e.g. acid sandy soils in low rainfall areas of the Mediterranean, but more commonly a broad range of genotypes of a promising species are sought.

2. THE ROLE OF PLANT COLLECTIONS AND ECOLOGY IN PRIORITY SETTING

More than 25 expeditions have been mounted by Australian plant collectors with one or more forage species as a major component. Programs such as that conducted by the University of Birmingham in liaison with the Aegean Agricultural Research Institute in Izmir, Turkey, are a first class example of a way to define the genetic resource base of forage legumes. Whilst not as detailed or comprehensive as these, ecogeographical observations accompanying past plant collection tours can, and have, opened up new pasture and grain options for Australian farmers. Plant collections, if done properly with soil and climatic analysis and site details, enable the definition of potential species and where they are most likely to fit if relocated from their centres of diversity.

Completely new forage species options can stem from the observations of the plant collector. Serradella was recognised by John Gladstones because if its performance on very poor acid soils in Portugal when he was searching for lupins and subterranean clover. The potential and subsequent release of *Medicago murex* arose from ecological observations of adaptation to acids soils in Sardinia (Francis and Gillespie 1981). More recently Nutt, Carr and Samaras (1996) observed the success of *Trigonella balansae* in the Greek Islands and it now shows great promise as a completely new pasture species for Southern Australia. In the same way *Biserrula pelecinus* was recognised as a coloniser of sandy acid sites and much more focused efforts at expanding the collection were made by Howieson and Loi (1994).

We are now in a position to superimpose refinements to the classical broad based expeditions like that of Donald and Miles in 1951 which crossed 10 Mediterranean countries. We have better defined agronomic needs and have a much better picture of where in the world we can collect appropriate plants. Good ecogeographical and recording systems are now available and provide a good guide in most cases as to where to focus our search for particular species. Where the targets can be defined by soil preference e.g. preference for granitic soils, they can be further focused now in regions of climatic similarity. CLIMAX and GIS systems can be used to further refine collection targets.

However, even without the modern data handling systems, Australia's Eric Crawford (1984) proposed a list of priority species for southern Australia using his wide knowledge of species ecology gained from numerous overseas collections, coupled with preliminary agronomic evaluations on his field station near Adelaide. His list (table 1) was prophetic but has not been reviewed.

Whilst past collections can define *primary centres* in the Mediterranean for collections of various species using data bases, collection experience and technology, the *secondary centres* of Mediterranean climate in other countries can be equally or more rewarding for some species. They should never be overlooked in collection planning. For example, until quite recently most subterranean clovers commercialised have been Australian naturalised strains. It could well be argued, as indeed it was, by Gladstones and Collins (1983) that we should exploit the naturalised and presumably already adapted varieties first. Interestingly many of the best adapted *Medicago polymorpha* varieties come from Chile where there is a high proportion of spineless types. Australia is also rich in naturalised *M polymorpha* but they are almost all very spiny types. South Africa has a rich population of *Ornithopus sativus* presumably a legacy of introductions from Portugal and the population has adapted to drier climates (Gillespie 1989). However a large proportion of the more productive free standing species have not become naturalised in Australia. There is still room for plant collection tours for many forage species which become a priority in the light of economic, sociological and environmental changes.

Table 1. National genetic resource banks - priorities for temperate pasture species (Crawford 1984)

Trifolium	*Medicago*	*Other Genera*
T. alexandrinum	*M. aculeata*	*Astragalus*
T. ambiguum	*M. intertexta*	*Hedysarum*
T. michelianum	*M. littoralis*	*Lotus*
T. cherleri	*M. murex*	*Onobrychis*
T. fragiferum	*M. orbicularis*	*Tetragonolobus*
T. hybridum	*M. polymorpha*	*Trigonella*
T. isthmocarpum	*M. rigidula*	
T. nigrescens	*M. rugosa*	
T. pallidum	*M. scutellata*	
T. pratense	*M. tornata*	
T. purpureum	*M. truncatula*	
T. repens	*M. turbinata*	
T. resupinatum		
T. semipilosum		
T. subterraneum		
T. squarrosum		
T. vesiculosum		

3. THE NEED FOR NEW SPECIES

3.1 The changing face of agriculture

Opportunities in agriculture and the effect of the imposed systems on the land itself are in a state of constant change. This changing pattern of land use means we must widen the reservoir of forage plant genetic resources. Changing farm practice and economic factors, such as the expense of producing seed of some dominant species like medics and subterranean clover, have been a major impetus to the search for new pasture species. Perhaps the most radical change in thinking for mediterranean Australia has been the realisation that free standing, easily harvestible seed crops could provide valuable alternatives to the annually regenerating pastures. These would not need to be very hard seeded, but must be easy to harvest and establish and provide a dense and dominant stand for rotation with cereals or oil seeds. CLIMA's first forage plant release, Cadiz pink serradella, was specifically tailored for such a strategy. For the purpose of the paper, it is apparent that a number of species could fit the model over a wide soil and rainfall range. Australian pasture collections have been so dominated by subterranean clover and medics in the past that for many, if not most other species eg *T vesiculosum,* there are too few accessions on which to base a full scale selection program. These provide targets for focused plant collections.

3.2 Environmental changes and niches

The edict for sustainable systems has in reality provided a stimulus for the search for new crop and forage options for different ecological niches or environmental challenges: Agroforestry, for example, is form of land use probably requiring a new forage species options. More advanced knowledge of species potential and problems arising with the expanding mix of grain crops, also supports the need for a wider range of forage species options.

A major priority is new species as alternatives to annual medics on fine textured strongly alkaline soils. A different set again is needed for neutral or mildly acidic fine textured soils like the red brown earths. The observation of our own collection missions are showing some useful relationships of species and soil type. Ecogeographical observations on collection trips such as that in Crete (Francis, Maxted, Kyriakakis and Bennett 1995) help build a picture of species adaptation (table 2).

Table 2. Frequency of major species groups on high pH soils

Species	Percentage of accessions (number of sites)
Trifolium repens	0 (10)
T. subterraneum	5 (21)
T. nigrescens	22 (15)
T. pallidum	40 (15)
T. resupinatum	47 (17)
T. lappaceum	50 (12)
T. angustifolium	50 (10)
Lathyrus annuus	45 (11)
Lupinus spp.	0 (14)
Vicia villosa	15 (11)
V. hybrida	42 (12)
V. sativa	51 (37)

In Australia we face major threats from rising water tables and increasing acidity. Deep rooted annuals or perennials are needed to help prevent it. Where there is already a problem, a series of waterlogging and saline tolerant species are needed.

It is possible, given the above guidelines to suggest groups of promising free standing species for which, partly because of our predilection with subterranean clover and medic, a lot more germplasm is needed. This list (table 3) can be expected to change progressively as we assess new germplasm and learn more of its potential, and as new challenges arise.

4. ADDITIONAL SPECIES OF PROMISE

There are many others which with more knowledge may make the list in time - On alkaline soils species in genera like *Astragalus, Scorpiurus, Psoralea, Tetragonolobus,* and *Coronilla* are some which can provide fresh options for grazing animal enterprises.

Table 3. Priority species for increased genetic stocks

Acid sandy soils	Distribution	other features
Ornithopus sativus	Mediterranean, Portugal	
Lupinus luteus	Mediterranean, Portugal	slightly bitter
Biserrula pelecinus	Mediterranean	deep roots
T incarnatum	S Europe, Caucasus	hay type
T. cernuum	W Mediterranean	waterlogging
Fine textured calcareous *		
T purpureum	E Mediterranean, Turkey	
T lappaceum	E Mediterranean Turkey	
T. alexandrinum	Mediterranean, W Asia	hay, waterlogging
T. dasyurum	E Mediterranean	
T. leucanthum	Mediterranean	
T. argutum	E. Mediterranean, SE Turkey	
Trigonella balansae	E Mediterranean, Cyprus	herbal properties
Hedysarum coronarium	Europe, West Med, Tunisia	biennial
Onobrychis viciifolia	Europe	perennial
Fine not calcareous		
T. pallidum	E Mediterranean, Crete	high pH
T. vesiculosum	E Europe, Central Asia	
T. clypeatum	E Mediterranean, Greece Is	rapid early growth
T. glanduliferum	E Mediterranean, Turkey	resistant to RLEM
T. michelianum	E Mediterranean, Turkey	salt waterlogging
T. squarrosum	W Mediterranean, Morocco	
T. spumosum	Mediterranean	
T. nigrescens	Mediterranean (common)	shady places
T. hirtum	Mediterranean, Cyprus	
Forages and hays		
L. clymenum	Mediterranean Morocco, Spain	moderate acid soils
L. tingitanus	Mediterranean Morocco, Spain	moderate acid soils
L. ochrus	Mediterranean, Morocco. Tunisia	low altitudes only
L. cicera	E Mediterranean mainly	dry areas
V. villosa	Mediterranean	moderate acid soil
V. sativa ssp. *macrocarpa*	Mediterranean	large seeds
V. pannonica	W Asia, E Europe	toxin free?
V. articulata	Mediterranean	wide soil range
Salinity / waterlogging		
Melilotus indica	Mediterranean to Asia	
M. officinalis	Europe, central Asia, China	biennial herbal
Lotus corniculatus	Native Europe, central Asia	perennial
L. sativus	S Europe and Asia, Ethiopia	low ODAP
T. resupinatum	Mediterranean	high pH soils
T. fragiferum	Mediterranean	
T. clusii	Mediterranean	
T. physodes	Mediterranean	
Agroforestry shady places		
T. clusii	Mediterranean	
T. boisseri	E Mediterranean, Greece, Crete	

Other Trifoliums with some adaptation to alkaline soil. *T. echinatum, T. stellatum, T. grandiflorum, T. angustifolium, T. resupinatum, T. cherleri, T. bullatum.*

On more sodic soils, *Melilotus species* have been little utilised in Australia nor have *Lotus* species. I believe this is very largely related to the dominance of the medic and subterranean clover programs or expenditure into lucerne research in the case of the perennials. This may be unfortunate in the light of the current thrust for tolerance of sodicity and rising water tables. The quaint situation is that some of the involved researchers in both annual and perennial *Medicago* have undertaken expensive screening for marginal differences in salt tolerance and waterlogging, neglecting species which probably represent a quantum leap in tolerance e.g. *Melilotus officinalis*, *Trifolium resupinatum* and *T. fragiferum*.

Others have been completely overlooked or forgotten. The potential value of species like *Lathyrus tingitanus* have been extolled by various authors (Oram and Brock 1972) but never followed up. *Anthyllus vulunaria* is not regarded as important today, but was once widely used in Europe (Cullen 1976). It is known to be widely distributed, but on the better soils has been displaced by food and energy crops. It is still likely to be useful on poorer land in drier areas.

Medicago lupulina is an interesting species, neglected, but deserving of study for its annual/ biennial character and adaptation to highly (possibly sodic) alkaline soils.

These are all species which justify at least some input. Post graduate studies are often an ideal way to get into a new species and it is to be hoped that better direction can be given by supervisors as it is one area where the special funds seem to allow a bit of blue sky - even in pasture research.

REFERENCES

Crawford E J (1984). The potential for further acquisitions of alternative pasture legumes with desirable characteristics. Proceedings Workshop on alternative legumes. Struan, South Australia, October 1984.

Cullen J. (1976) *Notes Royal Botanical Gardens, Edinburgh* **35**:1.

Francis C M and Gillespie D (1981). Ecology and distribution of subterranean clover and *Medicago* species in Sardinia. *CSIRO Plant Introduction. Revue* **13**:15.

Francis C M and Gladstones J S (1983) Exploitation of genetic resources through breeding *Trifolium*. Proceedings conference on Genetic resources of forage plants pp. 251. (Eds. J G McIvor and R A Bray), (CSIRO: Australia.)

Francis C M, Maxted N, Kyriakakis S, Bennett S (1995). Crete: Observations on the ecogeography of grain, forage and pasture legumes. CLIMA Occasional publication series No 11

Gillespie D J (1989) Report on pasture legume plant collection tour. Cape Province 1989. Western Australian Department of Agriculture.

Gladstones J S and Collins W J (1983) Subterranean clover as a naturalised plant in Australia. *J Aust. Inst. Agric. Sci.* **49**,:191

Harlan J R (1983). The scope for collection and improvement of forage plants. In 'Genetic Resources of Forage Plants'. (eds McIvor and Bray). (CSIRO: Australia).

Howieson J and Loi A (1994). Report on the distribution and preliminary evaluation of alternative pasture legumes and their associated root nodule bacteria collected from acidic parts of Greece (Serifos), Morocco, Sardinia and Corsica. CLIMA Occasional Publication No 4.

Ladizinsky G (1987). Pulse domestication before cultivation. *Economic Botany* **41**, 60.

Nutt B, Carr S and Samaras S. (1996). Collection of forage plants and grain legumes and associated rhizobia from selected Greek Islands. CLIMA Occasional Publication 13.

Oram R N and Brock K D (1972). Prospects for improving plant protein, yield and quality by breeding. *J. Aust. Inst. Agric. Sci.* **38**, 163.

CHAPTER 9

The host-rhizobia relationship

J.G. Howieson[1]
Co-operative Research Centre for Legumes in Mediterranean Agriculture (CLIMA), The University of Western Australia, Nedlands, WA 6907. Australia.
[1] *Current address: Centre for Rhizobium Studies, Murdoch University, Murdoch, WA 6150. Australia*

Abstract

Conserving and experimenting with the genetic resources of legumes are almost worthless pursuits if similar attention is not given to their microsymbiotic partners. The nitrogen fixing symbiosis between members of the Leguminosae and the four separate but related bacterial genera Rhizobium, Bradyrhizobium, Azorhizobium and Sinorhizobium is a specialised interaction which, when optimised, can contribute enormously to the world's protein production.

At the field level in the last decade there has been an expansion of interest in the adaptation of legumes and their root-nodule bacteria to the environment. In Australia, the impetus for this increased activity has been the changing environmental, economic and agronomic circumstances of rural agricultural production. In Europe, an increased emphasis upon low input forms of agriculture has been the motivation for exploring new legumes.

In the laboratory, our knowledge of the physiological interaction between legume and microsymbiont, and its molecular control, has seen a parallel expansion. This accumulation of understanding of the signalling process, which is now recognised as the first stage in nodulation, provides the opportunity to intervene and amend the recognition between legume and microsymbiont to the advantage of the agriculturalist. This advantage may be expressed at either the experimental or applied level.

This paper will discuss some of the recent advances in legume bacteriology, from the dual perspectives of the laboratory and the field, and their potential application to agriculture and the science which surrounds it.

1. INTRODUCTION

Nitrogen fixation by legumes, in association with root-nodule bacteria from the genera *Rhizobium, Bradyrhizobium* and *Sinorhizobium,* is pivotal to the low input - high output form of agriculture widely practised in southern Australia. Most of the productive legumes and their microsymbionts in this system have been introduced to Australia, either deliberately or accidentally, since European settlement. A corollary to this is that there exists, in centres of diversity outside Australia, a wealth of legumes that may one day have application to Australian agriculture. In recognition of this opportunity, there are currently several major research efforts to select alternative pasture and grain legumes with particular specialised agronomic traits (Carr 1998). To develop this wealth of germplasm efficiently requires knowledge of the microsymbiotic requirements of new legumes and of how these microsymbionts react to both the Australian edaphic environment and to the suite of organisms already resident in Australian soils. Without attention to this detail, legume development will be slow, if not futile in many regions of southern Australia.

Selection of optimal microsymbionts for new legumes

Legume material introduced to Australia may have one of several reactions to the local commercial, endemic and naturalised root-nodule bacteria. There may be no nodulation with any strains, as was the case for the monotypic genus *Biserrula pelecinus* (Howieson *et al.* 1995). At the other extreme, there may be nodulation followed by nitrogen fixation at near-

S. J. Bennett and P. S. Cocks (eds.), Genetic Resources of Mediterranean Pasture and Forage Legumes, 96–106.

optimal levels. Both of these scenarios are relatively uncommon. More frequently there is nodulation but with a mediocre and inadequate level of nitrogen fixation. A key role for the legume bacteriologist, then, is to supply legume introduction programs with knowledge of the likely reaction of their plant material to root-nodule bacteria in the larger agricultural environment, and where this reaction is imperfect, to provide appropriate inoculants.

1.1 Characteristics sought in inoculant strains of root-nodule bacteria

The legume bacteriologist seeks to provide agronomists with strains which have the following characteristics:

high nitrogen fixation with the target host legume
no deleterious effects on non-target hosts
adaptation to the edaphic environments for which the legume is primarily targeted
genetic stability in soil, culture and in storage
satisfactory growth and survival characteristics in inoculant carrier materials
These characteristics are dealt with in more detail below.

1.1.1 High nitrogen fixation.

The selection of rhizobial strains with a capacity for high nitrogen fixation becomes more complex as the number of legume species with which the particular strain can nodulate increases. In the simplest example, again for *B. pelecinus* where its specific rhizobia has no other known host in Australian agriculture, the selection of an appropriate inoculant strain can rapidly be achieved by screening a range of root-nodule bacteria for nitrogen fixation upon several genotypes of the legume. Techniques for screening vary from glass tube culture (Vincent 1970) to a sand medium in 1 kg pots held under axenic conditions (Howieson *et al.* 1995). Under normal circumstances, a group of strains can readily be identified which fix nitrogen well in association with the host genotypes. However, the ultimate success of such an approach, given that appropriate techniques are adopted, is directly related to the breadth of germplasm available for evaluation combined with the degree of specificity exhibited by the particular symbiosis under investigation. In the data for *B. pelecinus* presented in Fig. 1, strains WSM1542 and WSM1558 collected from Sardinia were highly effective in preliminary experiments with the three host genotypes. On the basis of this outcome WSM1558 was used as an experimental inocula in Western Australia in 1996. Despite fixing nitrogen well with a genotype from the Greek Island, Serifos, (Howieson and Loi 1994) in our initial experiments, WSM1558 has since been shown to be poorly effective with genotypes of *B. pelecinus* collected more recently from nearby islands in the Aegean Sea (Nutt *et al.* 1996). This outcome illustrates the importance of being continually aware of potential intraspecific variation in nitrogen fixation for any given symbiosis as new germplasm is introduced to breeding and selection programmes.

The task of selecting appropriate microsymbionts becomes more complex where several legume species are nodulated by a single rhizobial species. This is the scenario for the *Medicago* genus where, in Australia, there exists commercial pressures (based upon the economics of scale) to inoculate the species *Medicago sativa, M. tornata* and *M. littoralis* (and indeed hybrids between *M. littoralis* and *M. truncatula*) with a single strain of *R. meliloti*. Fig. 2 illustrates some of the research undertaken to select that particular strain, WSM826, in the late 1980's. It was clear at that time that available inoculant-quality strains such as U45 and CC169 were unsuitable for lucerne and for some genotypes of *M. littoralis*. A range of rhizobial germplasm was hence assembled from which improved strains eventually emerged. Examples of the variability of annual medics in their reaction to some of this rhizobial germplasm are given in Fig. 2.

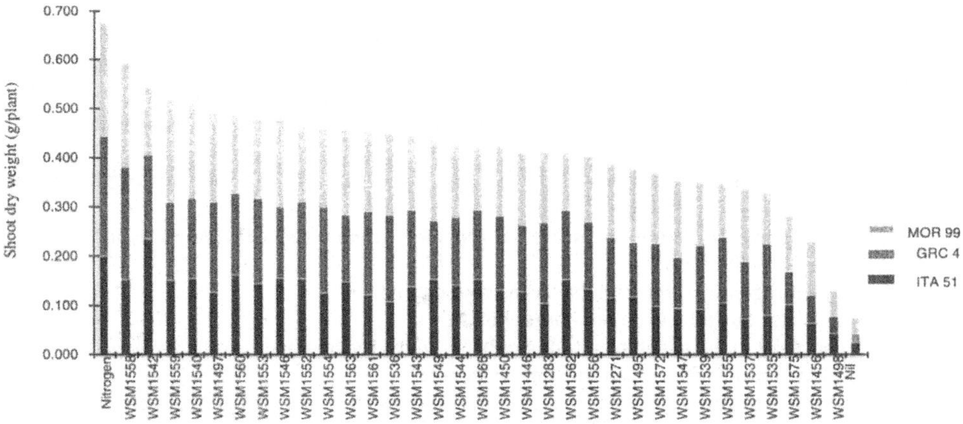

Figure 1. The dry matter yield of 3 genotypes of *Biserrula pelecinus* grown in N-free axenic sand culture following separate inoculation with 33 strains of root-nodule bacteria collected from that species (A. Loi, J. Howieson and S. Carr, unpublished data).

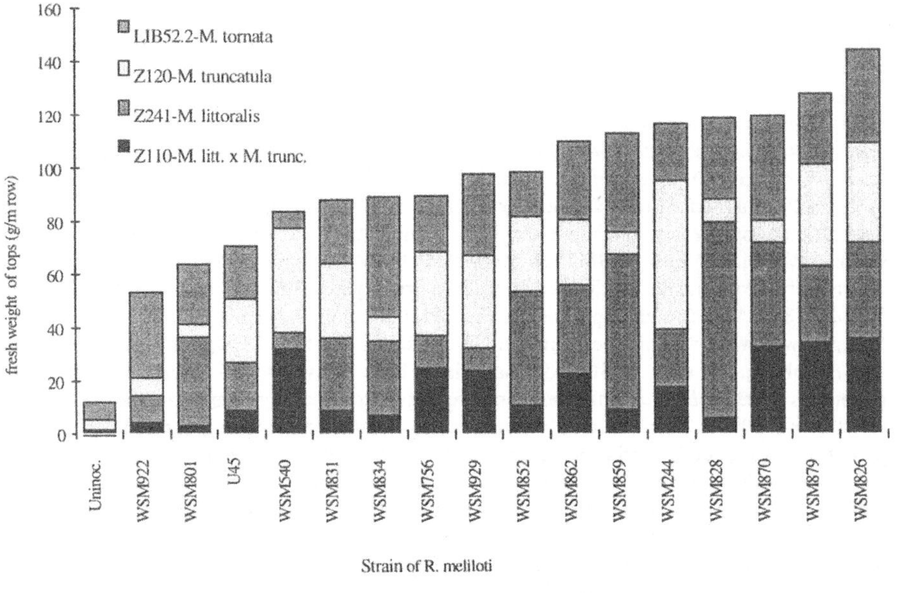

Figure 2. The yield of 4 species of *Medicago* inoculated separately with 16 strains of *Rhizobium meliloti.* Plants were grown in a field sand which was low in nitrogen status and free of *R. meliloti* (B. Nutt, P. Evans and J. Howieson, unpublished data.).

It is instructive to observe that interspecific hybridisation with *Medicago* species (a technique adopted to transfer important genes for insect resistance) yielded plant genotypes such as Z110 which were ineffectively nodulated by some strains (e.g. WSM852) which nonetheless fixed nitrogen adequately on representatives of the parent species.

Forage legumes in the *Lathyrus* and *Vicia* genera are nodulated by rhizobia from the species *Rhizobium leguminosarum* bv *viceae*. This rhizobial species additionally nodulates crop legumes from those two genera and the *Lens* genera. In Australian agriculture a single rhizobial strain (currently SU303) is used to inoculate all three genera, whether they be utilised for forage or grain production. *R.l viceae* is genetically unstable (see later) and in the search for improved inoculant strains substantial specificity in this symbiosis has been revealed. At a site in the south-west of Western Australia, 75 strains collected from *Vicia* and *Lathyrus* species growing in the Aegean Islands (Nutt *et al.* 1996) were screened for nitrogen fixation upon *V. faba*, with widely varying outcomes. The majority of strains which were poorly effective on *V. faba* in this experiment were collected from *Lathyrus* species whilst the majority of strains found to be highly effective on *V. faba* had been, in fact, isolated from this species (Table 1). As was evident for the *Medicago* genus, where specificity at the species level is encountered within a symbiosis, the task for selecting highly effective and broadly compatible microsymbionts becomes less trivial.

Table 1. The effect of host of origin on the effectiveness for nitrogen fixation of strains of R. leguminosarum bv viceae inoculated separately onto Vicia faba at a field site in south-western Australia (adapted from Carr *et al.* 1996)

	Percentage of maximum site yield of *V. faba*		
Host of origin	<50%	51-75%	>75%
Vicia faba	17	44	62
Vicia sp.	0	24	24
Lathyrus sp	83	27	14
unknown	0	5	0

1.1.2 Effect on non-target hosts.

A strong emphasis in the pasture programs of the Co-operative Research Centre for Legumes in Mediterranean Agriculture (CLIMA) has been upon diversifying the suite of pasture legumes available for agricultural exploitation in southern Australia (Howieson 1993, Carr 1996). The genus *Trifolium* is providing many potential "new" species, all of which are nodulated by *R.l trifolii*. However, not all *Trifolium* species are effectively nodulated by the strain of *R.l trifolii* currently used to inoculate *T. subterraneum*, the most important forage legume in southern Australia. Thus, in selecting alternative inoculants for these new species of *Trifolium* it has been of fundamental importance to be aware that organisms introduced for these legumes have the potential to also nodulate subterranean clover. As a matter of course, potential inoculants have been pre-screened for their ability to fix nitrogen optimally in association with subterranean clover. Those strains not fully effective on *T. subterraneum* are rejected irrespective of their reaction upon the alternative *Trifolium* species. Whilst this diminishes the potential germplasm base for selection of inoculant strains for the new clovers, if the germplasm base is initially broad enough, suitable strains are likely to be found. Fig. 3 illustrates data collected whilst seeking inoculants for *T. vesiculosum*, *T. purpureum*, *T. glanduliferum* and *T. incarnatum*. The commercial inoculants WU95 and TA1 did not fix nitrogen optimally in association with *T. glanduliferum* and several others of these new species (Howieson *et al.* 1996) and could not be considered as inoculants.

Fortunately, strains such as WSM1328 and WSM409 were broadly effective at nitrogen fixation with the range of clover species under investigation, including *T. glanduliferum* and subterranean clover, and their further development to commercial inoculant status has proceeded.

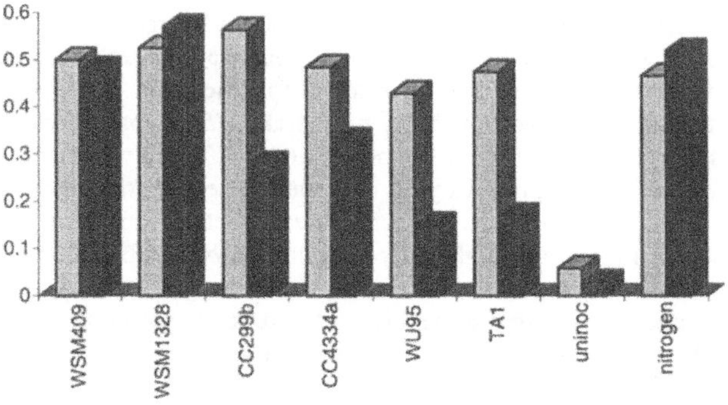

Figure 3. The yield (g/pot) of *T. subterraneum* (light bars) and *T. glanduliferum* (dark bars) grown in a N-free medium when inoculated separately with 6 strains of *Rhizobium trifolii.* The commercial inocula for clovers at the time of these experiments were WU95 and TA1.

1.1.3 Adaptation to the Edaphic Environment.

Pasture legumes, through their capacity to regenerate, form the basis of a sustainable and low cost ley-farming system on large areas of agricultural land in southern Australia. For the ley-farming system and derivatives of it (Reeve and Ewing 1994) to maintain viability and productivity, symbiotic nitrogen fixation accruing from the pasture phase must be optimal in every season that the pasture is allowed to regenerate. This will only occur if sufficient root-nodule bacteria survive and are present in high numbers in the soil at the start of each growing season, when seedlings are most susceptible to infection (Bhuvaeswari *et al.* 1980, Evans *et al.* 1988). The capacity to colonise soils in sufficient quantity to provide prompt nodulation is one of the definitive criteria in our selection of inoculant quality strains of rhizobia.

We move through three phases in the evaluation of persistence qualities in rhizobia.

- Initially, strains are sourced from regions of the world which share edaphic and climatic homology with the proposed target niche for the legume in Australia,
- Strains are then characterised for cultural parameters, such as pH tolerance *in vitro*, growth in broth, stability, and finally
- Strains are assessed for persistence in representative target soils.

These phases are substantial and logistically demanding, particularly when adaptation to the edaphic environment must be assessed at a range of sites over a range of seasons. A fuller description of the first and last phases is provided by Howieson (1995a), whilst the parameters in the second phase are discussed by Howieson (1995b).

1.1.4 Genetic stability

Root-nodule bacteria in the hands of the rhizobiologist are very often subjected to conditions quite alien to their normal *in situ* habitat. Whereas rhizobia are thought to have evolved as saprophytes in soil, in rhizospheres and then into microsymbionts (Parker 1957), in laboratories they are cultured on rich media, stored on the same at low temperature, and often desiccated by vacuum drying to facilitate prolonged storage. Historically, the species *Rhizobium leguminosarum* has reacted to such treatment by mutating in relatively high frequency to a non-infective form (although we cannot say that perhaps the same does not happen *in situ*). This is particularly noticeable in the biovar *viceae* and the propensity for such a transformation can represent an annoying and costly characteristic. For example, elite strain WSM937 was taken through the comprehensive edaphic adaptation assessment outlined above prior to becoming aware of genetic instability. Steps are now taken to ensure that elite strains are stable in the laboratory environment and detection of instability usually means strains are not pursued as potential inoculants.

1.1.5 Growth and survival characteristics in the manufacturing process

Rhizobial inoculants produced in Australia must conform to high standards of quality control. In particular, it is considered essential that commercial inoculants be routinely prepared with 1-3 billion rhizobial cells per gram of peat carrier material. For this level to be attained, a strain must be capable of growth to approximately 5 billion cells per ml in batch culture, followed by 2 or 3 generations once introduced into the carrier. The reasons for this stringency are that the range of sub-optimal conditions encountered in storage of inoculant, in the process of inoculation or during nodulation can be partially overcome by the application of high numbers of rhizobia to the seed at the outset. For example, Fig. 4 shows the cost in terms of reduced dry matter production of *Medicago polymorpha* sown into an acidic environment (pH 4.8) following inoculation with strain WSM688 produced in a peat carrier containing either 100 million or 1 billion cells per gram of peat (Howieson, unpublished data).

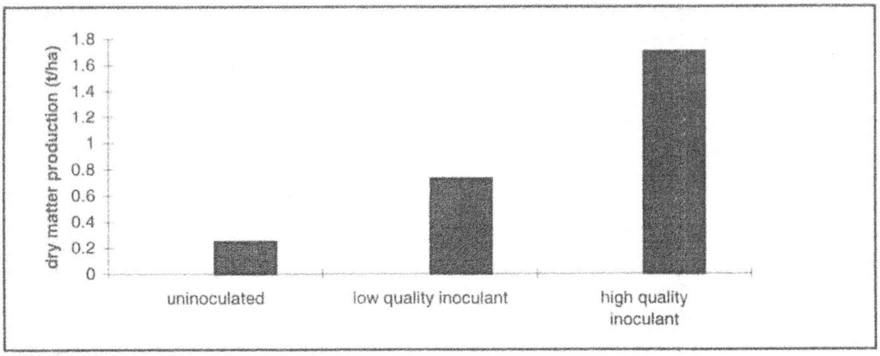

Figure 4. Production from M. polymorpha cv Santiago 70 days after sowing in the field at Muresk (Northam, WA) following different inoculation treatments (low quality = 100 million cells/g; high quality = 1 billion cells/g).

The production loss caused by delayed and reduced nodulation in the treatment with the low quality inoculant was despite seed being sown immediately after inoculation into a moist seed bed. Thus, rhizobia which grow well under the conditions of the manufacturing process, which multiply in the carrier material and which survive well once inoculated onto legume seed are sought when selecting inoculant quality strains of root-nodule bacteria.

2. THE ROLE OF THE LEGUME BACTERIOLOGIST IN STRATEGIC RESEARCH

The application of molecular techniques to studies of rhizobial ecology began in earnest in the late 1980's and early 1990's. This was precipitated by the use of rhizobia as a model organism in molecular biology by several groups around the world which has lead to a sophisticated analysis and understanding of the function of genes required for nodulation, amongst others. There have been two major uses of this new technology. Firstly, by linking introduced genes to the regulators of some of the well characterised nodulation (*nod*) genes it is now possible to make novel investigations into key events in the early steps of nodulation. Reporter genes such as *lacZ* and *GusA* (Wilson *et al.* 1994) have been widely used in this manner. In our laboratory we are currently selecting appropriate *Bacillus thuringiensis* toxin genes to link to *nod*-genes in a strategy with which we expect to control *Sitona discoideus* damage in medic and lucerne pastures. Secondly, the advent of DNA amplification technology in combination with related genetic techniques (see Howieson and Rome 1996) has provided a means with which to genetically categorise different strains of rhizobia at the within species level. This undertaking has historically proven very difficult. Hence, processes such as legume root exudate-*nod* gene interactions, competition between rhizobia for substrates and nodulation, rhizobial population dynamics and population biodiversity can now be studied with the aid of molecular tools. These tools will become part of standard techniques for the rhizobial ecologist in the near future. I will now briefly consider two research areas to directly benefit from this new molecular technology; quantifying rhizobial biodiversity and manipulating competition between strains.

2.1 Understanding rhizobial biodiversity

In selecting improved inocula for the new legumes introduced into Australia, two questions are fundamental.These are;

- How genetically and phenotypically variable are root-nodule bacteria in their natural environments, and,
- How much variability do we need to capture to satisfy the requirements of new legumes introduced into modern agricultural systems?

By using standard techniques for host-strain matching (e.g. Vincent 1970) it has long been possible to investigate the variability in nitrogen fixation that arises when different genotypes of the host and microsymbiont are paired. This has given some indication of the biodiversity of root-nodule bacteria collected from within a given geographic region. However, the question has always remained whether two strains which give a similar reaction when matched with a particular host genotype are in fact genetically the same. By analysis of a portion of the bacterial DNA using the Polymerase Chain Reaction (PCR, Harrison *et al.* 1992, Richardson *et al.* 1995) it is now possible to fingerprint strains according to their genetic makeup. In the example in Fig. 5, five strains of rhizobia were collected from *Biserrula pelecinus* at one site (within a small field) in Sardinia, with a view to exploring the diversity of rhizobia associated with this species. The strains were examined for both their nitrogen fixation potential with three genotypes of the host legume as well as their PCR amplification profiles.

Based upon the phenotypic analysis (Fig. 5), it appeared that the five strains reacted differently on the three host genotypes and that therefore they might be considered to be separate entities. This would be a surprising result, given their original close proximity at collection. However, the agarose gel (Fig. 6) clearly shows that in fact all strains (arranged in the same order as for Fig. 5) were genetically distinct.

The power of the PCR technique in application to rhizobial ecology lies in its ability to rapidly distuinguish between strains in this manner. With continued application, we expect to be able to answer the second question which relates to the amount of variability in root-nodule bacteria that is required to be preserved for future activities in matching rhizobial strains to legume hosts and the target environment.

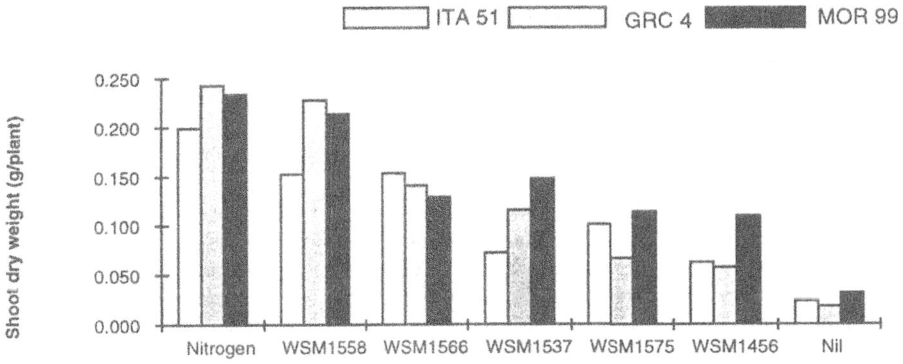

Figure 5. The yield of 3 genotypes of B. pelecinus grown in a N-free sand medium when inoculated separately with 5 strains of root-nodule bacteria isolated from this species (A. Loi, J. Howieson and S. Carr, unpublished data).

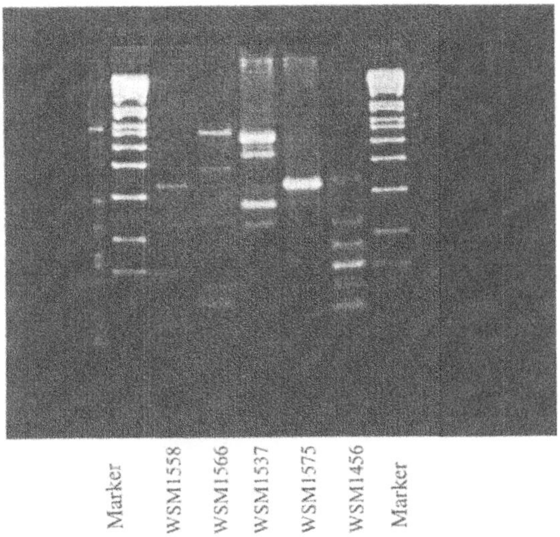

Figure 6. An agarose gel showing distinct DNA amplification banding patterns for the 5 strains of root-nodule bacteria from B. pelecinus in Fig. 5 (A. Loi and J. Howieson unpublished data).

2.2 Manipulating competition between strains

The availability of molecular techniques has provided several novel ways to experimentally address competition for nodulation.

(a) The major pragmatic hindrance to studying competition for nodulation between strains has been the time required to isolate strains from nodules at the completion of experiments and to then identify those strains in order to determine the outcome of the experiment. The availability of marker genes such as *GusA* and *LacZ* combined with the ability to insert them into the rhizobial genome at a point where no major bacterial functions are compromised, enables the researcher to visualise the outcome of competition experiments. Thus, following exposure to the appropriate substrate, nodules containing the transformed rhizobia are stained and nodule occupancy can be scored on the basis of the colour of the nodule.

(b) Investigation of the biochemical signalling that precedes nodulation, and the genetic control of these signals has provided the opportunity to intervene in the nodulation process at the recognition stage. One of the most researched genera in this regard is *Medicago*. In its symbiosis with *Rhizobium meliloti* there is evidence that plant genotypes release a suite of molecules which differentially induce rhizobial *nod* genes. For example, *M. littoralis* appears to produce betaines rather than flavonoid compounds in its seed coat, whereas *M. polymorpha* and *M. orbicularis* produce more flavonoids than betaines (Fig. 7). It has been revealed that the betaine compounds induce the gene *nodD2* in *R. meliloti* whereas the flavonoid compounds induce *nodD1* (Phillips *et al.* 1994). Knowing the different functions of these inducer molecules and the sites of their reception in rhizobia provides the oportunity to intervene in legume-rhizobial recognition because manipulation of the rhizobial *nod*-genes is relatively achievable in modern laboratories. So, where competition for nodulation by background rhizobia creates a challenge to optimal nitrogen fixation, there is the potential to create an exclusive combination of legume and rhizobia in the inoculant by tailoring the inoculant rhizobial *nod*-genes to recognise distinct inducer molecules. The steps required in such research entail:

- the selection of a legume genotype which produces exudates which are not powerful inducers of the *nod*-genes in the rhizobia dominating the soil, and then to
- engineer or select an inoculant strain to recognise these specific inducers and to synthesise the appropriate *nod*-factor.

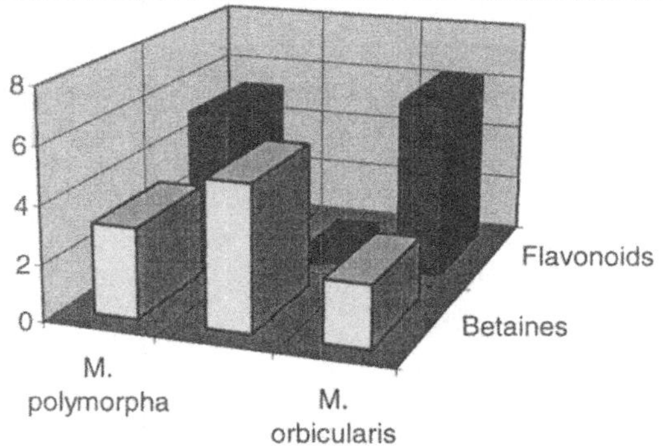

Figure 7. Nod-gene inducing compounds (expressed as B-glactosidase units) released from the seed coat rinses of three species of Medicago (adapted from Phillips *et al.* 1995)

(c) The creation of an exclusive recognition process is one way in which to manipulate competition for nodulation between rhizobial strains. In another approach, it has been shown that the inorganic nutritional environment of the rhizosphere can influence the attachment of rhizobia to the legume roots and root hairs. The concentration of phosphate in solution was found to condition the level of attachment of two contrasting rhizobial strains to *Medicago polymorpha* (Fig. 8).

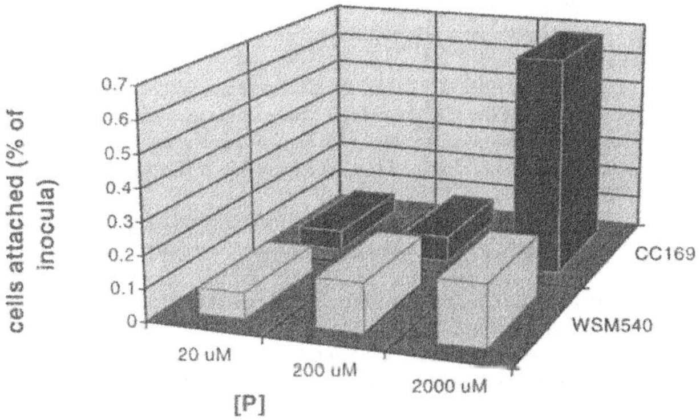

Figure 8. The attachment of two strains of R. meliloti to roots of M. polymorpha as influenced by the concentration of phosphate (P) in solution (Howieson *et al.* 1993)

Both WSM540 (acid tolerant) and CC169 (acid sensitive) attached at similar levels when the P concentration was 20 uM. Raising the concentration 10-fold to 200 uM saw WSM540 attach twice as efficiently as CC169, whereas raising it 100-fold favoured CC169. The knowledge that P concentration can have such a marked effect on attachment of rhizobial cells to legume roots might well be useful in a strategy to manipulate competition for nodulation. It would present no great pragmatic difficulty to pellet legume seed with soluble phosphate based compounds in order to change the P status of the rhizosphere.

3. CONCLUSIONS

There has been a pronounced movement towards the development of alternative pasture and grain legumes in Australia. Some of these legumes have microsymbiotic partners which are completely new to Australian agriculture whereas others interact with species of root-nodule bacteria already in the soil. To extract the full value from legume introduction and evaluation requires the selection of appropriate inoculant strains of rhizobia. Special qualities are sought in inoculant strains, particularly with respect to their potential to persist in the target environment and to be compatible with important existing legumes. The ecological research necessary to understand the reaction of rhizobia introduced to their new environment has been greatly expedited by the availability of new molecular tools. These tools will quickly become part of basic techniques in modern laboratories where the host-rhizobia relationship is researched.

ACKNOWLEDGEMENTS

GRDC provides substantial funding for the rhizobial research referred to in this paper. I thank Dr Steven Carr for useful comments in preparing this manuscript, and to Mr Ron Yates for excellent technical support in conducting the field experiments.

REFERENCES

Bhuvaneswari T.V., Turgeon B.G. and Bauer W.D. (1980). Early events in the infection of soybean (*Glycine max* L. Merr.) by *Rhizobium japonicum*. I. Localization of infectible root cells. *Plant Physiology* **66**, 1027-31

Carr S.J. (1998). A pasture ideotype for farming systems on the acid soils of the drier Western Australian wheatbelt. Proceedings Genetic resources workshop Perth, February 1998 (Ed. P.S. Cocks). CLIMA Occassional Publication (in prep.).

Carr S.J., Howieson J.G., Nutt B.J. and Yates R. (1996). Specificity in the *Rhizobium leguminosarum* bv *viceae - Vicia faba* symbiosis. Proceedings of the 11th Australian Nitrogen Fixation Conference: Perth.

Evans J., Hochman Z., O'Connor G.E. and Osborne G.J. (1988). Soil acidity and *Rhizobium*: their effects on nodulation of subterranean clover on the slopes of southern New South Wales. *Australian Journal of Agricultural Research* **38**, *605-18.*

Harrison S.P., Mytton L.R., Skot L., Dye M. and Cresswell A. (1992). Characterisation of *Rhizobium* isolates by amplification of DNA polymorphisms using random primers. *Canadian Journal of Microbiology* **38**, 1009-1015.

Howieson J.G. (1993) Alternative pasture legumes and rhizobial research in CLIMA subprogram P3. 'Proceedings of the Second National Alternative Pasture Legumes Workshop' (Eds D.L. Michalk, A.D. Craig and W.J. Collins) Primary Industries of S. Australia Technical Report 219.

Howieson, J.G., Ewing, M.A., Robson, A.D. and Abbott, L.K. (1993). External phosphate and calcium concentrations, but not the products of rhizobial nodulation genes, affect the attachment of *Rhizobium meliloti* to roots of annual medics. *Soil Biology and Biochemistry* **25**, 567-573

Howieson J.G., Carr S.J. and Yates R.J. (1996). Selection of effective, acid tolerant rhizobia for new *Trifolium* species adapted to acid, infertile soils in southern Australia. Proc. 4th Int Symp. on Plant-Soil Int. at low pH. Belo Horizonte: Brazil.

Howieson J.G., Loi, A. and Carr S.J. (1995.) *Biserrula pelecinus* L.- a legume pasture species with potential for acid, duplex soils which is nodulated by a unique root-nodule bacteria. *Australian Journal of Agricultural Research* **46**, 997-1009.

Howieson J.G and Loi A. (1994). The distribution and preliminary evaluation of alternative pasture legumes and their associated root-nodule bacteria collected from acidic parts of Greece (Serifos), Morocco, Sardinia and Corsica. *Agricola Mediterranea* **124**, 1-17.

Howieson J.G. (1995a). Characteristics of an ideotype acid tolerant pasture legume symbiosis in Mediterranean agriculture. Proceedings of the Third International Symposium on Plant Soil Interactions at Low pH, *Plant and Soil* 171, 71-76.

Howieson J.G. (1995b). Rhizobial persistence and its role in the development of sustainable agricultural systems in Mediterranean agriculture. *Special Volume Soil Biology and Biochemistry* **27** 603-610.

Howieson J.G. and Rome S. (1996). Rhizobial ecology - challenges and opportunities emerging from a marriage of molecular biology and applied science. 'In Recent Progress in Research on Symbiotic Nitrogen Fixation' (Ed. J.S. Pate) CLIMA Occasional Publication 14.

Nutt B.J., Carr S.J. and Samaras S. (1996). Collection of forage, pasture and grain legumes and their associated rhizobia from selected Greek Islands. CLIMA Occasional Publication 13.

Parker C.A. (1957). Evolution of nitrogen fixing symbiosis in higher plants. *Nature* **179**, 593-594.

Phillips, D.A., Dakora F.D., Sande E., Joseph C.M. and Zon J. (1994). Synthesis, release and transmission of alfalfa signals to rhizobial symbionts. *Plant and Soil* **161**, 69-80.

Phillips, D.A., Wery J., Joseph C.M., Jones A.D. and Teuber L.R. (1995). Release of flavonoids and betaines from seeds of seven *Medicago* species. *Crop Science*. **35**, 805-8.

Reeve, T.J. and Ewing, M.A. (1993) Is ley farming in Mediterranean zones just a passing phase? *In* 'Proceedings. XVII International Grasslands Congress' (ed M. Baker) pp2169-2177. (New Zealand Grassland Association: Palmerston North).

Richardson, A.E., Viccars, L.A., Watson, J.M. and Gibson, A.H. (1995). Differentiation of *Rhizobium* strains using the polymerase chain reaction with random and directed pimers. *Soil Biology and Biochemistry* **27**, 515-524.

Vincent J.M. (1970). 'A manual for the practical study of root-nodule bacteria'. IBP Handbook 15, (Blackwell Scientific Publications: Oxford).

Wilson K.J., Sessitch A., Corbo J., Giller K.E., Akkermans A.D.L. and Jefferson R.A. (1994.) Molecular markers as tools to study the ecology of microorganisms. pp149-56. In 'Beyond the Biomass: Compositional and functional analysis of soil microbial communities.' (eds R.K. Dighton and K.E. Giller) (John Wiley: U.K).

CHAPTER 10

The agronomic purpose of evaluation: Relating nurseries to field situations

Pier Paolo Roggero[1] and Claudio Porqueddu[2]

[1] Dipartimento di Biotecnologie agrarie e ambientali, Università degli studi, via Brecce bianche, 60131 Ancona, Italy

[2] C.N.R., Centro di Studio sui Pascoli Mediterranei, via E. De Nicola, 07100 Sassari, Italy

Abstract

The most successful Mediterranean pasture legumes varieties were obtained from the exploitation of genetic resources collected in the Mediterranean basin. The selection programs are often based on extensive germplasm collections followed, initially, by seed multiplication in a nursery, where preliminary evaluation of the bio-agronomic characteristics is made. This crucial phase can be effectively utilised, if the evaluation characters are chosen to test the performances of the germplasm and an initial selection is made for new varieties in the field. Emphasis has been given in the past to climatic and edaphic adaptation, but experience has shown that the successful introduction of a new variety also depends on the crop management. This is closely related to the type of farming system and the biotic competition (weeds, pests, diseases) in which it is grown. To date, an ideotype for new varieties of annual self-reseeding legumes cannot ignore the increasing requirement of a low-input oriented agriculture, that in many developed countries includes the use of these forage crops for alternative, non-forage purposes.

In this paper, the ecological characteristics of the Mediterranean rangelands are taken as a model for setting up the breeding objectives and as a guideline for choosing relevant morphological and agronomic characters to be assessed at the nursery level. Examples of the agronomic purpose of germplasm evaluation are given from recent and past screenings of annual clovers and medics, and field experienced introductions of annual pasture legumes varieties in deteriorating swards of the Mediterranean basin, both for forage or other non-productive uses.

An ecological analysis of germplasm collection sites, including the recent management history of the ecosystem and other relevant biotic indicators can provide useful information, that correlated with agronomic traits, assists in the selection of accessions to keep in the following phases of selection, thus improving the efficiency of the agronomic screening.

In testing for varieties to be grown under strong limiting environmental conditions, the climatic and edaphic environment of the nursery should be as similar as possible to the environment where the material is to be grown. The most limiting environmental factors, temperature and soil moisture, should be accurately monitored over the biological cycle of the collected plants, for extrapolation of the data collected in the nursery (e.g. flowering dates and duration) to a wide range of field situations. The single plant or single row forage and seed yields are not sufficiently correlated to the yields at the plant community level, but observation of some key yield components that are less influenced by the management, such as seed size, hardseededness, pod morphology and winter growth can contribute to the identification of promising genotypes from nursery data.

Keywords Adaptation, Mediterranean basin, pasture ecology, diversification, sustainability

1. INTRODUCTION

The first step in the agronomic evaluation of collected germplasm is the establishment of spaced plant or row nurseries, with the dual objective of obtaining sufficient seed for subsequent agronomic screening and of acquiring some preliminary data on the agronomic value of the collected populations (Crawford 1983). The nursery phase can be effectively utilised to advance selection if the breeding objectives have been clearly defined and if the

S. J. Bennett and P. S. Cocks (eds.), Genetic Resources of Mediterranean Pasture and Forage Legumes, 107–119.

observed agronomic characters are maintained in the subsequent phases of selection and in the field.

The current concern over the use of annual pasture legumes to improve the sustainability of intensive farming systems, or for use in non-productive uses in the Mediterranean climate areas is highlighted in this paper to indicate an alternative exploitation of collected genetic resources that may not be suitable for productive uses.

The ecological characteristics of the Mediterranean rangelands are taken as a model for determining breeding objectives, for guidance in the choice of relevant agronomic characters to be assessed at the nursery level, and to match the collection missions to specific sites.

The objective of this paper is to give some remarks on which agronomic traits are important to record and how their expression in nursery rows compares with field situations, in order to improve the efficiency of a breeding programme.

2. BREEDING OBJECTIVES, FIELD CONDITIONS AND NURSERY TRIALS

The evaluation of new germplasm during the initial phases of seed multiplication is a critical point in the selection of desired varieties. The risk that may occur is that valuable accessions that would perform well under field conditions are discarded as they are not successful under nursery and plot conditions. Many agronomic traits evaluated during these initial screenings are heavily influenced by the management of a nursery situation.

New improved varieties in the Mediterranean basin, when released in Australia were often found to be unsuccessful under farm conditions (Roggero *et al.* 1990). Many promising varieties following plot experiments aimed at reducing the typical feeding gaps of the mediterranean pastoral systems were not transferred into practice. Possible solutions to these problems depend amongst others on the correct approach adopted for agronomic evaluation. This is discussed in the following sections.

2.1 Screening germplasm collections for sustainable pasture systems

There is considerable concern in the scientific community about the sustainability of agricultural systems, and plant breeders and agronomists are rapidly adapting their scientific objectives to the new trend of the '90's (Sleper *et al.* 1991). The concept of sustainability in agriculture, depends on the socio-economical context in which it is practised. Some important features of sustainable agriculture in developed countries contrast with those of developing countries. However, there is a general consensus about the unsustainable economic, social and environmental consequences of external-input-oriented agriculture (Roggero *et al.* 1996). There is also a consensus on the need to achieve more efficient and rational use of labour, knowledge, and management skills, to replace off-farm inputs (Pretty and Howes 1993). The need to reduce off-farm inputs to reduce costs and enhance a farmer's income, should direct breeders and agronomists towards developing nutrient- and water-use efficient varieties, with a high persistence under sub-optimal conditions (Bramel-Cox *et al.* 1991). The high level of diversification of farming systems that is present in the Mediterranean basin is sufficient to identify opposing priorities in adjacent countries. For example, in the northern part of the Mediterranean basin, self-reseeding legumes are often recommended as species to be used for extra-productive purposes, whereas in the southern regions, the same species and varieties are used for rangeland improvement.

One important concept to emerge from this international debate, is that breeders should make an effort to adapt the variety to the environment instead of adapting the environment to the variety (Coffman and Smith 1991). One key point of this approach is to identify what is the "environment", as in most cases this is identified almost exclusively by its climate and soil type. Varieties may already be available that are adapted to components of a pasture system. However, other biotic and antrophic factors may also be important in the survival of selected varieties for a given environment.

Breeding strategies should be directed by the function of a sustainable pasture and farming system. Farm sustainability is negatively affected by fixed rules and schemes, as farmers always have to cope with erratic weather patterns and fluctuating prices. For these reasons, in many mediterranean-type areas of the world, crop diversification is a way for the farmer to regulate his income (Buddenhagen 1990).

The field conditions in which the released varieties are to be grown must be carefully considered when evaluating the agronomic characteristics of a germplasm collection following the nursery trials. If a flexible farm management scheme is required, then flexible, adapted material should be produced. However, it may not be possible to achieve this using only one species or variety (Francis 1987). It is common to use mixtures of varieties of the same or different species for pasture over-seeding in the Mediterranean (Bullitta *et al.* 1989; Olea *et al.* 1985). The use of mixtures can extend the growing season and help to reduce the damage caused by biotic adversities (Niks *et al.* 1993).

To account for these factors at the nursery level, evaluation of agronomic characters should be qualitative rather than quantitative and related to persistence and a flexible use rather than to yield. The assessment of useful morphological characters, such as pod spininess of medics (*Medicago*) and leaf markers should still be recorded.

2.2 The semi-arid Mediterranean pasture ecosystem: a model for breeding objectives

The overall biodiversity of a Mediterranean pasture ecosystems is very high and hence complex interactions of the biotic and abiotic components take place. The collection of selected biotic components (e.g. one legume species) from such complex ecosystems is often made regardless of its role in the ecosystem. This role is not necessarily the same as the role the species is going to have in a new ecosystem (i.e. an improved pasture). An example of this is the success or promising success of some introductions like *Ornithopus* species and *Biserrula pelecinus* L. in Australia (Howieson *et al.* 1996; Revell 1992), that in the Mediterranean basin are considered species of minor agronomic importance.

As annual species usually dominate grazed rangelands, the dynamics of the botanical composition of a pasture may vary considerably from year to year (Rivoira 1976). In wild Mediterranean plant communities it is common to find a number of annual legumes present, and for one species the intrinsic diversity is such that it can display a wide range of characters of agronomic importance, mostly related to reproduction. In a recent study, conducted over an area of about 1 ha in north-eastern Sardinia, Italy, 165 species were identified, among which 33 legumes (32 of which were annuals) were recorded. These were composed of 12 *Trifolium*, 5 *Vicia*, 4 *Lotus*, 4 *Medicago*, 3 *Lathyrus*, 2 *Ornithopus*, 1 *Lupinus*, 1 *Scorpiurus* and 1 *Tetragonolobus* species (Bullitta *et al.* 1990, Bagella and Roggero 1996). The ecological role of these species was found to vary from year to year depending on the selection pressure of the dominant ecological factor. There are a number of ecological factors that can influence the species and the genetic diversity of a pasture, and its dynamics with time. Apart from the rainfall and temperature regime, biotic factors like concomitant species, insects, diseases, grazing pressure, management, or abiotic factors such as fire, dynamics of macro-nutrients in the top soil layers or waterlogging can impose a strong selection pressure. Ecological factors, including agricultural management, have different effects on the separate species, according to their reproductive biology. For example *Trifolium nigrescens* is self-incompatible (Bullitta *et al.* 1989b), while many other annual pasture legumes are cleistogamous. Other variations in reproductive biology include hardseededness and seed dormancy of the seed bank, intraspecific variation in flowering time and duration, and seed burial depth amongst others. If we restrict the analysis to only one species, for instance an annual pasture legume, variations in its genetic population structure may develop over time, depending on the occasional selection pressure of the concomitant biotic and abiotic factors (Piano *et al.* 1993). Knowledge of the population structure of the collected species is therefore important when determining its breeding program, particularly when the reproductive biology of the collected species has not been sufficiently investigated. Intraspecific variation of the reproductive features of the germplasm collected at one site can be viewed as a "security mechanism" for survival, that is not necessarily of agronomic value, but is essential to the persistence of the species in that particular ecosystem. At each collection site, duration of flowering time is often determined by the date of flowering of different genotypes, and this variation is also related to the germination pattern of the seed bank in that year.

During agronomic evaluation of the germplasm, starting at the nursery level, it is important to estimate such variability and to relate it to the ecological conditions of the site of origin.

2.3 Relating bio-agronomic characteristics of accessions to the ecological data of the collection site

The relationship between environmental indicators of a collection site and the bio-agronomic characteristics of the collected germplasm can be observed during the nursery phase. This type of observations aids the classification of the germplasm collection. It is important to be aware that many morphological and agronomic characters are related to earliness of flowering and that earliness is frequently the agronomic variable most closely related to the climate of the collection site. To isolate the direct and indirect correlations appropriate statistical analysis is required (e.g. path coefficient regression analysis; Bullitta *et al.* 1989b). Multivariate statistical procedures are nowadays an easily accessible tool for the definition of homogeneous clusters of collected accessions. Within each cluster only a few accessions that are representative of each cluster can be kept in the subsequent dense stand evaluation, (Loi *et al.* 1995, Porqueddu *et al.* 1996).

From the literature there is clear evidence of the selection pressure made by the climatic and edaphic factors, while less evidence is recorded about the effects of the biotic factors on the population structure of a species (Piano and Francis 1992, Piano *et al.* 1993a). This is mainly due to the lack of information about the biotic components of ecosystems where the accessions are collected. A good step towards a better understanding of the biotic component is to classify the collection sites by means of biotic indicators, such as the floristic, geobotanical and phytosociological analysis (Biondi 1994, Rieley and Page 1990).

3. SUCCESSFUL PASTURE IMPROVEMENT: THE FINAL TASK OF THE AGRONOMIC EVALUATION

Annual self-reseeding legumes have been extensively used for pasture improvement in the Mediterranean basin for over thirty years. Most of the varieties available in the market are Australian selections originating from the Mediterranean basin. In many cases the agronomic performance of these varieties under plot and field trials are satisfactory in the first few years after sowing, but subsequently, the contribution of the introduced species (the sown variety) to the forage yield often decreased as native species become more competitive (Bullitta *et al.* 1990, Cerrasco *et al.* 1991, Sitzia *et al.* 1996). In Spain, the reason for this lack of persistence was attributed to climatic factors such as low winter temperatures, spring rainfall, time of the break of season (Olea *et al.* 1995). When a natural seed bank of pasture legumes is present in the pastures of the Mediterranean basin, fertilisation of that pasture without over-seeding may be sufficient to obtain satisfactory agronomic results, particularly if it is repeated for some years (Bullitta *et al.* 1989, Yañez *et al.* 1991). This can be partly explained by the relatively low solubility of phosphate (P) fertilisers in the soil and to the progressive enrichment of the seed bank in the years following the first fertilisation as a consequence of an improved seed yield. In fact it is likely that the seed bank of annual pasture legumes in a natural unfertilised pasture is not quantitatively sufficient to guarantee a prompt response to the P soil enrichment, but that the "domestication" of the structured wild populations under less oligotrophic conditions is often agronomically more reliable than the introduction of selected uniform varieties (Yañez *et al.* 1991). The role of the introduced legume varieties, in these cases, is only to cover the yield gap of the first few years following the initial fertiliser application. Therefore, wild ecotypes of pasture legumes have a "plastic" population structure, that allows persistence under sub-optimal conditions, and a satisfactory production when environmental conditions become more favourable.

When spaced plants derived from a collection are grown in a nursery, they represent the expression, under artificial conditions, of single genotypes that are part of a complex community whose agronomic value is based on the inter and intraspecific diversity. Therefore if the aim is to achieve a structured community, characterised by complex biological interactions, that can guarantee the persistence and productivity of the most valuable species in a given environment, we must be aware that this will be hard to achieve with an artificial system based on the use of a single uniform variety. On the other hand, it would also be expensive to produce seed of a multiline "structured" variety.

4. EVALUATING ADAPTATION AND PERSISTENCE OF PASTURE LEGUMES FROM NURSERY TRIALS

4.1 Soil and climate adaptation

The most important attribute of a new pasture or forage legume variety is its adaptation to the climate and soil type. For this reason the agronomic evaluation of any collection of species should be conducted in the ecosystem where it is expected to perform. Where nurseries are concerned, if the climate (rainfall and temperature) and soil characteristics (soil fertility, texture and pH) are similar to those of the target environment, adaptation of the genotypes under evaluation is an obvious attribute. However, there is the need to maximise seed yield in the early phases of multiplication, and therefore the collected accessions are often screened under the absence of biotic competition for water and nutrients.

In legumes an example of the marked difference between the site of origin and the nursery is given by the effect of the Rhizobium strain x soil type interaction on N uptake and hence on production (Howieson 1995). If we assume that the variation of plant production is explained to a certain extent by the symbiont strain features, we cannot ignore that the overall agronomic result in the nursery is affected by the success of the symbiosis (Howieson and Rome 1996).

4.2 Persistence

Many factors are associated with persistence and, because of this, the problem is genetically complex. Plant characters important for persistence in one environment, e.g. drought tolerance, may not be necessarily in another environment where another character is more important for survival, e.g. resistance to pests (Caradus and Williams 1989). A low level of domestication, especially in pasture plants, is often a positive attribute for persistence, e.g. shattering pods, abundant hard seeds, small seeds and special morphological characteristics for propagation (Jaritz 1987). It would be of great practical value to predict persistence during the preliminary screening in nursery rows. By doing so the selection process would be much quicker than by conventional methods or by subjecting mixtures to natural selection (Cocks 1992 1993).

In annual pasture legumes the persistence attributes include a number of mechanisms that are related to a species reproductive biology and ecology (Ehrman and Cocks 1996). Outbreeding species are often characterised by a high within population variability that can provide an ecological advantage under sub-optimal conditions, but the multiplication and breeding strategies of outbreeding species is complex and requires the nurseries to be isolated against undesirable cross pollination of adjacent accessions (Moreno-Gonzáles and Cubero 1993).

4.3 Time of flowering and maturation period

Considerable importance is placed on the time of flowering and seed ripening in relation to the length of the growing season for which a new cultivar is being selected (Crawford 1983).

The capacity to flower early and to keep on flowering and maturing pods is a distinct advantage in regions with an extremely variable spring climate. For example, variation in flowering time is considered a key factor in the adaptation of cluster clover (*T. glomeratum*), the most widespread and ecologically most successful clover in southern Australia (Smith *et al.* 1995). Regions with late spring frosts may benefit from late-flowering species with the capacity also to respond to late spring rainfall, while dry regions would take advantage from early maturing types.

Flowering time is highly correlated to the climatic characteristics of the collection site, and is also correlated to other relevant seed and forage yield performances (Loi *et al.* 1995 1996). This suggests that of all possible selection factors, spring temperature and rainfall are important in the adaptation of an annual pasture legume to a given environment. Therefore, populations with high intraspecific variation in flowering pattern could be subjected to relatively rapid genetic drift when introduced to a new environment, particularly if the spring rainfall limits the reproduction of later maturing types (Esquinas-Alcázar 1993). However, a wide within-accession variation of flowering time can be related to a relatively low selection pressure at the collection site if the environment is less limiting, as shown for Sardinian subterranean clover populations (Piano *et al.* 1996).

A long duration of the flowering phase and the ability to maintain flowers and ripe legumes at the same time, may be an important persistence mechanism against adverse abiotic factors or a flexible agricultural management. In nature, long flowering duration is maintained through a mixture of early and late genotypes, and the seed production and hardseededness of both genotypes is sufficient to guarantee the persistence of the seed bank. Data on the date of flowering collected in a nursery can provide valuable preliminary information about earliness if the reproductive mechanisms of the collected species are known, the nursery environment (light, photoperiod and temperature) is monitored and the ecological data of the collection sites are available.

Interactions with pasture management and the affect of climate on the reproductive development and growth of many mediterranean pasture and forage legumes is still unknown. Recent studies have shown that mathematical models developed for the reproductive behaviour of the pea (Pisum sativum) are not easily transferable to other pasture and forage legumes (Lelievre 1996).

4.4 Seed production and its components

Legume seed yield, namely the number of pods produced per unit area, can be analysed by its components the number of pods per plant, the number of seeds per pod and the seed size in relation to plant density. Some seed yield components can be accurately evaluated during nursery trials (Cocks 1990b). Pod number is one of the most important seed yield components, but the number produced are often effected by management, photoperiod and temperature rather than showing variation in relation to genotype. It is usually higher in outbreeding species. In the case of persistence, seed number is more important than seed weight. Therefore the production of numerous small seeds represent an ecological and agronomic advantage for pasture species, as the competition between the reproductive and vegetative growth is limited.

Seed size influences the persistence of pasture legumes in several ways. For example, larger-seeded species will emerge from greater depths than smaller species (Carter and Challis 1987), but are less likely to survive ingestion by sheep (Thompson *et al.* 1990). In the past plant breeders have selected pasture legumes on the basis of the "seedling vigour", which was related to seed size, larger-seeded species having larger seedlings (Crawford *et al.* 1989).

A dense population of prostrate, small seedlings should have a better survival rate and maintain an adequate photosynthetic area than a sparser stand of large seedlings. Such seedling populations would be obtained by small rather than large seeded strains (Francis and Gladstone 1974). Seedlings from small seeds will emerge only if the seeds are near the surface, which is normal in a permanent pasture, but may not be so in an arable farming system (Puckridge and French 1983). The choice between small or large seeded species or varieties will more frequently be in favour of the latter because of better establishment in the absence of specialised equipment to prepare the seedbed (Jaritz 1987) and because the seed grower would benefit from higher seed weights.

Seed size is considered a relatively stable parameter (Harper *et al.* 1970), though this is probably the result of using population means. Several studies on medics has shown that, within plants, variation in seed size often depends on the time of production and on the flower position within the plant (Cavers and Steel 1984, Cocks 1990a).

4.5 Hardseededness and hard seed breakdown

The softening of hard seed of several annual medics and subterranean clovers is a 2-stage process; the first occurs at a rate that increases with increasing temperature and the second one requires diurnal temperature fluctuations. These differ between species (Taylor 1993). It is possible that in many circumstances the level of hardseededness may be less important than the pattern in which it is lost (Carr 1996).

In a three year experiment on the hardseededness and pattern of hard seed breakdown on 35 Sardinian accessions of burr medic (*Medicago polymorpha*) and *M. polymorpha* cv. Circle Valley under field conditions (Porqueddu *et al.* 1996) three groups were identified. Marked differences were observed among them in terms of pod dimensions, spine length and seed weight (table 1). Seeds of one group, identified as *M. polymorpha* var. *polymorpha*, were found to be 96% hard, similar to that of cv. Circle Valley. The other two groups were identified as *M. polymorpha* var. *vulgaris* and they had a markedly different pattern of hard seed breakdown.

Table 1. Pod and seed characteristics of three groups of Sardinian burr medic (*Medicago polymorpha* L.) accessions (from Porqueddu *et al.* 1996)

Characters scored	*M. polymorpha* var. *polymorpha*	*M. polymorph* var. *vulgaris* (I)	*M. polymorpha* var. *vulgaris* (II)	*M. polymorpha* cv. *Circle Valley*
No. of coils	3.0	2.5	2.1	4.0
Length of spine (mm)	0.9	1.6	1.4	0
1,000 seeds weight (g)	3.1	2.3	2.2	3.2
Soft seeds* (%)	4	9	16	5

* end of the summer.

Hardseededness of the same Sardinian *M. polymorpha* accessions varied, according to the weather of the previous spring (Fig. 1). The variation recorded among the Sardinian accessions of burr medic suggest that hard seed breakdown and hardseededness are subjected to climatic selection pressures, but also within the same accession a dynamic pattern between years is expected. This may be an explanation for the fluctuating species composition of

Mediterranean rangelands, where persistence is a long-term based concept, not necessarily synonymous with a constant forage yield.

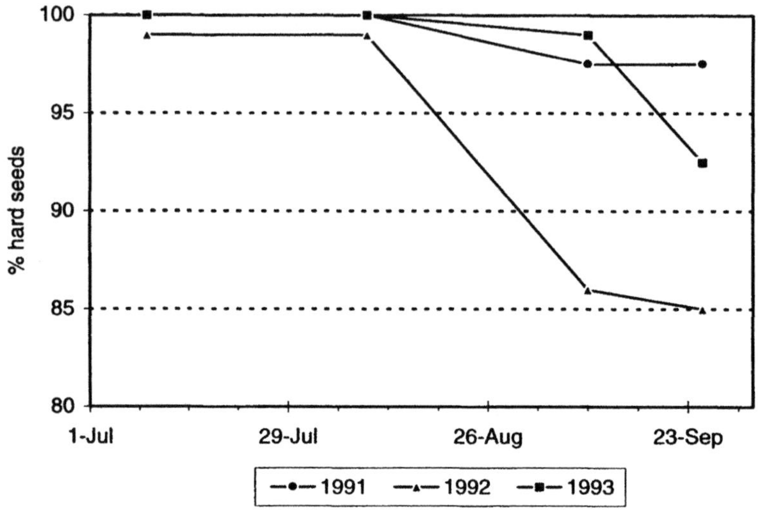

Figure 1. Variation in the pattern of hard seed breakdown of the same *M polymorpha* accessions in relation to the year of seed production (modified from Porqueddu *et al.* 1996)

4.6 Pod morphology and mechanisms of self-reseeding

Small pods are less likely to be selected by grazing sheep in summer than large pods. Small pods also have the advantage that they are easily incorporated into the top soil, although they are subject to greater predation by ants. Pod spines are an important seed dispersal mechanism although they are not essential to persistence. Among the morphological characteristics of the pod, the ones related to pod burial have an important evolutionary advantage in areas that have always been grazed (Piano and Pecetti 1995), while many other morphological attributes may have taxonomic importance with no ecological relevance. Pod burial has a greater effect on increasing emergence levels than any difference in pod structure (Blumenthal and Ison 1996, on murex medic genotypes). Nevertheless when tested in the field, pod burial by sheep trampling during summer grazing improved emergence in a genotype, characterised by lighter, more open pods, that were more easily trampled, than another with heavier, closed pods (Table 2).

Table 2. Effect of summer grazing on the percentage of germinable seeds at the end of the summer. (from Blumenthal and Ison 1996)

Genotype	Grazed	Ungrazed
Medicago murex CD26	43.3	18.8
Medicago murex CD53	18.2	21.4
Trifolium. subterranean cv. Dalkeith	41.8	48.8

Pods characteristics are also important when determining the harvesting technique. For instance the seed harvest of *Serradella* and *Biserrula* does not require vacuum machinery, thus reducing costs and the negative impact on the soil surface.

4.7 Persistence attributes of perennial pasture legumes

Perennial pasture legumes have not been extensively investigated and used in the semi-arid and arid mediterranean areas under rainfed conditions. Most of the perennial forage and pasture legume varieties available in the seed market are not suitable for Mediterranean drylands (Roggero *et al.* 1990). Nevertheless, perennial legume herbs are frequently found in the semi-arid and arid rangelands of the Mediterranean basin and the agronomic potential of this wild germplasm should be exploited for productive and non-productive uses.

The persistence of perennial legumes under mediterranean conditions is heavily dependent on mechanisms of drought tolerance or escape during the summer. Several perennial legumes occur in the natural Mediterranean rangelands, even when heavy grazing pressure annuals are dominant. However there is the possibility of exploiting the genetic resources of perennial legumes, particularly of those species that are able to escape summer drought, to actively re-grow at the season break, to survive heavy grazing pressure and to associate vegetative propagation with self-reseeding ability.

Sulla (*Hedysarum coronarium* L.) is an example of a short lived perennial that is of particular interest for the pastoral areas of the southern part of the Mediterranean basin, where heavy clay and calcareous soils occur (Sulas *et al.* 1995). Another interesting genus of legumes that is suitable for sandy and acid soils is the genus Lotus. It is frequently found in the Mediterranean arid rangelands, but its genetic potential has yet to be exploited.

The physiological mechanisms that allow the persistence of perennial legumes in dry environments are worth further investigation.

4.8 Agronomic attributes of pasture and forage legumes for extra-productive purposes.

During the last decade, in the more developed mediterranean countries, forage and pasture legumes have been increasingly used in non conventional farming systems, where extra-productive targets predominant, such as organic farming (Piano and Talamucci 1996), vineyards and orchards (Caporali *et al.* 1993), firebreaks (Piemontese *et al.* 1996, Masson and Gintzburger 1986), and cover-crops (Etienne and Vleix 1995, Pardini *et al.* 1995a). Specific features are required for specific uses. For swards to be grazed in the forest, shade tolerance, earliness of flowering and a low amount of dead residual is required. For firebreak lines, low sward height and a high regeneration ability are required. For vineyards, earliness, aptitude to survive without utilisation and rapid establishment in autumn are required. Frost tolerance is required in many hill mediterranean regions (Pardini *et al.* 1995b). The non conventional uses cannot be ignored when defining the breeding tasks of Mediterranean forage and pasture legumes, as it is foreseeable that in the more developed Mediterranean countries seed demand for new varieties adapted to specific uses will increase in the incoming years.

5. CONCLUSIONS

The role of pasture and forage legumes in the mediterranean farming systems is evolving according to the socio-economic requirements of the different regions, and to the increasing concern about the eco-compatibility of agricultural practices. Breeding tasks, and hence the agronomic purposes of germplasm evaluation, are currently being directed towards the diversification of genetic resources in view of a flexible utilisation and a sustainable cropping system. A knowledge of the complex relationships that occur in the field between biotic and abiotic components of the ecosystem is an essential step in the identification of breeding tasks. The selection criteria adopted in the earliest phases of the agronomic evaluation should be based on a range of indicators of agronomic value to reduce the risk of loosing valuable variation in characters that are not sufficiently expressed under nursery conditions.

Under sub-optimal and variable environmental conditions, like in many pastoral farming systems of the mediterranean environment, the maintenance of a high level of inter- and intraspecific genetic diversity is essential to achieve satisfactory and persistent pasture swards (Piano *et al.* 1991). Francis (1987) advocated the use of mixtures of genotypes, in the case of cereals under similar environmental conditions, and multiline varieties have been described as a possible strategy to cope with adaptation and resistance to biotic adversities (Borlaug 1981). Pests and diseases are rarely considered to be a problem in the rangelands of the Mediterranean basin, as they often rely on natural communities for their forage production, whose resilience against adversities is based on a dynamic biotic diversity. The complex ecological relationship that exists between pests, pathogens and related predators and parasites in the grazing ecosystems of the Mediterranean basin, can not be transported and established in a distant ecosystem when a new plant species is exported. However, the identification and introduction of the most relevant predators and parasites of the major pests and the use of other agronomic, management and breeding strategies should be investigated and developed to replace the present extensive use of chemicals.

The existing genetic diversity of pasture and forage legumes of the Mediterranean basin has yet to be fully exploited for agricultural and environmental uses (Piano and Talamucci 1996). The identification and selection of new species, and the development of cropping systems with a high level of diversification in space and time represent two important security measures against the possible arrival of new biotic adversities and environmental problems, that would present an unsustainable cost for farmers and society.

ACKNOWLEDGEMENTS

The authors wish to thank Prof. Giuseppe Rivoira and Prof. Fabio Veronesi for the helpful critical reading of this paper.

REFERENCES

Bagella, S., Roggero, P. P, (1996). Effetti del fuoco e della concimazione sulla diversità floristica di un cisteto. *Giornale Botanico Italiano* **130**, 1-499.

Biondi, E. (1994). The phytosociological approach to landscape study. *Annali di Botanica* **52**, 135-141.

Blumenthal, M. J. and Ison, R. L. (1996). Plant population dynamics in subterranean clover and murex medic swards 3. Effect of pod burial, summer grazing and autumn cultivation on emergence. *Australian Journal of Experimental Agriculture* **36**, 533-538.

Borlaug, N. E. (1981). Increasing and stabilizing food production. In 'Plant Breeding II'. (Ed. K. J. Frey.) pp. 467-492. (The Iowa State University Press:Ames.)

Bramel-Cox, P. J., Barker, T., Zavala-Garcia, F., Eastin, J. D. (1991). Selection and testing environments for improved performance under reduced-input conditions. In 'Plant Breeding and sustainable agriculture: considerations for objectives and methods.' (Eds. D. A. Sleper, T. C. Barker, P. J. Bramel-Cox.) pp. 29-56. (CSSA Special publication no. 18: Madison.)

Buddenhagen, I. W. (1990). Legumes in farming systems in Mediterranean climates. In 'The role of legumes in the farming systems of the Mediterranean areas'. (Eds. A. E. Osman, M. M. Ibrahim and M. A. Jones) pp. 3-29. (ICARDA: Printed in the Nederlands).

Bullitta, P., Bagella, S., Caredda, S., Roggero, P. P. (1990). Fire and shrub control in the Sardinian Mediterranean environment. Abstracts of the IV International Rangeland Congress, Montpellier, 22-26 April 1991, p. 395.

Bullitta P., Bullitta S., Roggero P. P. (1989). Agronomic methods to increase pastureland production in Mediterranean marginal areas. Proceedings of the XVI International Grassland Congress, Nice, 4-11 October 1989: 1591-1592.

Bullitta S., Roggero P. P., Viegi L., Veronesi F. (1989). *Trifolium nigrescens* Viv. subsp. *nigrescens*, leguminosa foraggera utilizzabile nel miglioramento dei pascoli in Sardegna: I. Prime acquisizioni su popolazioni naturali sarde. *Rivista di Agronomia* **3**, 289-296.

Caporali, F., Campiglia, E., Paolini, R. (1993). Prospects for more sustainable cropping systems in Central Italy based on subterranean clover as a cover crop. Proceedings of the XVII International Grassland Congress, New Zealand and Australia, 2197-2198.

Caradus, J. R. and Williams, W. M. (1989). Breeding for legume persistence in New Zealand. In 'Persistence of forage legumes'. (Eds. G. C. Marten, A. G. Matches, R. F. Barnes, R. W. Brougham, R. J. Clements, G. W. Sheath), pp. 523-540. (American Society of Agronomy, Crop Science Society of America, Soil Science Society of America: Madison, USA).

Carr, S. J. (1996). A pasture ideotype for farming systems on the acid soils of the drier Western Australian wheatbelt. Centre for legumes in Mediterranean Agriculture (in press).

Carrasco, C. L., Paredes, J., Verdasco, P., Olea, L. (1991). 'Mejora de pastos mediante fertilizacion e introduccion de especies de la campana de oropesa'. (Consejeria de Agricultura de la Junta de Comunidades de Castilla-La-Mancha: Toledo).

Carter E. D. and Challis S. (1987). Effect of depth of sowing medic seeds on emergence of seedlings. Proceedings 4th Australian Agronomy Conference, Victoria, p. 182.

Cavers, P. B. and Steel, M. G. (1984). Patterns of changes in seed weight over time in individual plants. *American Naturalist* **124**, 324-335.

Cocks, P. S. (1993). Legumes from the Mediterranean basin: A continuing source of agricultural wealth for southern Australia. Cooperative research Centre for Legumes in Mediterranean Agriculture, Technical paper no. 1.

Cocks, P. S. (1992). Plant attributes leading to persistence in grazed annual medics (*Medicago* spp.) growing in rotation with wheat. *Australian Journal of Agriculture Research* **43**, 1559-1570.

Cocks, P. S. (1990a). Dynamics of flower and pod production in annual medics (*Medicago* spp.) II in sward at low and high density. *Australian Journal of Agriculture Research* **41**, 923-931.

Cocks, P. S. (1990b). Dynamics of flower and pod production in annual medics (*Medicago* spp.) in spaced plants. *Australian Journal of Agricultural Research* **41**, 911-921.

Coffman, W. R. and Smith, M. E. (1991). Roles of public, industry, and international research center breeding programs in developing germplasm for sustainable agriculture. In 'Plant Breeding and sustainable agriculture: considerations for objectives and methods'. (Eds. D. A. Sleper, T. C. Barker, P. J. Bramel-Cox). pp. 1-10. (CSSA Special publication no. 18: Madison, USA).

Crawford, E.J. (1983). Selecting cultivars from natural occurring genotypes: Evaluating annual *Medicago* species. In 'Genetic resources of forage plants'. (Eds. McIvor, J. G. and Bray, R. A.) pp. 203-215. (CSIRO: East Melbourne).

Crawford, E. J., Lake, A. W. H., Boyce, K. G. (1989). Breeding annual *Medicago* spp. for semiarid conditions in southern Australia. *Advances in Agronomy* **42**, 399-437.

Ehrman, T. and Cocks, P. S. (1996). Reproductive patterns in annual legume species on an aridity gradient. *Vegetatio* **122**, 47-59.

Esquinas-Alcázar, J. T. (1993). Plant genetic resources. In 'Plant Breeding: Principals and prospects.' (Eds. M. D. Hayward, N.O. Bosemark and I. Romagosa). pp. 33-51. (Chapman and Hall: London).

Etienne, M. and Vleix, J. (1995). 'Installation d'élevages ovins sur grandes coupres'. (FAO Meeting on Sylvopastoral systems: Avignon).

Francis C. M. and Gladstones J. S. (1974). Relationship among rate and duration of flowering and seed yield components in subterranean clover (Trifolium subterranean). *Australian Journal of Agriculture Research* **25**, 435-442.

Francis, C. M. (1987). Domestication of indigenous species: Incorporation of useful characters into pasture and forage crops. In 'Legume genetic resources for semi-arid temperate environments'. Proceedings of an international Workshop on Genetic resources of cool-season pasture, Forage and food legumes for semi-arid temperate environments. (Eds. A Smith and L. Robertson.) pp. 394-420.(Cairo, Egypt 19-24 June 1987: ICARDA, Syria).

Harper, J. L., Lovell, P. H., Moore, K. G. (1970). The shapes and sizes of seeds. *Annual Review of Ecology Systematics* **1**, 327-356.

Howieson, J. G., Loi, A., Carr, S. J. (1996). *Biserrula pelecinus* L., a legume pasture species with potential for acid, duplex soils which is nodulated by unique root nodule bacteria. *Australian Journal of Agricultural Research* **46**, 997-1007.

Howieson, J. G. (1995). Rhizobial persistence and its role in the development of sustainable agricultural systems in Mediterranean environments. *Soil Biol. Biochem.* **27**, 603-610.

Howieson, J. G. and Rome, S. (1996). Rhizobial ecology - challenges and opportunities emerging from a marriage of molecular biology and applied science. In 'Recent progress in research on symbiotic nitrogen fixation' (Ed. J. S. Plate). pp. 14, 24. (Cooperative research Centre for Legumes in Mediterranean Agriculture, Occasional pubbblication no.?.)

Jaritz, G. (1987). General principles in the evaluation of pasture and forage legumes. In 'Legume Genetic Resources for Semi-arid Temperate Environments'. Proceeding of an International Workshop on Genetic resources of cool-season pasture, Forage and food legumes for semi-arid temperate environments. (Eds. A. Smith and L. Robertson). pp. 237-257. (Cairo, Egypt 19-24 June 1987).

Lelievre, F. (ed.) (1996). 'Maitrise de la reproduction des légumineuses, pour régulariser leur production et améliorer leur place dans l'agriculture européenne'. (Final report of the EU RandD AIR project CT92-0279).
Loi, A., Porqueddu, C., Veronesi, F., Cocks, P. S. (1995). Distribution, diversity and potential agronomic value of *Medicago polymorpha* in Sardinia. *Journal of Agricultural Science, Cambridge* **124**, 419-426.
Loi, A., Cocks, P. S., Howieson, J. G., Carr. S. J. (1996). Morphological characterisation of Mediterranean populations of *Biserrula pelecinus* L. *Plant Breeding* (in press).
Masson, Ph. and Gintzburger, G. (1986). 'Enherbement des vignobles en climat Méditerranéen: perspectives d'utilisation del légumineuses annuelles'. (Symp. Int. Culture Vigne: Montpellier). pp. 145-152.
Moreno-González, J. and Cubero, J. I. (1993). Selection strategies and choice of breeding methods. In 'Plant Breeding: Principals and prospects.' (Eds. M. D. Hayward, N.O. Bosemark and I. Romagosa). pp. 281-313. (Chapman and Hall: London).
Niks, R. E., Ellis, P. R. and Parlevliet, J. E. (1993). Resistance to parasites. In 'Plant Breeding: Principals and prospects.' (Eds. M. D. Hayward, N.O. Bosemark and I. Romagosa). pp. 422-447. (Chapman and Hall: London).
Olea, L., Paredes, J., Verdasco, P. (1985). Mejoras de pastos con trebol subterraneo: I influencia del clima en la persistencia. Anales de Istituto Nacional de Investigaciones Agrarias, Serie Agricola 28, 1-41
Pardini, A., Piemontese, S., Argenti, G., Staglianò, N., Talamucci, P. (1995a). Mantenimento di bande parafuoco con trifoglio sotterraneo pascolato ai fini della riduzione della fitomassa combustibile e delle perdite di suolo in un'area forestale toscana. *Rivista di Agronomia* **29** (3 suppl.), 427-433.
Pardini, A., Piemontese S., Staglianò, N. (1995b). Scelta di cultivar di trifoglio sotterraneo (*Trifolium subterraneum* L.) in funzione di diverse destinazioni produttive ed extra-produttive in ambiente mediterraneo. *Rivista di Agronomia* **29** (3), 267-272.
Piemontese, S., Staglianò, N., Pardini, A., Talamucci, P. (1996). Impieghi extraproduttivi di alcune leguminose autoriseminanti in sistemi multiuso. *Rivista di Agronomia* 30, (in press).
Piano, E. and Talamucci, P. (1996). Annual self-regenerating legumes in Mediterranean areas. Proceedings of the 16th General Meeting of the European Grassland Federation, Grado, Italy, September 1996: 865-878.
Piano, E. and Pecetti, L. (1995). L'interramento delle strutture riproduttive in trifoglio sotterraneo (*Trifolium subterraneum* L.; T. brachycalycinum Katzn. et Morley). I. Variabilità per la capacità di interramento. *Rivista di Agronomia* **4**, 582-588.
Piano, E. and Francis, C. M. (1992). The annual species of *Medicago* in the Mediterranean region. Ecogeography and related aspects of ploant introduction and breeding. Proceedings of the X International Conference of the EUCARPIA Medicago spp. Group, Lodi, 15-19 June 1992, Istotuto Sperimentale per le Colture Foraggere, Lodi, I, 373-385.
Piano, E., Pecetti, L., Carroni, A.M. (1996). Climatic adaptation in subterranean clover populations. *Euphytica* **92**, 39-44.
Piano, E., Pecetti, L., Sardara, M., Missio, A., Carroni, A. (1993a). Repationship between adaptive features and environment of origin in subterranean clover. In 'Management of Mediterranean shrublands and related forage resources'. pp. 24-27. (FAO REUR Technical series 28).
Piano E., Spanu F., Pecetti L. (1993b). Structure and variation of subterranean clover populations from Sicily, Italy. *Euphytica* **68**, 43-51.
Piano, E., Pusceddu, S. and Spanu, F. (1991). Observations on the distribution and ecology of annual medics and subterranean clovers in Sicily (Italy). Proceedings of the IV International Rangeland Congress, Montpellier, France, 405-409.
Pretty, J. N., Howes, R. 1993. Sustainable agriculture in Britain; recent achievements and new policy challenges. Research Series Sustainable Agriculture Programme, International Institute for Environment and Development, London, UK, pp. 2, 74.
Porqueddu, C., Loi, A., Cocks, P. S. (1996). Hardseededness and pattern of hard seed breakdown in Sardinian populations of *Medicago polymorpha* under field conditions. *Journal of Agriculture Science, Cambridge* **126**, 161-168.
Puckridge , D. W. and French, R. J. (1983). The annual legume pasture in cereal-ley farming systems of Southern Australia. *Agriculture Ecosystems and Environment* **9**, 229-267.
Revell, C. K. (1992). New yellow serradella varieties for low rainfall pastures. *Journal of Agriculture, Western Australia* **33**, 121-127.
Rieley, J, Page, S. (1990). 'Ecology of plant communities'. (Longman Scientific *and* Technical: Singapore).
Rivoira, G. (1976). Foraggicoltura asciutta in ambiente mediterraneo. *Rivista di Agronomia* **1**, 3-22.
Roggero, P. P., Rosales M., Bellon S. (1996). Sustainable feeding systems based on the use of local resources. *Annales de Zootechnie* **45**, suppl., 105-118.
Roggero P. P., Porqueddu C., Bullitta S., Veronesi F. (1990). The choice of suitable species and varieties for the rainfed Sardinian forage systems. Proceedings of the 6th Meeting of the F.A.O. Sub-network on Mediterranean pastures and fodder crops, Bari, 17-19 ottobre 1990, 77-81.
Sitzia, M., Sulas, L., Porqueddu, C., Fois, N. (1996). Dairy sheep farming system in Sardinia: comparison between rotational grazing and continuous stoking. Proceeding of the 16th European Grassland Federation, Grado, 15-19 September 1996, Italy.

Sleper, D. A., Barker, T. C., Bramel-Cox, P. J. (1991). 'Plant Breeding and sustainable agriculture: considerations for objectives and methods'. pp. 29-56. (CSSA Special publication no. 18: Madison, USA.

Smith, F. P., Cocks, P. S., Ewing, M. A. (1995). Variation in the morphology and flowering time of cluster clover (*Trifolium glomeratum* L.) and its relationship to distribution in Southern Australia. *Australian Journal of Agricultural Research* **46**, 1027-1038.

Sulas, L., Porqueddu, C., Roggero P. P., Bullitta, P. (1995). The role and potential of sulla (Hedysarum coronarium L.) in the Mediterranean dairy sheep farming system. Proceedings of the 5th International Rangelands Congress. Salt Lake City, Utah, USA, 23-28 July 1995, pp. 543-544.

Thompson, E. F., Rihawi, S., Cocks, P. S., Osman, A. E., Russi, L. (1990). Recovery and germination rates of seeds of Mediterranean medics and clovers offered to sheep at a single meal or continuously. *Journal of Agricultural Science, Cambridge* **114**, 295-299.

Yañez, F., Alcolado, V., Paredes, J., Verdasco, P., Lopez-Carrasco, L., Olea, L. (1991). 'Mejora de pastos de secano en el S.O. de la provincia de Ciudad Real'. pp. 1-19. (Consejeria de Agricultura, Junta de Communidades de Castilla La Mancha: Toledo).

CHAPTER 11

Evaluation of the feeding value of pasture legumes

S K Baker and R A Dynes
1 CSIRO Division of Animal Production, Floreat Park, Western Australia 6014
2 CSIRO Centre for Mediterranean Agricultural Research, Floreat Park, Western Australia 6014
3 Cooperative Research Centre for Legumes in Mediterranean Agriculture, The University of Western Australia, Nedlands, Western Australia 6907

Abstract The distinction between nutritive value and feeding value of forages for animal production is an important one, since variation in voluntary feed intake accounts for a large proportion of the variation that is observed in the feeding value of forages, while nutritive value is largely independent of voluntary feed intake. Some of the limitations in using estimates of dry matter digestibility as indices of nutritive value, and some alternative approaches to evaluation of the feeding value of forages including pasture legumes are discussed. For example, an index can be calculated from production responses of grazing sheep which reflects differences among cultivars of subterranean clover, *Trifolium subterraneum*, in their feeding values, expressed in terms of efficiency of production of wool. Weston (1982 1985 1996) quantified the extent to which the amount of forage that an animal consumes falls short of the amount that the animal needs to consume to meet its capacity to use energy as the constraint to forage intake (FCC). Feeding value of forages and some biomechanical and chemical constraints to voluntary feed intake, including the presence of plant secondary compounds, are considered in this context.

Keywords nutritive value, *Trifolium subterraneum*, wool growth, feed intake, constraint to intake

1. INTRODUCTION

Profitability in a grazing enterprise comes from the sale of animal products (wool, meat, milk) and in a ley cropping enterprise from the sale of crops, usually cereals or pulses, because pastures and forages otherwise have no intrinsic value for a producer, unless they are cut and sold for fodder. It would follow therefore that the most desirable pasture would be one that is sustainable and that will maximise profitability in the grazing and cropping enterprises that it supports. It thus is axiomatic that breeding and selection programmes for pasture or forage legumes should be for improved pasture 'quality' rather than for improved legume 'quality' *per se*. Evaluation of the quality of pastures for animal production in terms of pasture yield or of some estimate of the digestibility of dry matter alone are of limited value, and some alternative approaches are discussed. The discussion is focused primarily on legumes and, where possible, on comparisons among genotypes or cultivars of legumes in a mediterranean environment.

2. FEEDING VALUE

Two terms are commonly used to describe the 'quality' or value of a pasture or forage for animal production: nutritive value and feeding value (Ulyatt 1973; Purser 1980; Ulyatt 1980). Nutritive value refers to the response(s) in animal production per unit of voluntary feed intake, and is a function of digestibility of nutrients and the efficiency with which the nutrients are used for maintenance or production. Feeding value refers to animal production responses when feed available does not limit voluntary feed intake, and is a function of voluntary feed intake and nutritive value. The distinction between nutritive value and feeding value is an important one, and Ulyatt (1973) has estimated that variation in voluntary feed

S. J. Bennett and P. S. Cocks (eds.), Genetic Resources of Mediterranean Pasture and Forage Legumes, 120–131.

intake accounts for at least 50% of the variation that is observed in feeding value of forages. Thus yield of wool, meat or milk from a pasture will be the product of voluntary feed intake, the digestible energy content of the intake and the efficiency of utilisation of absorbed nutrients.

Animal production depends on the supply of energy from the feed, provided that all other nutrients (protein, minerals, vitamins, water) are present in adequate amounts and in appropriate balances. In general, the efficiency of energy utilisation in ruminants increases as metabolisable energy intake increases, and in grazing ruminants efficiencies are most likely to be greatest when voluntary feed intake is at or near *ad libitum*. There are a number of excellent reviews of the value of pastures for animal production in temperate, tropical and mediterranean environments, including those by Ulyatt (1973), Purser (1980), Ulyatt (1980), Minson (1982), Norton (1982), Wilson (1982), Jones and Wilson (1987), and Minson (1987 1990).

Across a range of forages, an overall positive correlation between voluntary feed intake and digestibility can be demonstrated (Blaxter 1961; Minson *et al.* 1964; Blaxter *et al.* 1966; Troelsen and Campbell 1969), and there was a view that if it was possible to show that increased voluntary feed intake always was related to increased digestibility, then breeding for improved digestibility of pasture plants should improve voluntary feed intake (Cooper *et al.* 1962). There have been notable successes in selection for increased digestibility (reviewed for example, by Minson (1990) and Buxton and Casler (1993)), and they are likely to be reflected in improvements in nutritive value. However the relationship between digestibility and voluntary feed intake is empirical rather than mechanistic, and it is not a precise one; digestibility accounts for only about 60% of the variation in voluntary feed intake of forage. Thus, voluntary feed intake of forages of the same digestibility can differ. Further, it is a common misconception that digestibility is a character of the forage, the implication being that digestibility is an absolute index of nutritive value. It has been demonstrated for a wide range of legumes and grasses that forages of similar digestibility can differ in nutritive value, that forages of similar nutritive value can differ in feeding value, and that the digestibility and thus nutritive value of a forage can be altered by altering particle size (for instance, see examples by Ulyatt (1971 1973) or Minson (1990)).

In forages, cell contents are usually readily available and completely digestible, and as a consequence most variation in the feeding value of forages appears to be related instead to the chemical, structural and biomechanical characters of the cell walls. For example, among 42 genotypes of subterranean clover there was a strong positive relationship between digestibility *in vitro* of stem material from dry, mature plants and the non-cell-wall content of the stems (and presumably a negative relationship with cell-wall components), which accounted for about 60% of the variation in *in vitro* digestibility and was unrelated to maturity (time to flowering) (Taylor *et al.* 1987). This suggests that there is potential for selection in subterranean clover, and perhaps in other legumes in a mediterranean environment, to improve nutritive value, since in other studies environmental factors (rainfall, temperature, shading and fertiliser treatments) appear to have a greater effect on plant yield than on the nutritive or feeding values of forages. Environmental influences on nutritive value or feeding value are reflected in changes in the protein, soluble carbohydrate or structural carbohydrate composition of the forages, but they appear to be associated principally with changes in plant morphology, notably the ratio of leaf to stem (Wilson 1982; Buxton and Casler 1993; Rossiter *et al.* 1994). Leafiness thus can be a good index of nutritive value or feeding values, but this is not always the case.

3. EVALUATING FEEDING VALUE OF PASTURES

Marked seasonality of animal production in mediterranean environments is well documented and in his review Purser (1980) characterised pastures in a typical farm enterprise in the mediterranean regions of southern Australia in terms of the quality and quantity of feed available. Feeding value of a pasture can be estimated only when feed availability is not limiting - a more likely circumstance in spring and summer in this environment than in autumn and winter when, under practical conditions, feed availability is a major limitation to animal production. Comparative feeding value among pastures can be assessed in grazing situations where the amount of forage on offer does not limit voluntary feed intake, but this is a condition that is not always met in stocking-rate comparisons. In addition, if a *cause* of differences in feeding value among pastures is sought then the relative contributions of the components of feeding value are needed, and reliable measurements of voluntary feed intake and components of nutritive value (selection, nutrient intake, nutrient digestibility and nutrient utilisation) are required.

Because reliable measurements of voluntary feed intake or the components of nutritive value (selection and efficiency of utilisation of absorbed nutrients) were not available, Purser and Southey (1987) used rates of liveweight change and wool growth to calculate an index which reflects the value of pastures for wool growth. They argued that if liveweight change is an indicator of metabolisable energy intake in the absence of more reliable estimators (White *et al.* 1979), and if live weight and liveweight change account for a high proportion of the observed variation in wool growth (Allden 1979; White *et al.* 1979), then live weight, liveweight change and rate of wool growth together would provide an estimate of the supply and efficiency of utilisation of nutrients from pasture for wool growth that would be independent of voluntary feed intake. Thus the coefficient of the efficiency of wool production that they calculated can be regarded as an estimator of the nutritive value of a pasture.

In the same way, if voluntary feed intake is not constrained by the availability of forage, then an index or coefficient calculated from live weight and liveweight gain during a period when the animals are in positive liveweight gain, would provide a measure of the feeding value of the pasture, and conversely a measure of the extent to which production responses are limited by some factor other than the availability of forage. Using this approach the feeding value of eight cultivars of subterranean clover, *T. subterraneum*, have been estimated using unpublished data from Nicholas (Table 1). In an early report of this study (Anon 1972) it was shown that there were significant differences amongst the cultivars in the liveweight responses by sheep grazing the plots. There is a 2-fold range in the coefficients of efficiency of wool production, indicating that there were substantial differences in the efficiency with which these eight cultivars were used for wool production and thus in their feeding value. In addition a poor efficiency of utilisation for wool production tended to be reflected in poor profitability of wool production (annual wool production per ha) (kg greasy/ha, $r = 0.31$; $p = 0.08$). These differences cannot be explained by differences in maturity (time to flowering), and there were no differences among cultivars or between stocking rates in seedling density soon after the break of season (Table 1), although seeding density was lower in 1974 than in 1975.

There was a negative correlation between rate of liveweight gain during the period of gain and annual fleece weight (kg greasy/head; $r = 0.65$; $p < 0.001$), and whereas the correlation of the clover content of the pasture late in spring (late September) with the rate of liveweight gain was positive ($r = 0.72$; $p < 0.001$; see also Thorn (1985)), the correlation with the coefficient of efficiency of wool production was negative ($r = -0.74$; $p < 0.001$). This would suggest that pastures with a low content of clover were used more efficiently for wool production that those with a high content of clover, and pastures which promoted greatest liveweight gain were used least efficiently for wool production.

In comparisons such as this the comparative feeding value of pastures can be assessed and the pastures ranked in order of merit. However it provides no information about how well the pastures supply nutrients to meet the animals' need for maximum production, about how much better production could be, or about the factors that might limit production.

Table 1. Wool production over 2 seasons (1974 and 1975), pasture measurements, and coefficient of efficiency of wool production from pastures of each of 8 cultivars of subterranean clover, *T. subterraneum* grown at North Bannister, Western Australia, and stocked with sheep at either 7.4 or 9.9 sheep/ha (2 plots per stocking rate). The pastures were sown in 1968 and the plots had been stocked continuously since then. In 1975 one plot at each stocking rate was sown with Clipper oats. From unpublished data provided by D Nicholas.

Cultivar	Midland	Uniwager	Daliak	Seaton Park	Woogenellup	Dinninup	Dwalganup	Geraldton	Standard error
Coefficient of efficiency of wool production[1]	0.40[2]	0.37	0.35	0.34	0.33	0.28	0.25	0.21	0.008
Annual wool production									
kg greasy/head	5.93	5.69	5.63	5.84	5.65	5.43	5.13	5.07	0.071
kg greasy /ha, on plots stocked at:									
7.4 sheep/ha	43.1	43.3	42.5	44.7	43.1	41.7	40.6	41.0	1.93
9.9 sheep/ha	59.7	54.6	54.5	55.8	54.1	51.7	47.1	45.6	1.93
Plant density in May (plants/dm^2)	451	663	612	424	345	465	549	651	149.3
Pasture in September:									
pasture on offer (t DM/ha)	2.9	2.1	2.8	2.9	4.0	3.2	2.2	1.5	0.31
clover content (% of DM)	61	33	54	59	76	100	84	66	7.7
Days from sowing to flowering[3]	-[4]	103	97	110	130	113	83	97	-
Grain yield (t/ha) in 1975 on plots that had been stocked at:									
7.4 sheep/ha	3.85	4.18	3.87	4.22	4.05	3.92	3.69	3.87	-
9.9 sheep/ha	5.23	5.01	5.09	5.13	4.96	4.85	4.48	4.73	-

[1] Calculated according to Purser and Southey (1987) using annual fleece weight, initial live weight, and liveweight gain during the period of sustained gain in live weight (Figure 1). Coefficients decrease as efficiency of production of wool decreases.
[2] Values are least squares means, pooled across stocking rates except as indicated; variation between years is used as a covariate. Values followed by the same letter differ ($p < 0.05$)
[3] Indicative values only, from Collins *et al.* (1984) for cultivars grown in Perth.
[4] Not available

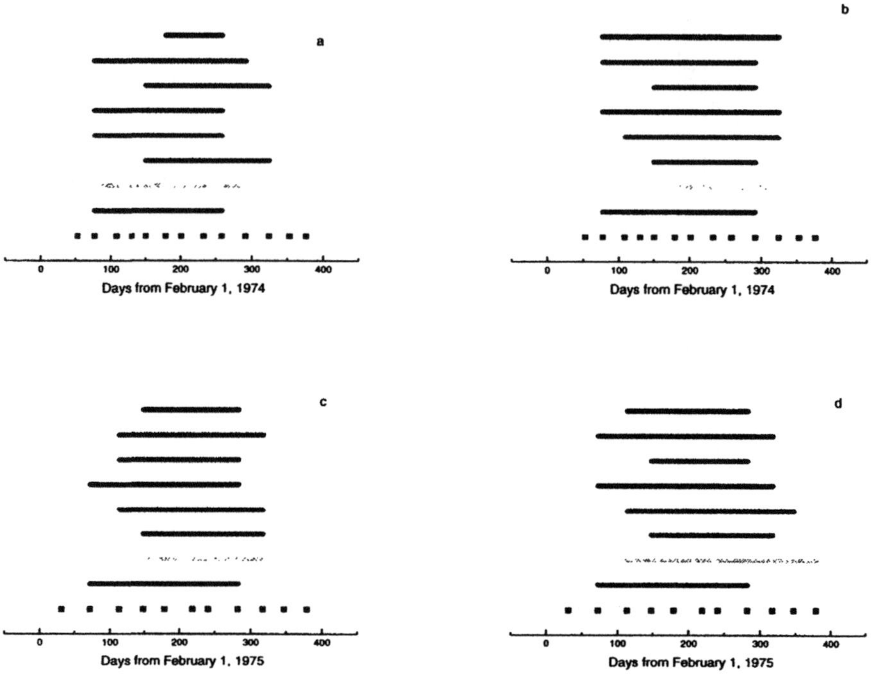

Figure 1. Length of periods of sustained liveweight gain in 1974 on plots stocked at 7.4 (a) and 9.9 (b) sheep/ha, and in 1975 on plots stocked at 7.4 (c) and 9.9 (d) sheep/ha. Bars from top to bottom correspond to: Geraldton, Uniwager, Dwalganup, Daliak, Seaton Park, Dinninup, Woogenellup, Midland. Points (■) indicate dates when sheep were weighed.

1. CONSTRAINTS TO INTAKE

The quantities of forage diets selected by grazing ruminants or consumed indoors usually fail to provide sufficient nutrients to meet the animals' needs for maximum production. A wide range of factors may contribute to limit nutrient intake and utilisation, and includes factors that limit the availability of pasture species, such as sparse distribution of pasture or high grazing pressure, or characters that limit intake of the forage, such as low palatability, poor adequacy of essential nutrients, high resistance to comminution by chewing during mastication and rumination, poor digestibility, or high contents of water, electrolytes or deleterious secondary plant compounds. Weston (1982 1985 1996) proposed that in forage-fed ruminants voluntary feed intake is regulated in an interplay of the rate of clearance of dry matter from the rumen and the amount of useful energy that is available to the animal, relative to the animal's capacity to use that energy. In this conceptual model forage diets generally fail to provide sufficient energy to meet the capacity of the animal to use energy due to a number of constraints, including the resistance of forage organic matter (OM) to removal from the rumen, low diet palatability, difficulty in forage harvesting and prehension, and environmental stresses. Thus an energy deficit relative to the capacity to use dietary energy intake usually exists. He quantified this deficit as forage consumption constraint (FCC), calculated as the difference between the quantity of forage that the animal actually consumes and the amount of that forage that the animal needs to consume to meet its capacity to use energy when any constraints are absent. In general, for weaner sheep (Weston 1996) and adult sheep (Weston and Davis 1991), the constraint to forage intake tends to decrease as the energy content of the diet increases.

2. PALATABILITY

Animals show discrimination among forages in their feeding behaviour but the mechanism(s) are not well understood. If a forage is rejected by an animal then clearly that forage will be of reduced feeding value compared to the animals' requirements, and there can be a paradoxical situation where the nutritive value of a forage can be high, but if unpalatability constrains intake and the forage is not eaten, then the forage is of low feeding value. Palatability constraints are defined as any characteristics of the feed which inhibit intake of the forage whether the forage is offered alone or as a choice. These constraints fall into two broad categories; chemical constraints and physical, or biomechanical, constraints. Chemical constraints include a reduction in the capacity to supply essential nutrients, changes in the taste or odour of the forage and increases in secondary plant compounds. Physical or biomechanical constraints include the resistance of the forage material to shear, compression or comminution, and are likely to be reflected in increased resistance of the plant material to mastication or rumination.

Aspects of feeding behaviour can be learned, from other members of the flock or herd or from experience, and ruminants use the senses of sight, smell, taste and touch in their choice of feeds. The animals' responses to chemical constraints to intake are used to study this discriminatory behaviour in situations where there is a choice of feeds and the animal is permitted to 'select' a feed that is more palatable than others (reviewed recently by Provenza 1995 1996). The extent to which biomechanical characters of forages influence palatability is less well studied. Doyle (1988) showed that low palatability of mature subterranean clover hay constrained intake and this was accompanied by a relatively low digesta load in the gut. Similarly Dynes *et al.* (1994) found that feeding spring-harvested, air-dried subterranean clover (predominantly cv Dinninup and Seaton Park) with an OM digestibility of 70% to weaner sheep resulted in low voluntary feed intake, relatively low digesta load in the gut and a failure to gain weight. Voluntary feed intake increased gradually over several weeks but failed to respond to additional protein or energy supplements, suggesting that some aspect of palatability of subterranean clover was limiting intake in these weaners.

Weston's conceptual model may provide a framework in which to assess the extent to which various factors may act as constraints to forage intake. If the factor(s) that constrains intake can be identified, then amelioration of that limitation should be reflected in a reduced constraint to intake. For example, this can be demonstrated using data from Ozanne *et al.* (1976) where FCC is reduced as the amount of phosphorus supplied through phosphate fertilisation of the pastures, based on subterranean clover increases, except where an equivalent amount of phosphorus was supplied to the sheep as a mineral supplement (Table 2). It is of interest to note that the pasture of highest feeding value (when FCC was least) was fertilised at rates in excess of those required for maximum dry matter yield of the pasture. Similarly Hemsley and Downes (1986) improved the feeding value for sheep of milled, dried tagastaste by providing supplements of macro and trace minerals, and this was reflected in a progressive decrease in FCC (Table 2).

Table 2. Effects on constraint to forage intake (FCC) of phosphorus fertilisation of pastures based on subterranean clover (*T. subterraneum*), of the condensed tannin content of lotus (*Lotus pedunculatus*), and of tagastaste with supplementary minerals

Effect of phosphorus (P) fertilisation of pasture, from Ozanne *et al.* (1976)			Digestibility (%)	Voluntary feed consumption (g/d)	Constraint to forage intake (FCC) (g OM/BW$^{0.75}$/d)[1]
Experiment 1	P intake (g/d)	Liveweight gain (g/d)	Dry matter		
Low P fertilisation	0.78	-23	42.4	1114	78
Medium P fertilisation	1.17	13	40.5	1167	81
High P fertilisation	2.41	113	39.7	1418	70
Experiment 2					
Low P fertilisation	0.73	-18	44	1042	79
Medium P fertilisation	1.23	16	39.9	1120	67
High P fertilisation	3.27	124	49	1420	52
Low P fertilisation, plus P supplement	3.88	33	38.6	1293	87
Effect of condensed tannins in *Lotus pedunculatus* fed freshly-cut, from two experiments reported by Barry and Duncan (1984)					
			Organic matter		
High tannin lotus			65.9	1187	23
Low tannin lotus			72.7	1240	11
High tannin lotus, no PEG[2]			54.7	951	55
High tannin lotus, with PEG[2]			62.7	1204	26
Effect of minerals supplements in sheep fed dried and chopped tagasaste, from Hemsley and Downes (1986)					
	Wool growth (g clean/d)	Liveweight gain (g/d)	Dry matter		
Milled oaten hay (90) + lupin grain (10)	8	60	64	1080	48
Tagasaste (100) + NaCl (0.25)	7	120	67	1380	28
Tagasaste (100) + macro minerals (3.0)	11	330	67	1500	22
Tagasaste (100) + macro and trace minerals (3.0)	12	310	68	1860	2

[1] Calculated according to Weston and Davis (1991).

[2] Polyethylene glycol sprayed onto the feed.

1. BIOMECHANICAL CHARACTERS OF FORAGES

In his study of 14 forages, Weston found that the fibre fractions determined by proximate analysis were related more to FCC than to voluntary feed intake (Weston and Davis 1991). They showed that the relationship of FCC with either the biomechanical characters or the fibre composition of the forages were better than the corresponding relationships involving intake. In particular there were strong positive correlation coefficients between energy required to comminute and FCC (r = 0.96, Weston and Davis 1991) and between energy required to shear and FCC (r = 0.94, Baker unpublished data), compared with a correlation coefficient of -0.79 between voluntary feed intake and FCC. Thus the energies required to comminute or to shear forage material are good predictors of FCC when no factors other than the biomechanical characters limit voluntary feed intake. If FCC is greater for a forage than would be predicted by either of these relationships, this would suggest that a factor or factors, other than the rate at which clearance of dry matter from the rumen, may be constraining intake and thus would reflect a feeding value of that forage that is lower than anticipated.

For penned sheep fed dry mature subterranean clover of different genotypes (Taylor *et al.* 1989), voluntary feed intake decreased as the energy required to shear the material increased (Baker *et al.* 1994) (Table 3). However when FCC for these clovers was compared with FCC predicted from the relationship between FCC and energy required to shear, the discrepancy between FCC and predicted FCC was greatest for sheep fed Mt Barker and Mt Helena A, and indeed these fell outside the 95% confidence interval about FCC and energy required to shear the 14 forages of Weston and Davis (1991). This suggests that some factor(s) other than the resistance of the plant material to comminution or shear constrained intake of these two subterranean clovers.

Table 3. Voluntary feed intake, digestibility of dry matter and constraint to forage intake (FCC) of dry, mature subterranean clovers, *T. subterraneum*, fed to adult wether sheep in pens.

Subterranean clover	Dry matter intake (g/d)[1]	Digestibility of dry matter (%)[1]	Energy required to shear (MJ/m^2)[2]	Constraint to forage intake (FCC) (g OM/kg $BW^{0.75}$/d)[3]	Predicted constraint to forage intake (FCC) (g OM/kg $BW^{0.75}$/d)[4]
Spencers Brook	381	48.6	23.8	96	105
Collie A	499	49.9	22.8	88	100
Mt Helena A	605	52.3	14.5	77	54
69S30.5.4.1	600	55.6	15.9	70	62
Mt Barker	732	58.0	12.3	61	42
Standard error	46.6	1.73	0.54	2.4	-

[1] From Taylor *et al.* (1989).
[2] From Baker *et al.* (1994).
[3] Calculated according to Weston and Davis (1991).
[4] Estimated from the relationship between the energy required to shear and constraint to forage intake (FCC) for the 14 forages used by Weston and Davis (1991) (Baker, unpublished data).
[5] Not applicable

2. SECONDARY PLANT COMPOUNDS

There is an extensive literature on secondary plant compounds in legumes and grasses, and their effects on ruminants and non-ruminants has been reviewed recently by a number of authors, including Hegarty (1982), Adams (1987), Barry and Blaney (1987), Lloyd-Davies (1987), Mangan (1988), Adams (1995), Cheeke and Palo (1995), Craig (1995), Dominguez-Bello (1995), Hammond (1995) and Reed (1995). Secondary plant compounds can be toxic to animals, can have subclinical effects that result in losses in animal productivity, and can enhance plant persistence in a sward through adverse effects on palatability. The effects of secondary plant compounds usually are dose-dependant. Like the plant characters that influence nutritive and feeding values of forages, the concentrations of many secondary compounds are influenced by environmental factors including effective rainfall, temperature optima, fertiliser treatments, and grazing pressure, and the concentrations of these plant secondary compounds can differ significantly amongst genotypes.

Deleterious secondary plant compounds may reduce voluntary feed intake by depressing any one of many of the pathways which regulate intake (for example, see Forbes 1995) and if they adversely affect any aspect of digestion or utilisation of the forage then the nutritive and feeding values of the forage will be reduced. Adverse effects of secondary plant compounds may be immediate or they can be cumulative over months or even years. In ruminants there are a number of detoxification mechanisms in many of the organs exposed to deleterious secondary plant compounds, including the mouth, reticulorumen, intestines, liver and kidney. In addition, the concentration and potency of some secondary plant compounds can be enhanced by the microbial activity in the reticulorumen.

Whereas some secondary plant compounds reduce the nutritive and feeding values of a forage, others enhance its nutritive and feeding value, and can depend on the amount of the secondary plant compound that the animal consumes. For example low concentrations of condensed tannins can enhance animal productivity by reducing the degradation of plant proteins in the reticulorumen. However, FCC for sheep fed lotus containing high concentrations of condensed tannins was twice as high as FCC for sheep fed lotus containing low concentrations of condensed tannins. The constraint to intake caused by the condensed tannins could be reduced by spraying polyethylene glycol onto the feed (Barry and Duncan 1984; Table 2). Similarly, the presence of other naturally occurring secondary plant compounds in pasture legumes could be exploited for effective grazing management through effects on FCC, either by changing voluntary feed intake, or by changing potential intake via changes in digestibility of the forage. So far no mechanisms exist to reduce FCC in the long term and, although using unique compounds in legumes to stimulate intake will depend on future research, recent research indicates the potential for this approach.

Naturally occurring phytooestrogens are a common feature of subterranean clovers, and manipulation of the concentrations of these oestrogens may be beneficial in producing lean animals since excessive fatness reduces carcass value and limits feed intake. When β-oestradiol is infused intravenously there is a depression in voluntary feed intake that is dose-related (Weston and Davis unpublished data) and it is greater in fat animals than in thin animals. Pastures containing sufficiently high concentrations to limit feed intake in spring, but low concentrations in summer and autumn, may have a role in management of wool production and quality. If fat deposition can be reduced in spring, then because a lean animal has a higher energy demand than a fat animal, it will consume more feed in summer and autumn when the feeding value of the pasture is poor. This in turn would enhance wool quality and the utilisation of this mature, dry pasture.

3. PROSPECTS FOR THE FUTURE

More often than not, nutritive value of pasture legumes and forage legumes have been inferred from *in vitro* studies or proximate analysis of these materials. These methods have considerable advantages; the amount of material required is small, and relatively large numbers of accessions can be screened. They also have a number of disadvantages, among them that proximate analyses and estimates of *in vitro* digestibility can only give an idea of relative nutritive value; they offer no information about feeding value. The paucity of information about feeding value of legumes in mediterranean environments stems partly from the methodological difficulties that are entailed in determining with accuracy voluntary feed intake and selection of pasture species and plant parts by grazing animals. For this the use of alkanes in the cuticular waxes of plants offers promise, and their use to determine intake, and selection of pasture species as well as plant parts is being evaluated (Dove and Mayes 1991; Baker and Klein 1994; Dove 1994; Dove and Mayes 1996; Dove *et al.* 1996).

Further research is needed to determine the role of palatability in performance of animals grazing legumes. Of importance also is the understanding and exploitation of differences in palatability within or between species and cultivars which could be exploited in the establishment of persistent pasture plants, especially in mixed swards, which have a high feeding value during both green and dry phases of the pasture season. This would rely on initial screening of agronomically promising lines using digestibility together with biomechanical properties, such as energy required to shear the material, to estimate FCC as an index of potential feeding value. Plants of potentially high feeding value could then be screened for known secondary plant compounds, and the most promising lines then would be used in larger-scale evaluation where the material either is fed to animals in pens, or is grazed and alkanes are used to determine intake, selection and digestibility, which in turn would be used to compare actual FCC with FCC estimated from biomechanical characters alone.

REFERENCES

Adams, N.R. (1987). Oestrogenicity of subterranean clover pastures and infertility in ewes. In 'Temperate pastures: their production, use and management.' (Eds J.L. Wheeler, C.J. Pearson, and G.E. Robards.) pp. 457-9. (Australian Wool Corporation, and Commonwealth Scientific and Industrial Research Organisation: Australia.)

Adams, N.R. (1995). Detection of the effects of phytooestrogens on sheep and cattle. *Journal of Animal Science.* **73**, 1509-15.

Allden, W.G. (1979). Voluntary feed intake, diet composition and wool growth. In 'Physiological and Environmental Limitations to Wool Growth.' (Eds J.L. Black and P.J. Reis.) pp. 61-78. University of New England Publishing Unit: Armidale.)

Anon (1972). Different pasture plants have different grazing values. *Journal of Agriculture, Western Australia* **13**, 123-4.

Baker, S. K. and Klein, L. (1994). Potential to use alkanes in plant cuticular waxes to discriminate plant parts of subterranean clovers eaten by ruminants. *Proceedings of the Australian Society of Animal Production.* **20**, 419.

Baker, S.K., Klein, L. and Purser, D.B. (1994). Physical characters of forages as constraints to voluntary intake. In Dove, H. 'The Plant-Animal Interface Revisited." *Proceedings of the Australian Society of Animal Production.* **20**, 55-64.

Barry, T.N. and Blaney, B.J. (1987). Secondary compounds of forages. In 'The Nutrition of Herbivores.' (Eds J.B. Hacker and J H Ternouth.) pp. 91-119. (Academic Press: Australia).

Barry, T.N. and Duncan, S.J. (1984). The role of condensed tannins in the nutritional value of *Lotus pedunculatus* for sheep. 1. Voluntary intake. *British Journal of Nutrition.* **51**, 485-91.

Blaxter, K.L. (1961). The utilization of the energy of grassland products. VIII International Grasslands Congress, 479-84.

Blaxter, K.L., Wainman, F.W., and Davidson, J.L. (1966). The voluntary feed intake of food by sheep and cattle in relation to their energy requirements for maintenance. *Animal Production.* **8**, 75-83.

Buxton, D.R. and M.D. Casler. (1993). Environmental and genetic effects on cell wall composition and digestibility. In 'Forage Cell Wall Structure and Digestibility.' (Eds H.G. Jung, D.R. Buxton, R.D. Hatfield, and J. Ralph.) pp. 685-714. (American Society of Agronomy: USA.)

Cheeke, P.R. and Palo, R.T. (1995). Plant toxins and mammalian herbivores: co-evolutionary relationships and antinutritional effects. In 'Recent Developments in the Nutrition of Herbivores.' (Eds M. Journet, E. Grenet, M-H. Farce, M. Thériez, and C. Demarquily.) pp. 437-56. (INRA Editions: Paris.)

Collins, W.J., Francis, C.M., and Quinlivan, B.J. (1984). Registered cultivars of subterranean clover - their origin, identification and potential use in Western Australia. Bulletin 4083. (Western Australia Department of Agriculture: Australia.)

Cooper, J.P., Tilley, J.M.A., Raymond, W.F., and Terry, R.A. (1962). Selection for digestibility in herbage grasses. *Nature* **195**, 1276-7.

Craig, A.M. (1995). Detoxification of plant and fungal toxins by ruminant microbiota. In 'Ruminant Physiology: Digestion, Metabolism, Growth and Reproduction.' (Eds W.V. Engelhardt, S. Leonard-Marek, G. Breves, and D. Giesecke.) pp. 271-288. (Ferdinand Enke Verlag: Stuttgart.)

Dominguez-Bello, M.G. (1995). Detoxification in the rumen. In 'Rumen Microbiology. Satellite Symposium of the IV International Symposium in the Nutrition of Herbivores.' (Ed. G. Fonty.) pp.45-9 (INRA: Paris.)
Dove, H. (1994). Plant cuticular wax alkanes - a new technique for estimating diet selection and intake in the grazing animal. In Dove, H. 'The Plant-Animal Interface Revisited.' *Proceedings of the Australian Society of Animal Production.* **20**, 55-64.
Dove, H. and Mayes, R.W. (1991). The use of plant wax alkanes as marker substances in studies of the nutrition of herbivores: a review. *Australian Journal of Agricultural Research.* **42**, 913-52.
Dove, H. and Mayes, R.W. (1996). Plant wax components: a new approach to estimating intake and diet composition in herbivores. *Journal of Nutrition.* **126**, 13-26.
Dove, H., Mayes, R.W. and Freer, M. (1996). Effects of species, plant part, and plant age on the n-alkane concentrations in the cuticular wax of pasture plants. *Australian Journal of Agricultural Research.* **47**, 1333-47.
Doyle P.T. (1988) Utilisation of clover diets by sheep. III. Palatability of a mature subterranean clover hay. *Australian Journal of Agricultural Research.* **39**, 891-8.
Dynes, R.A., Weston, R.H. and Peter, D.W. (1994). Role of gut load in the intake of subterranean clover in weaner sheep. Proceedings of the *Australian Society of Animal Production.* **20**, 420.
Forbes, J.M. (1995). 'Voluntary Food Intake and Dietary Selection in Farm Animals.' (CAB International: Wallingford, UK.).
Hammond, A.C. (1995). Leucaena toxicosis and its control in ruminants. *Journal of Animal Science.* **73**, 1487-92.
Hegarty, M.P. (1982). Deleterious factors in forages affecting animal production. In 'Nutritional Limits to Animal Production from Pasture.' (Ed J.B. Hacker.) pp. 133-50. (Commonwealth Agricultural Bureaux: Slough.)
Hemsley, J.A. and Downes, P.A. (1986) Nutritional value of tagasaste: effect of mineral supplements on voluntary feed intake, liveweight gain and wool growth in merino wethers. In 'Proceeding of Tagastaste Workshop.' pp. 3-5. (Western Australian Department of Agriculture: Manjimup.)
Jones, D.I.H. and Wilson, A. (1987). Nutritive quality of forage. In 'The Nutrition of Herbivores.' (Eds J.B. Hacker and J.H. Ternouth.) pp. 65-89. (Academic Press: Australia.)
Lloyd Davies, H. (1987). Limitations to livestock production associated with phytooestrogens and bloat. In 'Temperate pastures: their production, use and management.' (Eds J.L. Wheeler, C.J. Pearson and G.E. Robards.) pp. 446-56. (Australian Wool Corporation, and Commonwealth Scientific and Industrial Research Organisation: Australia.)
Mangan, J.L. (1988) Nutritional effects of tannins in animal feeds. *Nutrition Research Reviews.* **1**, 209-31.
Minson, D.J. (1982). Effects of chemical and physical composition of forage eaten upon intake. In 'Nutritional Limits to Animal Production from Pastures.' (Ed J.B. Hacker.) pp. 167-82. (Commonwealth Agricultural Bureaux: Slough.)
Minson, D.J. (1987). Estimating the nutrient value of forage. In 'Temperate pastures: their production, use and management.' (Eds J.L. Wheeler, C.J. Pearson, and G.E. Robards.) pp. 415-22. (Australian Wool Corporation, and Commonwealth Scientific and Industrial Research Organisation: Australia.)
Minson, D.J. (1990). 'Forage in Ruminant Nutrition.' (Academic Press: London.)
Minson, D.J., Harris, C.E., Raymond, W.F., and Milford, R.J. (1964). The digestibility and voluntary feed intake of S22 and H.I. ryegrass, S170 tall fescue, S48 timothy, S215 meadow fescue and Germinal cocksfoot. *Journal of the British Grasslands Society.* **19**, 298-305.
Norton, B.W. (1982). Differences between species in forage quality. In 'Nutritional Limits to Animal Production from Pasture.' (Ed J.B. Hacker.) pp. 89-110. (Commonwealth Agricultural Bureaux: Slough.)
Ozanne, P.G., Purser, D.B., Howes, K.M.W and Southey, I. (1976). Influence of phosphorus content on feed intake and weight gain in sheep. *Australian Journal of Experimental Agriculture and Animal Husbandry.* **16**, 353-60.
Provenza, F.D. (1995) Role of learning in food preferences of ruminants: Greenhalgh and Reid revisited. In 'Ruminant Physiology: Digestion, Metabolism, Growth and Reproduction.' (Eds W.V. Engelhardt, S. Leonard-Marek, G. Breves, and D. Giesecke.) pp. 233-47. (Ferdinand Enke Verlag: Stuttgart.)
Provenza, F.D. (1996) Familiarity and novelty in animal diets: implications for management. *Proceedings of the Australian Society for Animal Production.* **21**, 12-6.
Purser, D.B. (1980). Nutritional value of mediterranean pastures. In 'Grazing Animals.' (Ed F.H.W. Morley.) pp. 159-80. (Elsevier Scientific Publishing Company: The Netherlands.)
Purser, D.B. and Southey, I.N. (1987). Variations in the ability of pasture to support wool growth. In 'Temperate pastures: their production, use and management.' (Eds J.L. Wheeler, C.J. Pearson, and G.E. Robards.) pp. 359-61. (Australian Wool Corporation, and Commonwealth Scientific and Industrial Research Organisation: Australia.)
Reed, J.D. (1995). Nutritional toxicology of tannins and related polyphenols in forage legumes. *Journal of Animal Science.* **73**, 1516-28.
Rossiter, R.C., Taylor, G.B., and Klein, L. (1994). Environmental effects, in particular of rainfall, on the digestibility of dry, mature subterranean clover. *Australian Journal of Experimental Agriculture.* **34**, 25-32.
Taylor, G.B., Rossiter, R.C., Klein, L. and Collins, W.J. (1987). Influence of genotype on in vitro digestibility of dry, mature, subterranean clover. In 'Temperate pastures: their production, use and management.' (Eds J.L.

Wheeler, C.J. Pearson, and G.E. Robards.) pp. 432-4. (Australian Wool Corporation, and Commonwealth Scientific and Industrial Research Organisation: Australia.)

Taylor, G.B., Rossiter, R.C., Klein, L. and Collins, W.J. (1989). The intake and digestibility of dry mature subterranean clovers. XVI International Grasslands Congress, 809-10.

Thorn, C.W. (1985). Pasture management including the use of herbicides. In 'Pastures and their Management.' (Eds A.C. Dunlop and C.W Thorn.) pp. 40-60. (Australian Society of Animal Production, Western Australian Branch: Perth.)

Troelsen, J.E. and Campbell, J.B. (1969). The effect of maturity and leafiness on the intake and digestibility of alfalfas and grasses fed to sheep. *Journal of Agricultural Science, Cambridge.* **73**, 145-54.

Ulyatt, M.J. (1971). Studies on the causes of differences in pasture quality between perennial ryegrass, short-rotation ryegrass, and white clover. *New Zealand Journal of Agricultural Research*: **14**, 352-62.

Ulyatt, M.J. (1973). The feeding value of herbage. In 'Chemistry and biochemistry of herbage.' (Eds G.W. Butler and R.W. Bailey.) pp. 131-378. (Academic Press: London).

Ulyatt, M.J. (1980). The feeding value of temperate pastures. In. 'Grazing Animals.' (Ed F.H.W. Morley.) pp. 125-41. (Elsevier Scientific Publishing Company: The Netherlands.)

Weston, R.H. (1982). Animal factors affecting voluntary feed intake. In 'Nutritional Limits to Animal Production from Pasture.' (Ed J.B. Hacker.) pp. 183-98. (Commonwealth Agricultural Bureaux: Slough.)

Weston, R.H. (1985). The regulation of feed intake in herbage-fed ruminants. *Proceedings of the Nutrition Society of Australia.* **10**, 55-62.

Weston, R.H. (1996). Some aspects of constraint to forage consumption by ruminants. *Australian Journal of Agricultural Research.* **47**, 175-97.

Weston, R.H. and Davis, P. (1991). The significance of four forage characters as constraints to voluntary feed intake. In 'Proceedings of the III International Symposium on the Nutrition of Herbivores.' p. 31. (Malaysian Society of Animal Production: Serdang.)

White, D.H., Nagorka, B.N, and Birrell, H.A. (1979). Predicting wool growth of sheep under field conditions. In 'Physiological and Environmental Limitations to Wool Growth.' (Eds J.L. Black and P.J. Reis.) pp. 139-61. (University of New England Publishing Unit: Armidale.)

Wilson, J.R. (1982). Environmental and nutritional factors affecting forage quality. In 'Nutritional Limits to Animal Production from Pastures.' (Ed J.B. Hacker.) pp. 111-31. (Commonwealth Agricultural Bureaux: Slough.)

CHAPTER 12

The influence of storage conditions on seed viability

E.H.Roberts
Department of Agriculture, The University of Reading, Earley Gate, PO Box 236, Reading RG6 6AT, UK

Abstract So far as is known, all pasture and forage legumes show orthodox seed behaviour: that is over a wide range of storage conditions longevity increases with decrease in moisture content and decrease in temperature in a predictable manner. Although the patterns of loss of viability are similar for all orthodox seeds, for any given set of storage conditions different species may show considerable differences in longevity. Legume seeds in general store well. *Ex situ* conservation of forage and pasture legumes should therefore present few serious problems providing accessions are of initial good quality and are then stored under the conditions recommended by FAO and the International Plant Genetic Resources Institute, viz. at 3-7% moisture content (depending on species) and at -18°C or less. Under such conditions it should not be necessary to regenerate stocks for a century or more. If the accessions are of poor quality (initial low viability), or if inferior storage conditions are used, then not only may viability drop sooner to unacceptable levels but there may also be some genetic alteration due to genetic selection within accessions and an accumulation of mutations.

1. GENERAL CHARACTERISTICS OF SEED STORAGE BEHAVIOUR

Three categories of seed storage behaviour are now recognised - orthodox, recalcitrant (Roberts 1973a) and intermediate (Ellis *et al.* 1990b 1991a 1991b). Essentially orthodox seeds are those which can be dried to low moisture contents, 5% or less, without damage and whose longevity increases predictably over a wide range of conditions with a decrease in moisture content and temperature. So far as is known, all arable, pasture and forage species, including leguminous species are orthodox. Indeed almost all the Leguminosae are orthodox, except for just a few tree species from the humid tropics which are known to be recalcitrant (Hong *et al.* 1996). Recalcitrant seeds cannot be dried, even slightly, without some injury, or cooled much below 10°C (tropical species) or 0° (temperate species) without damage. Typically, even under ideal conditions (moist and cool temperatures under aerobic conditions) they can only be stored for a few weeks, months or, exceptionally, years. Intermediate seeds have an optimum temperature for storage of about 10°C and optimum moisture content of about 10% (Ellis *et al.* 1990b 1991a 1991b). Under these conditions a large proportion of seed may remain viable for several years. But so far no leguminous seeds have been classified as intermediate (Hong *et al.* 1996).

2. ORTHODOX SEEDS

Although the lifespan of orthodox seeds decreases with an increase in moisture content over the range within which crop seeds are typically handled, there are upper and lower moisture content limits beyond which this relationship no longer holds. When expressed as percentage moisture content, the upper limit differs between species depending on their hygroscopic characteristics which in turn depend on the chemical composition of the seeds. But in all cases, the upper critical moisture content appears to coincide with an equilibrium relative humidity of about 90% at 20°C, or a water potential of about -14MPa (Roberts and Ellis 1989; Zewdie and Ellis 1991): the corresponding moisture content varies from about

S. J. Bennett and P. S. Cocks (eds.), Genetic Resources of Mediterranean Pasture and Forage Legumes, 132–140.

15% (e.g. lettuce) to 18% (e.g. soya bean) in oily seeds, but in starchy seeds, such as cereals, it is much higher, typically about 26%. When seeds are stored under aerobic conditions the trend of decreasing longevity with increasing moisture content is reversed once seeds reach the upper critical moisture content, so that longevity increases again. It reaches a maximum when the seed is fully imbibed and longevity may then be prolonged for many years or decades - providing, of course, that they do not germinate. If fully imbibed seeds are stored under anaerobic conditions, however, they may only survive for very short periods - a few days, perhaps.

Arising from a hypothesis which was originally developed by Villiers (1975) it is postulated that sub-cellular damage occurs at a rate which increases with water potential and temperature. However at the upper critical moisture content the water potential (about -14 MPa) becomes sufficient for metabolic activities sustained by respiration to repair some of the damage which would otherwise accumulate. Repair is most effective when the seeds are fully imbibed. Intermittent wetting is sufficient providing the dry gaps are not long enough for a significant amount of damage to accumulate to an irreparable level.

Although moist storage has been considered as offering a method for long-term seed storage, the main problem is that sustained viability only occurs when seeds are fully imbibed, and in crop species it has not, so far, been possible to devise methods for imposing dormancy to prevent loss of fully imbibed seeds through germination. Accordingly the emphasis has been on drying seeds. But how far should they be dried?

The increase in longevity with decrease in moisture content ceases at the lower critical moisture content. Below this point, there is in most cases no further improvement or, more rarely, either only slight increases or slight decreases in longevity. Many of the reports of lethal damage at extremely low moisture content turn out not to be due to drying but to rapid imbibition of liquid water which can occur if seeds are very dry when they are set to germinate. For this reason, very dry seeds need to be conditioned by exposure to humid air for a day or two before they are set to germinate. Providing this precaution is taken lethal damage due to drying in orthodox seeds is rare.

The lower critical moisture content has now been determined at ambient temperature for at least 16 species (Ellis *et al.* 1988 1989 1990c). Again, although the value varies according to species, it seems to be related to the chemical composition of the seeds - oily seeds having a lower value than non-oily seeds. Some examples of lower critical moisture contents for leguminous crops are: groundnut (*Arachis hypogaea*) 2.0%; chickpea (*Cicer arietinum*) 5.3%; cowpea (*Vigna unguiculata*) 5.6%; common bean (*Phaseolus vulgaris*) 5.7%; pea (*Pisum sativum*) 6.2%; and mung bean (*Vigna radiata*) 6.3% (Ellis *et al.* 1989 1990a 1990c). These values are in equilibrium with about 10% relative humidity at ambient temperature. This is equivalent to about -350 MPa, a water potential at and below which water can be considered to be chemically inactive (Roberts and Ellis 1989). In practical terms it is difficult to dry below the critical values using a reasonably short and benign drying treatment; but in any case there is little point in trying to do so since longevity will not be improved.

3. THE ORTHODOX SEED VIABILITY EQUATION

The upper and lower critical moisture contents mark the limits within which the seed viability equation (Ellis and Roberts 1980a 1980b) is applicable. The temperature boundaries are less clear, but certainly the domain spreads from at least -20° to +90°C. Accordingly the viability equation is satisfactory for dealing with most practical problems. It was developed from an earlier version (Roberts 1960) which was unable to cope with variations in the initial quality of accessions and was inaccurate under more extreme conditions. The new version overcame these difficulties. It has been applied to many different species by several independent laboratories, over four of the ten super-orders of plants (Dickie *et al.* 1990), and has not so far been found wanting. It therefore may not be unreasonable for practical purposes to consider it of universal application to all orthodox seeds.

The viability equation is derived from two components. The first of these recognises that an accession of orthodox seeds stored under any given set of conditions within the domain in which it applies shows a normal distribution of lifespan of the individual seeds. As a consequence, survival curves can be described by negative normal cumulative distributions. Conveniently these s-shaped curves can be converted to straight lines if the percentage viability values are transformed to probits. The slope of the curve is then given by the reciprocal of the standard deviation (σ) of the distribution. Hence the survival curves can be written as;

$$v = K_i - p/\sigma \qquad (1)$$

where v is probit percentage viability, p is storage period (days) K_i is the intercept constant at time zero at the beginning of storage, and σ is the standard deviation of individual seed lifespans. K_i is also described as the seed-lot constant. Its value depends on the genotype and the environment to which the seeds were exposed after they matured on the mother plant and until the accession was stored. Clearly the value cannot be altered by storage since it describes the condition of the seed at zero time before storage starts. Accordingly all combinations of storage conditions can only affect the value of σ, since there is no other coefficient available for the storage conditions to influence (see Fig. 1).

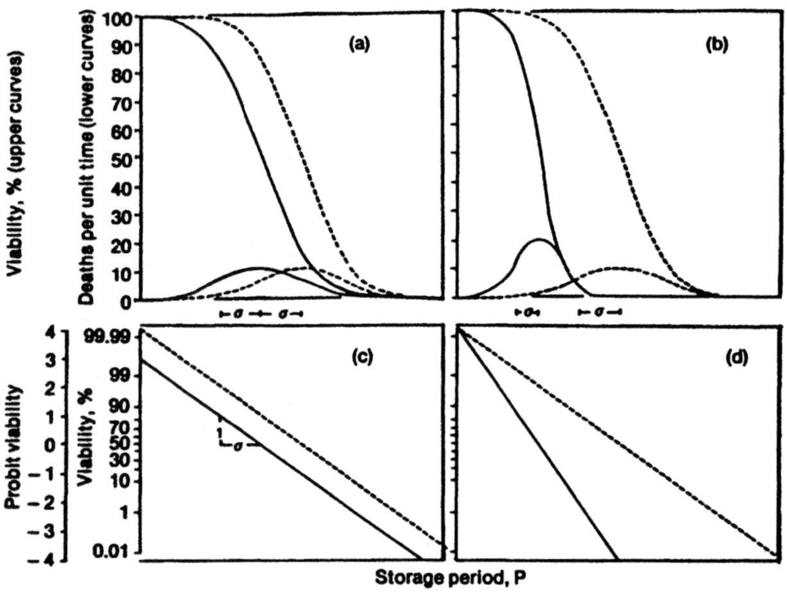

Figure 1. Fig. 1. Graphs (a) and (b) show typical seed survival curves, which are cumulative normal distributions of negative slope, under conditions when the temperature and moisture content remain constant during storage. The frequency distribution of life-spans which give rise to these survival curves are also shown. The standard deviation, σ, is indicated below each curve. Graphs (c) and (d) show the same survival curves as (a) and (b), respectively, when percentage viability is transformed to probit values; these curves are described by Eq. [1] and the slope, $1/\sigma$, is indicated on one of the curves in (c). Graphs (a) and (c) show two survival curves representing two different seed lots of the same species stored under identical conditions which, therefore, have identical slopes but where the seed lot constant, K_i, of one seed lot (---) has a greater value than the other seed lot (—). Graphs (b) and (d) show two survival curves of the same seed lot (therefore the same K_i value), stored in two different environments so that the slope, $1/\sigma$, in the less deleterious environment (---) has half the value of the other environment). From Roberts (1986).

The second component of the viability equation results from the observation that both moisture content and temperature independently affect the slope ($1/\sigma$) of the survival curve according to the following equation:

$$\log \sigma = K_E - C_W \log m - C_H t - C_Q t^2 \qquad (2)$$

where m is moisture content (% wet-weight basis), t is temperature (°C) and K_E, C_W, C_H and C_Q are constants. All these constants appear to retain the same values within a species and so, in contrast to the seed lot constant K_i, they only have to be determined once for a species, and they are therefore termed the species constants. Equations (1) and (2) may be combined by substituting the right half of equation (2) for σ in equation (1) and this results in the single viability equation

$$v = p/10^{K_E - C_W \log m - C_H t - C_Q t^2} \qquad (3)$$

It is fortunate that the species constants only need to be determined once in any species because considerable tedious experimentation is required to determine their values. Not surprisingly, therefore, the number of species for which values of the coefficients are available is limited (only 12 so far) and no forage or pasture legumes are included in this number. The task may turn out to be easier in the future, however, since there is now evidence that the values of C_H and C_Q may not differ significantly between species (Dickie *et al.* 1990). Furthermore if equilibrium relative humidity or water potential were to be used instead of percentage moisture content, then there is some evidence that the relative effect of water on seed viability may also not differ much between species (Roberts and Ellis 1989). If that turns out to be the case then it may be that only K_E varies between species and this will describe, under any given set of storage conditions, whether the species is inherently 'good' or 'poor' for storage.

Be that as it may, it is already clear that most leguminous species store relatively well. This being so then it could be expected that the majority of forage and pasture legumes should not require regenerating for more than a century when kept in long-term stores - providing seed viability is initially high, and providing the seed stores are run according to the 'preferred' conditions recommended by the FAO and the International Plant Genetic Resources Institute (IPGRI), *viz* in sealed containers at 4 - 7% moisture content (depending on species) and at -18°C or less (Anon 1994). 'Acceptable' conditions are defined as sub-zero temperatures and at similar moisture contents. Providing they are dried to these values, working collections could be stored for many decades even at ambient temperatures. Although no precise predictions can be made for leguminous pasture and forage species, in general some indication of what could be expected can be readily obtained by comparing what would be predicted under various conditions for soya bean (a relatively poor species for storage) and chickpea (which survives better). It would be expected that most forage and pasture species would fall somewhere within this range. The range can be calculated by referring to the nomographs in Figs. 2 and 3, which provide a rapid visual method of undertaking a variety of calculations based on equation (3), incorporating the appropriate species constants for chickpea and soya bean, respectively.

One complication in the Leguminosae is that many species, especially forage and pasture legumes, produce hard seeds, i.e. seeds which have coats which are impermeable to water. This has a number of consequences. Many wild annual or ephemeral species (e.g. many weeds) are capable of surviving in the soil because they remain moist, at least intermittently, and can therefore repair the sub-cellular damage which would otherwise accumulate. This behaviour also requires efficient dormancy mechanisms, otherwise the seeds would germinate prematurely. Other species have adopted a different strategy. Hard seed coats enable the seeds to produce their own low-moisture-content environment which can ensure extended longevity in a different way, i.e. by reducing the rate of accumulation of damage. The imposed low moisture content also acts as a dormancy mechanism which prevents premature germination and spreads its onset, often over several years, by gradual break-down of integrity of the hard seed coats (Russi *et al.* 1992a 1992b; Quinlivan 1968).

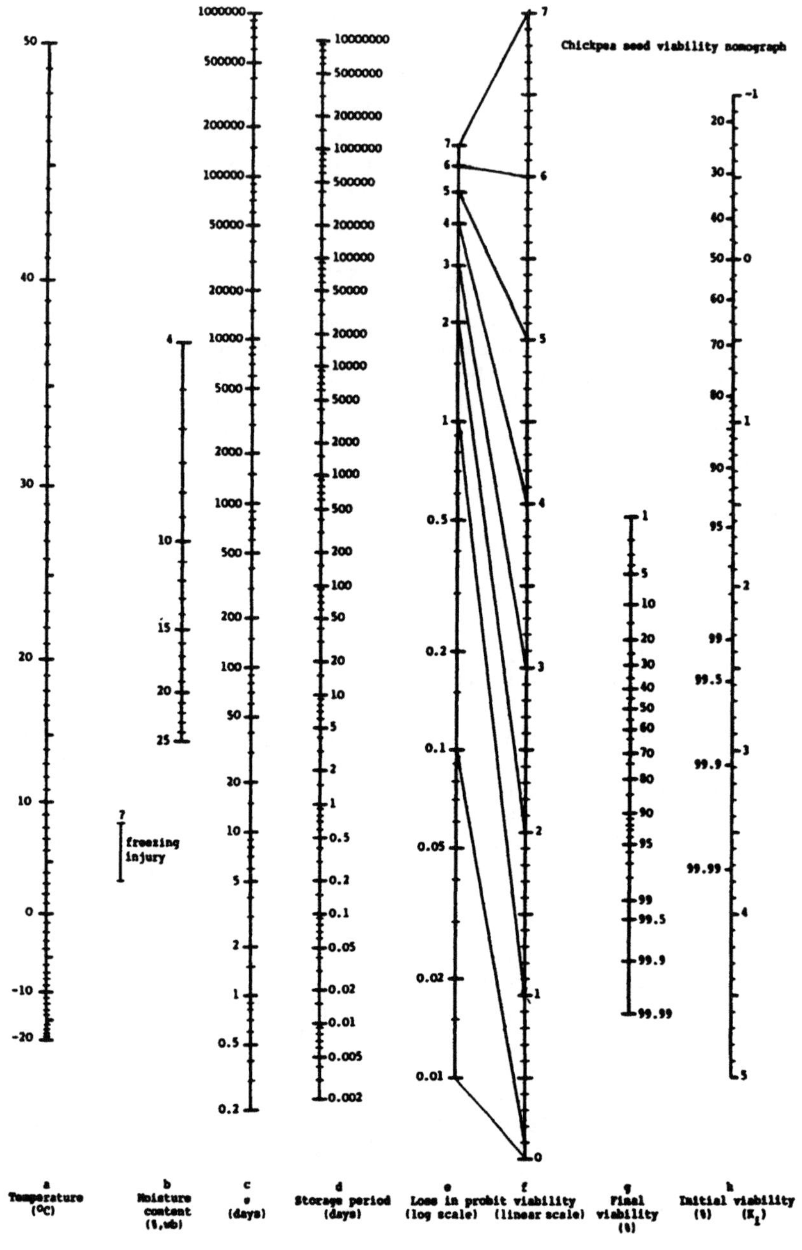

Figure 7. Seed viability nomograph for chickpea (*Cicer arietinum*). See Box 1 for a description of its use (From Ellis 1988).

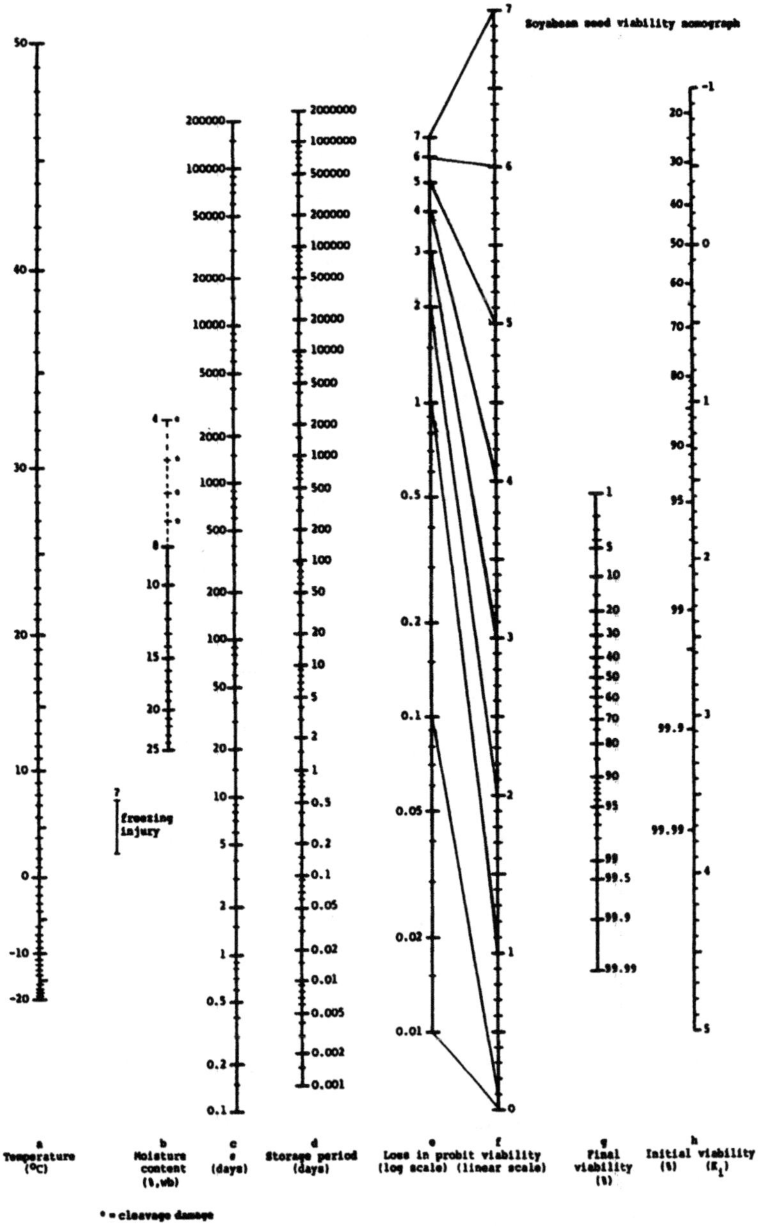

Figure 3. Seed viability nomograph for soya bean (*Glycine max*). See Box 1 for a description of its use (From Ellis 1988).

None of these characteristics presents insuperable problems for genetic conservation, but they do mean that special techniques need to be used to remove the effects of hard seed coats during viability monitoring or when accessions are regenerated (Hong *et al.* 1996; Ellis *et al.* 1985a 1985b).

Box 1. Seed viability nomographs

The seed viability nomographs, illustrated in Figs.2 and 3, can be used to solve various problems based on eq. 3 for chickpea (Fig.2) or soya bean (Fig.3). These species have been chosen as examples from the Leguminosae to represent species that show relatively good storage (Fig.2) or poor storage (Fig.3).

A straight line across identified values of temperature and moisture content on scales a and b respectively, projects the value of the standard deviation of seed longevity in days on scale c. Thus it can be seen, for example, that in chickpea (Fig.2) the standard deviation (σ) would be expected to be 300 days when seeds at 16% moisture content are stored at 20°C. Accordingly, the time taken to fall though one probit of germination (from 98 to 84% viability, or from 84 to 50% viability, for example) would be 300 days under these conditions.

Let us now assume that the initial viability is 99% (K_i = 2·32) and one is prepared to accept a drop in viability to 80%, then placing a straight line across scales h and g expresses this as a loss in probit viability of about 1.5 on scale f. This value may be directly transferred to a logarithmic scale (scale e). Putting a straight line between the value 1.5 on scale e and on the value 300 days (the value for viability to fall by one probit) on scale c, shows on scale d that it would take just over 400 days for viability to fall from 99% to 80% when the seed is stored at 20°C and 16% moisture content. Alternative storage conditions for achieving the same result can be found by using 300 days on scale c as a pivot and moving a ruler over scales a and b. For example storing at 29°C and 11% moisture content would be equally satisfactory.

4. LOSS OF SEED VIGOUR DURING STORAGE

Loss of viability in an individual seed is the final catastrophe resulting from the accumulation of sub-cellular deterioration that proceeds at a rate dependent on moisture content and temperature. There are many symptoms - disintegration of cell membranes and organelles, leakage of electrolytes and organic molecules from germinating seeds, slower germination, more morphological abnormalities in seedlings - all of which result in greater susceptibility to pathogens, and reduced ability to emerge from a seed bed especially under stressed conditions. The syndrome is commonly known as loss of vigour. Most if not all the symptoms are closely correlated with the percentage viability of an accession, providing appropriate transformations are carried out which take into account the normal distribution and other factors (Ellis and Roberts 1980c; Roberts 1986). Consequently accessions of low viability need careful handling when surviving seeds are set to germinate for regeneration.

5. GENETIC CHANGES DURING STORAGE

In addition to these symptoms of deterioration there are others changes which affect the genome - chromosome breakage and point mutations - which are of particular relevance to genetic conservation. Most of the research has been done on chromosome breakage since it is easier to quantify even though much of it is not of lasting significance since it disappears through diplontic selection during seedling growth (Roberts 1988). Some chromosome aberrations do however survive through to meiosis (Rao and Roberts 1989a). The main interest relates to the fact that visible chromosome aberrations are correlated with the induction of point mutations and these survive through to subsequent generations but are not expressed phenotypically until they start segregating in the second generation (Abdalla and Roberts 1969; Rao and Roberts 1989b). It has been calculated that 50% loss of viability may produce an equivalent load of mutations in the surviving seeds to treating fresh seeds with 10,000r of X-rays (Roberts 1973b).

The accumulation of chromosome aberrations in very dry seed for a given loss of seed viability is much greater than is the case for more moist seeds (Fig. 4). Nevertheless, since the *rate* of accumulation of the aberrations is least under very dry conditions, it is still an appropriate policy to store very dry seeds. But the quantitative relationships indicated in Fig. 4 also emphasise that it is particularly important to regenerate accessions before viability has fallen much, when seeds are stored under very dry conditions. Hence there are good reasons for maintaining a high regeneration standard, *viz*, regenerate as soon as viability falls to 85%.

In genetically heterogeneous accessions there is an additional argument for not allowing viability to fall too far since some genetic variation may be lost through selection against those components with lower values of K_i (Roberts and Ellis 1984).

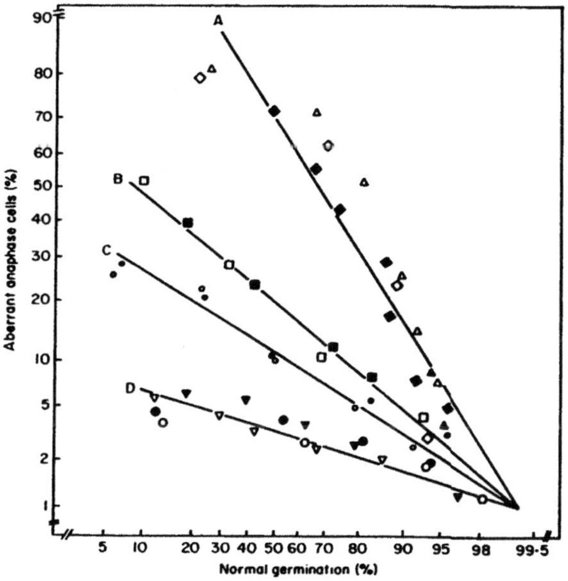

Figure 4. Relationships between the frequency of anaphase cells containing gross chromosome aberrations in the surviving seeds (probability scale) and percentage seed viability (probability scale) in lettuce (*Lactica sativa*) following storage for various periods at four moisture content ranges illustrated by different regression lines: A, 3·3 - 5·5%; B, 8·1%; C, 9·8%; D, 13·0 - 18·1% Individual storage conditions (temperature and moisture content) were as follows: 50°C, 3·3% (◆); 60°C, 3·3% (◊); 30°C, 5·5% (▲); 50°C, 5·5% (Δ); 35°C, 8·1% (■), 40°C, 8·1% (□); 35°C, 9·8% (small •); 40°C, 9·8% (small ◦); 35°C, 13·0% (▼), 40°C, 13·0% (∇); 30°C, 18·1% (large ●); 40°C, 18·1% (large O). Control before storage (**O**). From Rao *et al.* (1987).

6. FURTHER ADVICE

Additional practical advice on storage, germination tests, dormancy breaking removal of inhibition due to hard seed coats may be found in Ellis (1988), Ellis *et al.* 1985a 1985b) and in Hong *et al.* (1996). Advice on storage may be found in Cromarty *et al.* (1985).

REFERENCES

Abdalla, F.H. and Roberts, E.H. (1969). The effects of temperature and moisture on the induction of genetic changes in seeds of barley, broad beans and peas during storage. *Annals of Botany*, **33**, 153-167.

Anon (1994). 'Genebank Standards' (Food and Agriculture Organization of the United Nations, International Plant Genetic Resources Institute: Rome.)

Cromarty, A.S., Ellis, R.H. and Roberts, E.H. (1985). 'The Design of Seed Storage Facilities for Genetic Conservation'. Revised edition. (International Board for Plant Genetic Resources: Rome.)

Dickie, J.B., Ellis, R.H., Kraak, H.L. Ryder, K. and Tomsett, P.B. (1990). Temperature and seed storage longevity. *Annals of Botany* **65** 197-204.

Ellis, R.H. (1988). The viability equation, seed viability, nomographs and practical advice on seed storage. *Seed Science and Technology* **16**, 29-50.
Ellis, R.H. and Roberts, E.H. (1980a). Improved equations for the prediction of seed longevity. *Annals of Botany*, **45**, 13-30.
Ellis, R.H. and Roberts, E.H. (1980b). The influence of temperature and moisture on seed viability period in barley (*Hordeum distichum* L.). *Annals of Botany*, **45**, 31-37.
Ellis, R.H. and Roberts, E.H. (1980c). Towards a rational basis for testing seed quality. *In* 'Seed Production' (Ed P.D Hebblethwaite) pp. 605-635. (Butterworth: London.)
Ellis, R.H., Hong, T.D. and Roberts, E.H. (1985a). 'Handbook of Seed Technology for Genebanks. Vol. I. Principles and Methodology.' (International Board for Plant Genetic Resources: Rome.)
Ellis, R.H., Hong, T.D. and Roberts, E.H. (1985b). 'Handbook of Seed Technology for Genebanks. Vol. II. Compendium of Specific Germination Information and Test Recommendations.' (International Board for Plant Genetic Resources: Rome.)
Ellis, R. H., Hong, T. D. and Roberts, E. H. (1988). A low-moisture-content limit to logarithmic relations between seed moisture content and longevity. *Annals of Botany* **61**, 405-408.
Ellis, R. H., Hong, T. D. and Roberts, E. H. (1989). A comparison of the low-moisture-content limit to logarithmic relations between seed moisture content and longevity in twelve species. *Annals of Botany* **63**, 601- 11.
Ellis, R.H., Hong, T.D., and Roberts, E.H. (1990a). Moisture content and the longevity of seeds of *Phaseolus vulgaris*. *Annals of Botany* **66**, 341-8.
Ellis, R.H., Hong, T.D., and Roberts, E.H. (1990b). An intermediate category of seed storage behaviour? I. Coffee. *Journal of Experimental Botany* **41**, 1167-1174.
Ellis, R.H., Hong, T.D., Roberts, E.H. and Tao, K.-L. (1990c). Low moisture content limits to relations between seed longevity and moisture. *Annals of Botany*, **65**, 493-504.
Ellis, R.H., Hong, T.D. and Roberts, E.H. (1991a). An intermediate category of seed storage behaviour? II. Effects of provenance, immaturity, and imbibition on desiccation-tolerance in coffee. *Journal of Experimental Botany* **42**, 653-657.
Ellis, R.H., Hong, T.D. and Roberts, E.H. (1991b). Effects of storage temperature and moisture on the germination of papaya seeds. *Seed Science Research* **1**, 69-72.
Hong, T.D., Linnington, S. and Ellis, R.H. (1996). 'Seed Storage Behaviour: A Compendium. Handbook for Genebanks No 4.' (International Plant Genetic Resources Institute: Rome.)
Quinlivan, B.J. (1968) Seed coat impermeability in the common annual legume pasture species of Western Australia. *Australian Journal of Experimental Agriculture and Animal Husbandry* **8**, 695-700.
Rao, N.K., Roberts, E.H. and Ellis, R.H. (1987). Loss of viability in lettuce seeds and the accumulation of chromosome damage under different storage conditions. *Annals of Botany* **60**, 85-96.
Rao, N.K. and Roberts, E.H. (1989a). Seed ageing and meiotic chromosomal abnormalities in lettuce. *Cytologia* **54**, 373-379.
Rao, N.K. and Roberts, E.H. (1989b). Characteristics and inheritance of seed-ageing induced mutations in lettuce (*Lactuca sativa* L.). *Plant Breeding* **102**, 306-310.
Roberts, E.H. (1960). The viability of cereal seed in relation to temperature and moisture. *Annals of Botany*, **24**, 12-31.
Roberts, E.H. (1973a). Predicting the storage life of seeds. *Seed Science and Technology*, **1**, 499-514.
Roberts, E.H. (1973b). Loss of viability in storage: chromosomal and genetical aspects. *Seed Science and Technology*, **1**, 515-527.
Roberts, E.H. and Ellis, R.H. (1984). The implications of the deterioration of orthodox seeds during storage for genetic resources conservation. *In* 'Crop Genetic Resources: Conservation and Evaluation'(Ed. J.H.W. Holden and J.T. Williams), pp. 18-37. (George Allen and Unwin: London.)
Roberts, E.H. (1986). Quantifying Seed deterioration. *In* 'Physiology of Seed Deterioration'. (Eds M.B. McDonald Jr. and C.J. Nelson) pp.101-23. (Crop Science Society of America: Madison.)
Roberts, E.H. (1988). Seed ageing: the genome and its expression. *In* 'Senescence and ageing in plants'. (Eds L.D. Noodén and L.C. Leopold) pp.465-98. (Academic Press: New York.)
Roberts, E.H. and Ellis, R.H. (1989). Water and seed survival. *Annals of Botany* **63**; 39-52.
Russi, L., Cocks, P.S. and Roberts, E.H. (1992a). Hard-seededness and seed bank dynamics of six pasture legumes. *Seed Science Research* **2**, 231-241.
Russi, L., Cocks, P.S. and Roberts, E.H. (1992b). Coat thickness and hard-seededness in some *Medicago* and *Trifolium* species. *Seed Science Research* **2**, 243-249.
Villiers, T.A. (1975). Genetic maintenance of seeds in imbibed storage. *In* 'Crop Genetic Resources for Today and Tomorrow' (Eds O.H. Frankel and J.G. Hawkes) pp.297-316. (Cambridge University Press: London and New York.)
Zewdie, M. and Ellis, R.H. (1991). The upper-moisture-content limit to negative relations between seed longevity and moisture in niger and tef. *Seed Science and Technology* **19**, 295-302.

CHAPTER 13

The characterisation and preliminary evaluation of *Medicago* and *Trifolium* germplasm

G.C. Auricht[1], J.M. Prosperi[2], R. Snowball[3] and S.J. Hughes[1]

[1]*SARDI, South Australian Research and Development Institute, GPO Box 397, Adelaide, SA 5001, Australia.*
[2]*INRA, Institut National de la Recherche Agronomique - SGAP (LRGALM) Domaine de Melgueil. F34130 Mauguio FRANCE.*
[3] *AgWA, Agriculture Western Australia, Baron-Hay Court, South Perth, WA 6151, Australia.*

Abstract

The characterisation and preliminary evaluation of annual pasture legume germplasm provides seed and data essential for the effective conservation of the resource, the basis for further research and the development of new and improved pasture cultivars.

This paper deals with the characterisation and preliminary evaluation of annual pasture legumes at the Australian *Medicago* Genetic Resource Centre, Adelaide, South Australia; the French Forage Genetic Resource Centre, Montpellier, France and; the Australian *Trifolium* Genetic Resource Centre, South Perth, Western Australia. These Centres play a major international role in the preservation of pasture legumes with a total of 50,000 accessions conserved between them. Of these some 35,000 accessions, or 70% are characterised.

Preliminary evaluation and characterisation of pasture accessions is a major focus of these Centres, all of which are closely associated with national pasture improvement programs. Characterisation of germplasm involves, the multiplication of seed, the collection of data important for efficient utilisation of the resource and taxonomic identification of each accession.

The three Centres were established in the past 30 years and were set up with the aim of assembling germplasm of predetermined taxa, as this is the basis for the development of improved cultivars. Characterisation methodologies vary at each Centre, with the Australian Centres growing accessions in the field as spaced plants and the French program growing plants in a glasshouse soil tank. The characteristics recorded also vary between programs.

Issues of concern for the Centres are discussed and include; funding for long term conservation; evaluation of backlogs of germplasm; ownership of germplasm and conditions of access; encouraging effective use of the collections; the lack of diversity of many new alternative pasture legume species and the handling of data associated with the collections.

1. INTRODUCTION

This paper discusses the principles and practice of preliminary evaluation and characterisation of the annual pasture legume germplasm maintained at three key temperate forage genetic resource centres. These are:

1. The Australian *Medicago* Genetic Resource Centre (AMGRC), Adelaide, South Australia.
2. The French Forage Genetic Resource Centre (FFGRC), Montpellier, France.
3. The Australian *Trifolium* Genetic Resource Centre (ATGRC), South Perth, Western Australia.

These Centres each play a major international role in the preservation of pasture legumes with a total of 50,000 accessions conserved between them. To date some 35,000 accessions or 70% of their collections have been characterised. This is being increased at a rate of about 2,500 new accessions each year.

Preliminary evaluation and characterisation of pasture accessions is a major focus of the Centres which are all closely associated with national pasture improvement programs. These

S. J. Bennett and P. S. Cocks (eds.), Genetic Resources of Mediterranean Pasture and Forage Legumes, 141–149.

programs utilise genebank germplasm and associated data for the development of new pasture cultivars and for a range of research topics from population ecology to molecular biology.

The definition and functions of characterisation and preliminary evaluation are addressed in this paper as are the methodologies employed for growing germplasm and recording data at each of the three Centres. In addition to looking at these activities this paper highlights issues of concern which may be addressed at the workshop.

2. HISTORY AND FUNCTION OF GENETIC RESOURCE CENTRES

The following brief outline of each Centre provides an introduction to the characterisation and preliminary evaluation activities they undertake.

2.1 The Australian Medicago Genetic Resource Centre

The AMGRC was established in the early 1960's when it was recognised that the production of annual medics could be improved through the deliberate introduction and evaluation of a diverse array of germplasm from the Mediterranean region. Annual medics had for a long time been an important part of the cereal growing system in the alkaline soils of southern Australia when used in rotation with a wheat or barley crop. The medics grown at that time were, however, based on a limited gene pool of accidentally introduced and naturalised medics.

The AMGRC now maintains the worlds largest annual *Medicago* collection with some 20,000 accessions, plus maintains a further 12,000 accessions of related genera. The Centre introduces an average of 1,500 new accessions each year and distributes seed of some 2,000 accessions globally to pasture research scientists. The major recipient of AMGRC germplasm has been the National Annual Medic Improvement Program, which has primary responsibility for the development of improved cultivars for Australia. This program is based in Adelaide with collaborators in four other states.

The AMGRC program currently grows 1,000 accessions each year for characterisation and preliminary evaluation and grows a further 300 accessions for seed regeneration. All data associated with the collection is maintained in one database which is accessible to collection users. The Centre currently has some 12,000 accessions still to be characterised.

The AMGRC maintains samples of seed of each accession in a base collection, stored at -20°C, and in an active collection, stored at 2°C. For many accessions a third sample of unthreshed pods is also maintained, as a working collection, at low humidity and 15°C.

2.2 The French Forage Genetic Resources Centre

The FFGRC is not organised as a separate institution but rather corresponds to a network which includes INRA[1] Forage plant breeding stations, private breeding and selection firms and GEVES[2] (a public institution which is in charge of characterisation and registration of cultivars).

INRA stations or private firms collect cultivated or wild populations of the main forage species, evaluate these populations, select the accessions that will be included in the genebank and increase seed stocks. INRA stations keep a duplication of their collections in a number of cold chambers. These are located at Montpellier for the Mediterranean *Medicago spp.*, Lusignan for the *Dactylis*, *Festuca* and some *Medicago sp.* and Clermont-Ferrand for the genus *Lolium*).

[1] Institut National de la Recherche Agronomique - National Institute of Agronomical Research
[2] Groupement d'Etudes des Variétés Et des Semences -

GEVES is in charge of the management of seed stocks, the control of germination, the publication of an genebank listing (Index Seminum 1989), development of the database, as well as dispatch of seed samples. In France, all the forage species accessions are gathered at the 'station GEVES - Le Magneraud -- 17700 Surgères'. GEVES is also in charge of the long term conservation of all the accessions which are stored at -20° C. Cold stores for the mid-term conservation are maintained at 3°C, and at less than 20% relative humidity.

Under the rules of the Economic Community and to avoid problems in relation to the Rio and UN conventions on genetic resources, the only official collections maintained in France are those of accessions originating in France and Corsica. In the case of annual medics this totals only 112 accessions. There are however, other collections held at INRA plant breeding stations such as the one at Montpellier. At this station there are 2800 accessions from 9 countries representing 32 *Medicago* species. Seeds of these lines, collected from outside France, may be available with restrictions.

The aim of the INRA-Montpellier *Medicago* Plant Breeding laboratory is to introduce wild genetic resources of fodder plants and supply these to breeding and basic research programs. The germplasm introduced by the Centre is mainly used within INRA. The INRA improvement programs are conducting research to develop persistent forage plants, both perennial or annual self-reseeding types, which are adapted to grazing, show good adaptation to variable cultivation practices and which show rapid establishment.

Collection activities are concentrated on the western part of the Mediterranean basin (North and South). These collection activities are related directly to the needs of collaborating research programs. Only populations of *Medicago* species are introduced into the genebank although some accessions of *Trifolium subterraneum* were included in the past.

The FFGRC has 2095 accessions representing 25 different species of annual *Medicago*, some 200 wild populations of perennial *Medicago* (lucernes) as well as 500 samples of cultivated material. A database provides information on the collection site and some agronomic descriptors.

2.3 The Australian Trifolium Genetic Resource Centre

The ATGRC was established around the same time as the AMGRC in the early 1960's. This Centre was founded for research into Mediterranean pastures suited primarily to acidic soils and includes material from several Australian led collection missions in the 1950's. The Centre now has extensive collections of pasture legumes comprising 11,000 *Trifolium* accessions (primarily *T. subterraneum* (7,800), but including another 100 species), 1,880 *Ornithopus* accessions. and 120 *Biserrula pelecinus* accessions (Snowball 1993 1996). In addition to these, the Centre has a backlog of some 3,000 accessions awaiting characterisation.

The collection, like the AMGRC and other Australian Centres, is maintained as a base collection stored at -20°C and an active collection stored at 2°C. Data associated with the collection is maintained on two separate databases, one for seed stocks and the other for passport data , preliminary agronomic and morphological descriptors.

The ATGRC grows some 600-800 accessions annually for characterisation and distributes some 800 accessions each year from its collection. The primary recipient of ATGRC germplasm has been the Australasian subterranean clover and Alternative Legume Improvement Program, ASCALIP, based largely in Perth but with collaborators in five other Australian States.

3. STAGES OF CHARACTERISATION AND PRELIMINARY EVALUATION

There are generally considered to be four broad types of data associated with the botanic or agronomic description of plant genetic resource collections.

These are:

1. Accession data
2. Characterisation data
3. Preliminary evaluation data
4. Further evaluation data

Accession data refers to genebank identification descriptors for a given accession. Characterisation data is distinguished as being that data relating to highly heritable plant characteristics which are expressed consistently in all environments. Examples of the descriptors collected are;

- Vegetative data such are plant habit, internode length and leaf markers (shape, position and size),
- Inflorescence and fruit details such as peduncle to petiole ratio, number of florets and direction of pod whorl
- Seed data;. size, shape and colour.

Preliminary evaluation data is collected by the genebank Curator during the characterisation year and is a comparative assessment of each accession in relation to other accessions or to a standard. It must be remembered that field collected data in this category relates to the season in which it was recorded and generally is from unreplicated plots. Care must be taken when comparing data across years. Examples the preliminary evaluation data are;

- Vegetative data such as seedling vigour and herbage yield, scored in comparison with standard controls
- Flowering and fruiting characteristics, such as number of days from germination to first flowering, and to first maturity, degree of pod spininess
- Seed data such as seed yield and pattern of hardseed breakdown.

The number and the scope of the descriptors recorded during the characterisation and preliminary evaluation phases depends on the resources of the Centre and is generally decided each year following discussion with key groups utilising the collection. Up to 40 descriptors are routinely recorded for these two data groups, in the AMGRC program.

Further evaluation data is data collected in subsequent tests and trials, which are generally replicated and often sown at a number of sites. This data is generally not collected as part of the Genetic Resource Centre activities, but by scientists and agronomists. Ideally the information they collect is passed back to the GRC to be added to the database and subsequently aid future researchers in their selection of accessions from a collection. A few examples of the characters recorded include:

- Field performance
- Response to abiotic stress e.g. drought, frost (Prosperi et al 1993), boron toxicity or phosphate deficiency
- Response to biotic stress e.g. pests such as aphids or redlegged earth mites, fungal pathogens or nematodes
- Biochemical composition e.g. isoflavones
- Rhizobial requirements.

However this list is infinite, and is determined by the researchers themselves.

4. FUNCTIONS OF CHARACTERISATION

There are three main reasons for growing accessions in a characterisation and preliminary evaluation phase:

1. It enables the multiplication of seed from accessions which have been introduced generally in small quantities, and with seed of unknown, often poor, viability.
2. It enables the collection of data which will enhance the value of the accession in the genebank and enable efficient utilisation of the resource in subsequent research.

3. It allows taxonomic identification of the accession.

The multiplication of original seed is clearly essential to ensure that high quality seed is available for long-term conservation. It also ensures that sufficient seed is available for distribution, generally as one gram samples. Larger quantities of seed may also result and where storage permits these can be stored and distributed by the genebank in larger quantities saving a breeding or selection program one year of seed multiplication.

Collection of data during the characterisation and preliminary evaluation phases, and its addition to the GRC database increases the use and value of material held in the gene bank, and enables a number of subsequent areas of research or data acquisition to be greatly improved. These include:

- Provision of data for further research, such as in population, taxonomic or ecological studies (Bonnin *et al*,1996). Correct taxonomic identification is critical if accessions are to be stored and utilised correctly. In addition, such taxonomic studies enable the identification of new taxa. For example *Medicago syriaca* (Small et al 1990) or *Medicago rigiduloides* (Groose et al 1996).
- Efficient utilisation of germplasm, for example in plant breeding or molecular biology, by allowing selection of lines with specific traits (Auricht and Howie 1996).
- Aid decisions on the selection of *in-situ* conservation areas, and
- In the case of the FFGRC, where not all accessions are entered into the collection, it provides a basis for the selection and inclusion of material in the genebank.

Careful identification of descriptors is the key to ensuring that the data collected is well utilised and is useful to researchers. The International Plant Genetic Resources Institute (IPGRI) has published a list of descriptors for annual *Medicago* based on the AMGRC procedures (IPBGR 1990). IPBGR has produced a number of such publications for a range of crop and pasture species, and these give a good indication, or initial starting point on which descriptors to collect and the number of states that should be available for each descriptor.

5. METHODOLOGIES

A characterisation year in each Centre follows the procedure listed below:

1. Selection of lines from the genebank backlog.
   Priorities are allocated to various accessions or taxonomic groups. In the case of the FFGRC; *Medicag truncatula, M. polymorpha, M. rigidula* and *M. scutellata* are evaluated as first priority, rare species or unidentified accessions as second and the remainder as third.
2. Determination of characteristics for recording (descriptors and descriptor states).
3. Selection of comparators (control lines as standards for preliminary evaluation).
4. Preparation of seed including scarification and counting.
5. Acquisition of appropriate *Rhizobium* for each species (lines are generally not inoculated at the FFGRC).
6. Germination of seedlings in controlled conditions.
7. Transfer of germinated seed to peat pellets.
8. Maintenance of plants for some six weeks in controlled conditions to ensure establishment.
9. Transplanting to the field[3].
10. Maintenance of plants.
11. Collection of data.
12. Identification of off-types (mixtures) in accessions.
13. Collection and pressing of herbarium specimens of selected accessions.

---

[3] The methodology at different Centres varies. The French Centre conducts the first stage of multiplication in soil tanks in plastic greenhouses. Only 30 to 40% of accessions go on to field evaluation.

14. Taxonomic identification or confirmation. This may involve associated laboratory work. For example, microscopic pollen studies to distinguish *Medicago rigiduloides* from *M. rigidula* (Groose 1996).
15. Harvesting of seed.
16. Cleaning of pods and seeds.
17. Recording of pod / seed characteristics.
18. Any post harvest tests such as determination of hardseed breakdown pattern in *Medicago* or *Trifolium*.
19. Storage of seeds.
20. The transfer of data onto the database.
21. Analysis of data and reports.

6. DATA HANDLING AND RESULTS

With up to 40 descriptors, containing up to 20 different states, being recorded for 1,000 accessions in a program such as the AMGRC, efficient and accurate data handling is essential. Accordingly a hand-held data-logger with programmable touch sensitive screen is used for recording many characters. This eliminates the need for paper, doubie handling of results and avoids transcription errors. Data from the data logger is down-loaded directly into the AMGRC database. For programs growing fewer accessions or recording fewer descriptors, paper records are used with data entered manually into the databases.

Once in the database the data becomes available for analysis. A report is produced by the collection Curators, which is available to utilisers of the collection. An Australian wide scheme is being developed for a national database linking all eight Australian Plant Genetic Resource Centres. This scheme aims to simplify access to data on the full range of germplasm available in Australia. It will be known as AGRIS, the Australian Genetic Resources Information System.

Other than direct access, scientists interested in genebank characterisation and preliminary evaluation data have access to data in printed reports or on computer disks. Published reports such as the AMGRC annual report (Hughes and Auricht 1996) contain full descriptions of genebank activities including summarised characterisation and preliminary evaluation data for each of the key species grown.

The FFGRC through INRA publish an index seminum giving passport and availability details for accessions in their collection. However, this does not contain any characterisation or preliminary evaluation data. The ATGRC does not currently produce an annual report but rather prepares progress reports for its key funding agencies.

7. ISSUES RELATING TO THE EVALUATION OF GENEBANK ACCESSIONS

This section discusses some of the issues which must be addressed by Curators when undertaking characterisation and preliminary evaluation and when considering access to the seed and data which results from the process. The issues are grouped under a number of headings and range from technical issues to funding concerns. They are listed here in note form.

7.1 Germplasm Management

- Reducing collection backlogs
- Lack of diversity of alternative legumes
- Handling new species (which descriptors, management etc.)
- Acquisition of appropriate *Rhizobium*
- Quarantine changes in Australia (complicating the introduction of medics at present)

- Number of plants to grow per accession (to protect against genetic erosion)
- At what level of difference should accessions from one site be split
- The real value of core collections.

7.2 Utilisation and Ownership of Collections

- Who owns and who has rights of access to grow-on plots, seeds and data
- International treaties such as the Convention on Biological Diversity (material from pre- and post 1992 should be handled differently)
- Unforeseen uses of germplasm (which descriptors to select on)
- Communication with users (the best ways of reporting data)
- Encouraging utilisation of the data and the collection thereby enhancing its value
- Relationship with subsequent breeding and evaluation research (shared resources and return of evaluation data to the genebank)
- Extrapolation of preliminary evaluation data to other sites.

7.3 Funds

- For characterisation, preliminary evaluation and long-term conservation
- Charging for seed or data
- Equity share for GRCs from commercialisation of germplasm.

7.4 Technical Issues

- Access to suitable land
- Weed and pest control - vertebrate and invertebrate
- Maximising efficiency of mechanical processes associated with growing the characterisation plots
- Efficient data recording and management.

The three Genetic Resource Centres discussed here are major repositories of *Medicago* and *Trifolium* germplasm and are of international importance. The Centres distribute germplasm widely, in many cases utilising data collected during the preliminary evaluation and characterisation phase to guide the selection of accessions for specific requests. Through their activities these Plant Genetic Resource Centres underpin a variety of pasture improvement programmes in many countries from Australia to Europe and the United States. In order to continue to serve this function the Centres require secure, long-term funding for all of their core activities - from collection to characterisation, storage and utilisation. In order to be most effective the Centres must resource and balance each of these activities. To do this they require support from Government, industry and the research programmes utilising their resources and data, to help secure the necessary funding.

The second major requirement for the Centres to fulfil their national and international responsibilities is clear policy advice and direction. A broadly based, active and supportive advisory committee can assist the Centres greatly in this area.

8. CONCLUSIONS

The three forage genetic resource centres discussed in this paper have all been established with the one major aim, to utilise collected germplasm for plant improvement in established programs. The GRC centres are all located in close proximity to their collections.

Whilst they serve an important international role in the conservation of wild diversity, the utilisation aspect has provided the impetus and often the funding for the collections. With this background, characterisation and preliminary evaluation is a key function of the Centres as it provides both seed and data necessary for utilisation of the collection, in addition to providing for long-term storage.

Characterisation is the step between the collection and introduction of new germplasm and the utilisation of the germplasm in research or plant improvement. As an activity it often lacks the excitement of a collection mission and the commercial outcome of a breeding program and accordingly can suffer from funding shortage.

The quantity of data recorded during characterisation and preliminary evaluation requires a compromise between the number of descriptors recorded and the number of lines grown. Over time the number of descriptors recorded has reduced significantly at the Australian Centres in order to enable more lines to be grown each year. Despite this the collections all have backlogs of material awaiting evaluation and new accessions arriving each year. This is not an argument to stop collecting, indeed as more species are being investigated collection is a priority (Auricht 1994). The backlogs are rather a reminder that funding does need to be directed towards characterisation as well as just collection and utilisation. This applies equally to long-term preservation of germplasm which is an activity which often lacks visibility but which is of fundamental importance to the future of agriculture. The three Centres discussed in this paper each contain a significant proportion of accessions which are unique to their collection and in many cases irreplaceable given the changing land uses and intensity of agriculture in the accessions' countries of origin.

Funding for the characterisation and preliminary evaluation and long term storage activities of the forage GRCs will ensure this material remains available for future pasture improvement.

REFERENCES

Auricht, G.C. (1994). The Scope and Availability of Alternative Pasture Legume Germplasm. In 'Alternative Pasture Legumes. Technical Report 219'. (Eds. D.L. Michalk; A.D. Craig and W.J. Collins.) pp. 195-198. (Department of Primary Industries South Australia: Adelaide).

Auricht, G.C. and Howie, J.H. (1996). A Pasture Ideotype for the Drier Southern Australian Wheatbelt. In 'Alternative Legumes Workshop'. (Ed. P.S. Cocks.) In press.

Bonnin, I., Huguet, T., Gherardi, M., Prosperi, J.M., and Olivieri, I. (1996). High levels of polymorphism and spatial structure in a selfing plant species, *Medicago truncatula* (Leguminosae), using RAPD markers. *American Journal of Botany*. **83**, pp. 843-855.

Chaulet E., Prosperi J.M. (1994). Genetic diversity of *Medicago truncatula* Gaertn collected in Algeria. In 'Proceedings of the 7th Genetic Resources section meeting of Eucarpia, March 1994'. Pp. 255-257. (INRA: Clermont-Ferrand).

Groose, R.W. (1996). Cold tolerant annual medics *Medicago rigidula* and *M. rigiduloides*. In 'Proc. 35th North American Alfalfa Improvement Conference, Oakloholma, USA'. In press.

Hughes, S.J. and Auricht G.C. (1996). 'Australian *Medicago* Genetic Resource Centre Annual Report 1994'. (SARDI Research Report Series No. 12: Adelaide).

INRA, (1989). 'Index Seminum'. (Institut National de la Recherche Agronomique: Montpellier, France).

IBPGR, (1991). 'Descriptor lists for annual *Medicago*'. (IBPGR: Rome, Italy).

Prosperi J.M. (1993). Selection of annual medics for French Mediterranean regions. In 'Introducing the ley farming system to the Mediterranean basin. Proceeding of an international workshop. June 26-30th 1989, Perugia, Italy'. (Eds. S. Christiansen, L. Materon, M. Falcinelli and P.S. Cocks). pp. 173-191. (ICARDA: Syria).

Prosperi J.M., Angevain M., Génier G., Mansat P. (1994). Growth rhythm and habit in alfalfa. Genetic study of erect x prostrate hybrids. In 'Proceedings of the FAO Pastures and Fodder Crops Network - EUCARPIA *Medicago sativa* group Meeting. September 4/8 1994'. (INRA: Lusignan, France).

Small, E. (1990). *Medicago syriaca* a new species. *Can. J. Bot.* **68**, 1473-1478.

Snowball, R., Foster, K.J. and Collins, W.J. (1992). Australian genetic resources of *Trifolium* and *Ornithopus* species. *WA Journal of Agriculture* **33**, pp. 103-108.

Snowball, R.S. (1993). Preliminary agronomic evaluation and characterisation of Persian clover (*Trifolium resupinatum* L.). *Australian Plant Introduction Review* **24**, pp. 10-41.
Snowball, R.S. (1996). The Australian germplasm of pink flowering serradella species. In 'Proceedings of the 8th Australian Agronomy Conference, Toowoomba'. (Australian Society of Agronomy: Victoria). pp. 506-509.
Toll, J.A. (1996). Processing of germplasm, associated material and data. In 'Collecting Plant Genetic Diversity' (Eds. L. Guarino, V. Ramanatha Rao, and R. Reid) pp.577-597. (CAB International: Wallingford, UK).

CHAPTER 14

Plant genetic resources programme in Turkey with special reference to forage legumes

Cafer Olcayto Sabanci
Aegean Agricultural Research Institute, Izmir, Turkey

Abstract. Turkey, as the primary or secondary gene centre of many plant species, is rich in terms of genetic and plant diversity plus contains the distinctive forms of primitive cultivars or landraces of many forage legumes among others. The Aegean Agricultural Research Institute is responsible for the collection and conservation of the existing plant genetic diversity in Turkey. The Institute has three main departments, documentation, systematic and taxonomy, and conservation, and eight plant departments dealing with plant collection, multiplication/regeneration, and evaluation.

There are many forage species in the natural flora including some important legume genera such as *Vicia, Trifolium, Medicago,* and *Lathyrus* Some species within these genera were originally domesticated in Turkey. Expeditions have been organised regularly since the establishment of the Institute, and over ten thousand seed samples of forage species have been collected so far.

1. INTRODUCTION

Knowledge of plant genetic resources and in particular genetic variability is important for the improvement of plant varieties and for programmes to increase food production. Primitive cultivars, old varieties, landraces and wild relatives of cultivated plants have great potential in broadening germplasm pool used for plant improvement.

Turkey, located between 36.00 and 42.00^{o} N and 26.00 and 45.00^{o} E, with annual rainfall ranging from 300 to 1300 mm, is the centre of diversity for many cultivated plants. Vavilov (1926) set up eight geographic centers for cultivated plants, and Turkey was placed into two of these centers of origin, the Near East and the Mediterranean. Turkey is also the domestication center of some crops (Harlan 1971), such as barley, einkorn, emmer, peas, lentils, flax, vetch and chickpeas. Other evidence of the amount of genetic diversity and variability contained in Turkey is shown by the separation of Turkey by Davis (1970) into three phytogeographic regions; Euro-Siberian, Mediterranean and Irano-Turanian, each composed of many ecological conditions. Within these regions more than 3000 endemic plant species have been identified (Dokuzoguz 1990).

There are many reasons for genetic resources being at risk of loss erosion. Changes in cultivation systems, replacements of landraces with modern cultivars, over-grazing of pasture and meadows, use of chemicals, erosion cause by flood or wind are all reasons for the risk of genetic erosion. Plant collection and conservation are thus important to prevent further genetic resources. Many countries have national gene banks set up for the conservation and use of their own genetic resources, plus there are international organisations such as the International Plant Genetic Resources Institute (IPGRI), Rome, Italy.

Turkey is one of the pioneer countries in genetic resources, establishing an institute in 1932, followed by the collection of nearly 50 thousand plants. In 1963, an institute was opened in Izmir and the national plant genetic resources project was started in 1978 and now Aegean Agricultural Research Institute (AARI) now has overall responsibility for the plant genetic resources of Turkey.

S. J. Bennett and P. S. Cocks (eds.), Genetic Resources of Mediterranean Pasture and Forage Legumes, 150–162.

2. NATIONAL PLANT GENETIC RESOURCES RESEARCH PROGRAMME

The Aegean Agricultural Research Institute (AARI) is the co-ordination centre of the national plant genetic resources programme. The objectives of this program are the collection, conservation and evaluation of the plant genetic resources and plant diversity in Turkey. There are three main units; documentation, systematic and taxonomy, and conservation, with eight plant groups; cereals, forages, food legumes, vegetables, industrial crops, ornamentals, medicinal and aromatic plants, and fruits.

Conservation activities include both *in-situ* and *ex-situ* conservation systems.

2.1 *In-situ* conservation

The *in situ* project started in 1994 and is an important part of the national plant genetic resources research programme, and has been applied to both agricultural and forest tree species in collaboration with the Ministry of Agriculture and Rural Affairs, the Ministry of Forestry, and the Ministry of Environment. The plants studied in this project are wild and weedy relatives, and landraces of cultivated crops such as cereals, forages, food legumes, industrial crops, vegetables, ornamentals, fruits, nuts and grapes, medicinal and aromatic plants. All are endemic species.

The objectives of the project are: 1) to identify and establish *in situ* conservation areas in Turkey for the protection of the genetic resources of wild species of Turkey, 2) To test and develop a new approach to the conservation of genetic diversity, and 3) To provide sustainable *in situ* conservation areas for wild genetic resources.

Project components are defined into the following components:

1. Site surveys and inventories of designated sites
2. Gene management zones
3. Data management
4. National plan for *in situ* conservation.

Three areas were chosen as the proposed gene management zones, containing a number of survey and inventory sites:

1. Kazdag mountains located in the north-west which represents the Euro-Siberian, Mediterranean and Irano-Turanian elements,
2. Ceylanpinar State Farm in south-east which represents the Mediterranean, and Irano-Turanian elements,
3. Anatolian Diagonal in South and Central Turkey which represents Mediterranean, Irano-Turanian and Euro-Siberian elements.

The Aegean Agricultural Research Institute is responsible for the project in Kazdag, and for part of the Anatolian Diagonal. From an agricultural point of view, chestnut (*Castanea sativa*) and plum (*Prunus divericata*), a close relative of cultivated plum, were chosen as target species for Kazdag, and nine potential sites for gene management zones have been selected after the initial surveys. Inventory studies were completed to determine clear ecogeographic ranges and environmental heterogeneity of the target species.

The Field Crops Central Research Institute in Ankara is responsible for the project in Ceylanpinar where species belonging to *Triticum, Aegilops, Hordeum* have been defined as target species. Initial survey studies have already started in the Anatolian Diagonal, where *Vicia* and other forage legumes are likely to be the target species.

Five forest trees (*Abies equi-trojani, Abies clicica* ssp. *issaurica, Pinus nigra, Pinus brutia,* and *Cedrus libani*) were selected as target species in the Kazdag and Bolkar mountains. Population samples and seed collection is complete and gene management zones will be specified following evaluation of the data.

A database will be created for data management using geographic information systems (GIS). A review of the *in situ* conservation project will provide the foundation for a national plan to facilitate the co-ordination and co-operation of gene management zones into other natural conservation strategies, and to outline an implementation plan for continuing the *in situ* conservation activities.

2.2 *Ex-situ* collections

Expeditions are organised each year by the eight plant groups of AARI to landraces, wild relatives of cultivated crops, other wild plant species, and endemic plants of their groups. Collection trips are also planned to meet the needs of other national programs. *Ex-situ* conservation is important for generative collections in seed genebank, and for vegetative collections in field genebanks. The seed bank built in 1972 in maintains earlier collected material. Central Field Crops Research Institute in Ankara retains duplicates of all collections. They have 43 deep-freezes for short- and medium-term storage. The collections are available for plant breeders and researchers both from Turkey an abroad.

2.3 Storage facilities

There are three storage rooms in the genebank at AARI, a long-term storage room held at -18° C for base collections, a mid-term storage room at 0° C for active collections, and a temporary storage room held at 4° C for working material.

2.4 Documentation

A standard format is used for passport data from each accessions per site. Storage and regeneration information are kept separately for each accession in storage format. There is a uniform database program of all data gathered from standard formats.

2.5 Characterisation and evaluation

Characterisation and evaluation activities are conducted separately by the plant groups in co-ordination with the national plant breeding programmes. Data obtained by the plant groups is returned with the seed samples. If material is distributed from the genebank, information is requested for feedback when research project is completed.

2.6 Regeneration/multiplication

Regeneration of the seeds in genebank takes place when their viability has dropped below 80%. Multiplication of the seeds occurs when the amount of seeds drops below a set level. Regeneration and multiplication sites are chosen with consideration for the ecological conditions of the areas where the original accessions were collected. For this reason a number of institutes and breeders are involved. Original and regenerated/multiplied material is kept separately in the genebank.

3. FORAGE COLLECTIONS

The Forage Crops Department has been active in the collection, multiplication, regeneration, characterisation and evaluation of forages for many years. Over ten thousand seed samples have been collected to date (Anon. 1979-1996). The collected species are listed in Table 1.

Table 1. AARI forage collections

Agropyron cristatum	*Koeleria christata*	*Ornithopus compressus*
Agropyron elongatum	*Lathyrus spp.*	*Panicum crus-galli*
Agropyron intermedium	*Lolium loliaceum*	*Phalaris paradoxa*
Arrhenatherum elatius	*Lolium perenne*	*Phalaris tuberosa*
Astragalus asterias	*Lolium rigidum*	*Phleum montanum*
Astragalus hamosus	*Lolium temulentum*	*Pisum sativum*
Astragalus lyridis	*Lotus collinus*	*Poa annua*
Astragalus radiata	*Lotus corniculatus*	*Poa bulbosa*
Astragalus suberosus	*Lotus edulis*	*Poa nemoralis*
Astragalus tridiatus	*Lotus gebelia*	*Poa pratense*
Biserulla pelecinus	*Lotus strictus*	*Sanguisorba minor*
Briza minima	*Lotus tenius*	*Sanguisorba officinalis*
Bromus inermis	*Lotus uliginosus*	*Trifolium spp.*
Bromus japonicum	*Lupinus angustifolius*	*Trigonella fisheriana*
Coronilla emeroides	*Lupinus pilosus*	*Trigonella foenum-graecum*
Coronilla emerus	*Medicago spp.*	*Trigonella lunata*
Coronilla orientalis	*Melilotus officinalis*	*Trigonella monantha*
Coronilla scorpoides	*Onobrychis aequidentata*	*Trigonella monspeliaca*
Coronilla varia	*Onobrychis armena*	*Trigonella polycarpa*
Dactylis glomerata	*Onobrychis caput-galli*	*Trigonella rostrata*
Festuca ovina	*Onobrychis christa-galli*	*Trigonella schlumbergeri*
Hippocrepis unisiliquosa	*Onobrychis sativa*	*Vicia spp.*
Hymenocarpus circinnatus	*Onobrychis viciifolia*	

The four main genera collected within the forages are *Vicia*, *Medicago*, *Trifolium* and *Lathyrus*. The collections of these genera are described below.

3.1 *Vicia* collections

Davis (1970) recorded 59 *Vicia* species in Turkey, six of them endemics. A new species, *Vicia eristaloides*, has since been identified by Maxted (1995). *Vicia* species is shown in Table 2.

Table 6. Vicia collections]

V. anatolica	*V. ervilia*	*V. laxiflora*	*V. noeana*
V. articulata	*V. faba*	*V. lutea*	*V. palaestina*
V. assyriaca	*V. galeata*	*V. melanops*	*V. pannonica*
V. barbazitae	*V. galilae*	*V. meyeri*	*V. peregnina*
V. bithynica	*V. grandiflora*	*V. mollis*	*V. sativa*
V. cassia	*V. hirsuta*	*V. monantha*	*V. seriocarpa*
V. cracca	*V. hybrida*	*V. montbrettii*	*V. tetrasperma*
V. cuspidata	*V. johannis*	*V. narbonensis*	*V. villosa*
V. eristalioides	*V. lathyroides*		

There are a few vetch species which are cultivated in Turkey. Common vetch (*V. sativa*) is widely grown and wild or weedy forms also exist in fields, waste places, meadows, rocky slopes, woodlands margins. There is a large collection of V. *sativa* in the genebank, and nearly 50 % of the accessions are landraces or cultivated crops. The distribution of common vetch is shown in Fig. 1. It has been collected from sea level to 2200 meters.

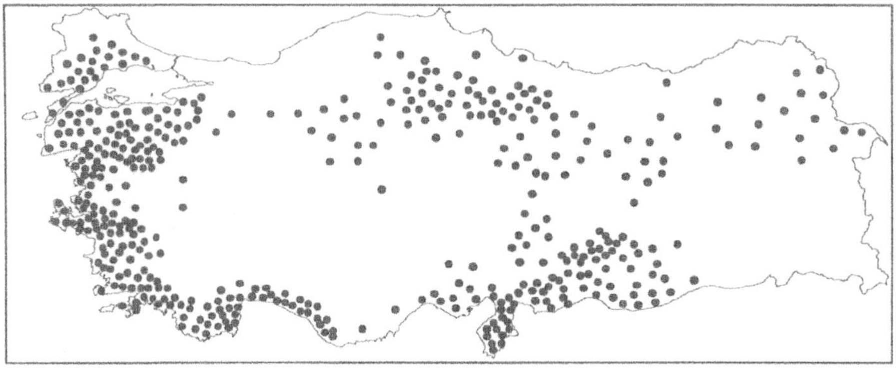

Figure 1. Vicia sativa collections

Narbon vetch (*V. narbonensis*) and Hungarian vetch (*V. pannonica*) are also cultivated. Narbon vetch is grown for seed production in north-west and south-east regions of Turkey(Fig. 2). Its wild or weedy forms can be found in fallow fields, woodlands or rocky slopes. Hungarian vetch was found in cultivated and fallow fields and on road sides (Fig. 3). Cultivated accessions were collected from north-west. The other samples collected from the other regions are mostly wild forms.

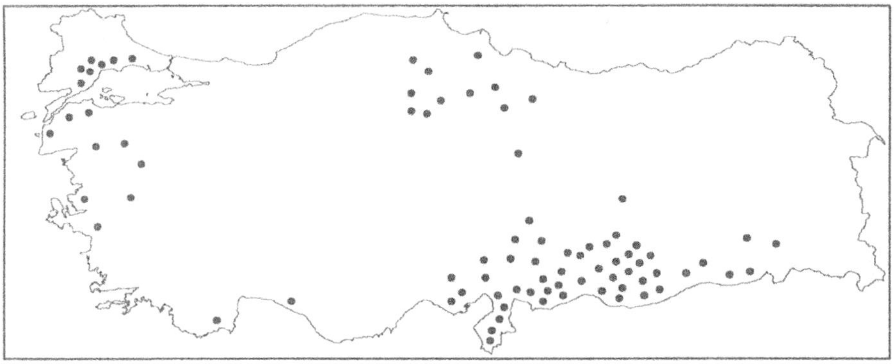

Figure 2. Vicia narbonensis collections

Bittervetch (*V. ervilia*) is another cultivated vetch. The collections are mainly from south-east of Anatolia, and a few accessions from north-west (Fig. 3). This species is a good example of genetic erosion occurring as a result of the replacement of landraces with high yielding common vetch varieties. The accessions held in gene bank came from expeditions in the early 80's. Bittervetch landraces have not been found in recent expeditions. Fig. 3 shows that bittervetch was collected mainly at higher altitudes from 600-1200 m.

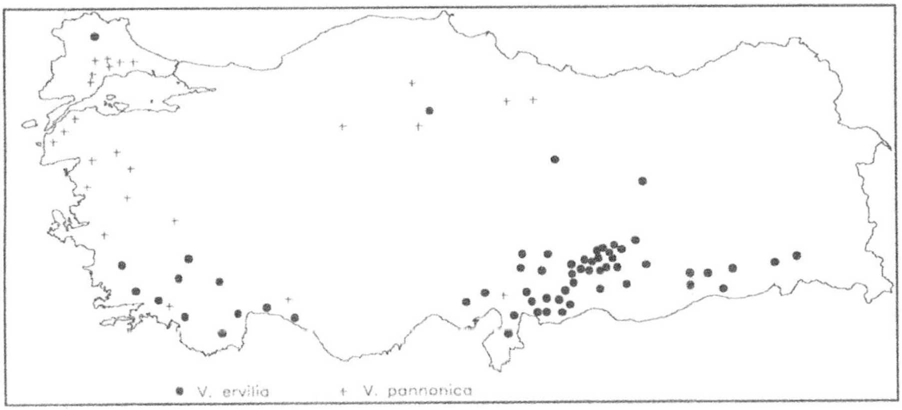

Figure 3. Vicia pannonica and *Vicia ervilia* collections

Hairy vetch (*V. villosa*) is cultivated to a much lesser extent. Two varieties were registered by AARI. This species is widespread throughout Turkey (Fig. 4), with collections mainly from coastal areas at altitudes of 100 to 1200 m. It has also be found at higher altitudes to 1700m.

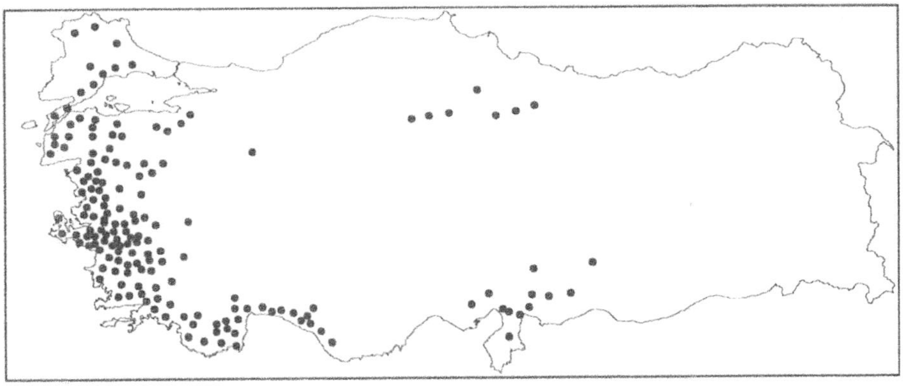

Figure 4. Vicia villosa collections

V. hybrida (Fig. 5) is a widespread species occurring mainly in the south and west regions. V. *peregnina* occurs mainly in the south-east (Fig. 6) and in a lesser extent in north-west. This contrasts with Davis (1975) who reported that it was a widespread species except north and north-east. *V. lutea* and *V. cuspidata* have also been collected from coastal areas (Figs. 7 and 8 respectively), although *V. lutea* exists in inner Anatolia.

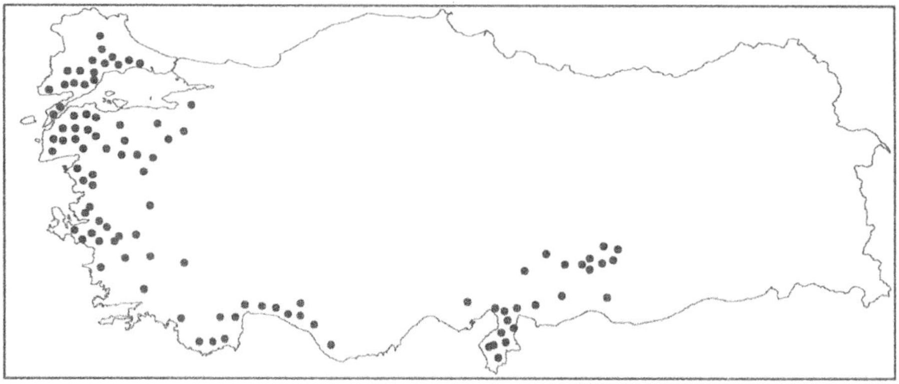

Figure 5. Vicia hybrida collections

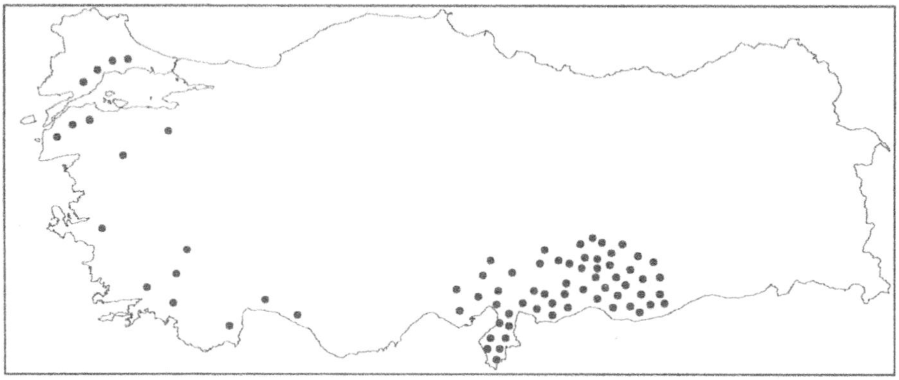

Figure 6. Vicia peregrina collections

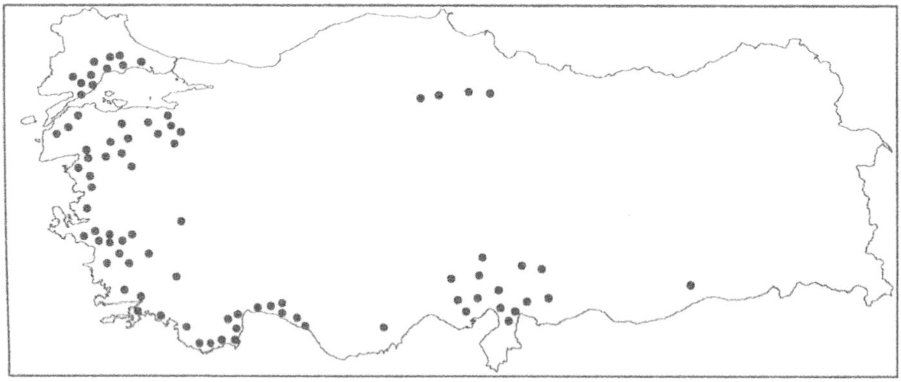

Figure 7. Vicia lutea collections

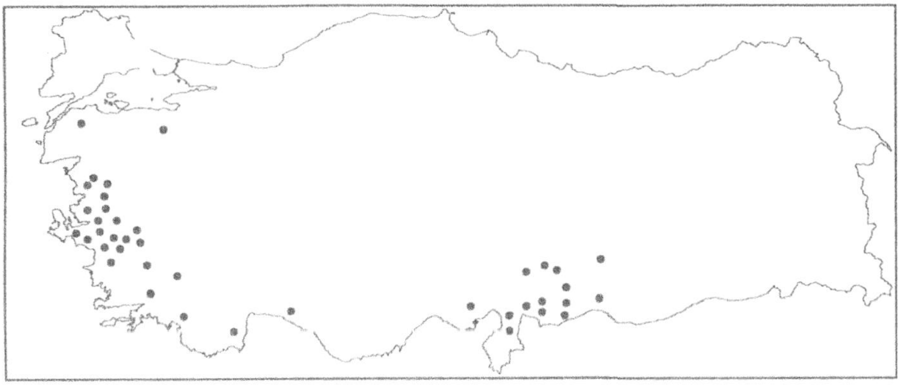

Figure 8. Vicia cuspidata collections

3.2 *Medicago* collections

Harlan (1951) stated that Turkey was the micro gene centres of many *Medicago* species. There are 29 different *Medicago* species in AARI collections (Table 3). Davis (1970) also listed *M. arborea, M. shepardii* (endemic), *M. doliata, M. rugosa* as present in Turkey, but they have not been found during any expeditions.

M. sativa is considered to have originated in eastern Anatolia, west and central Asia (Small 1982). It is a widespread species occurring over most of Turkey, but particularly in the far east and west (Fig. 9). It occurs from sea level to 2200 m. It is cultivated as a fodder crop in almost every part of the country and its wild forms exist in any grassy area.

Table 3. Medicago collections

M. aculeta	*M. falcata*	*M. minima*	*M. radiata*
M. arabica	*M. granadensis*	*M. murex*	*M. rotata*
M. blancheana	*M. intertexta*	*M. noeana*	*M. sativa*
M. coerulea	*M. laciniata*	*M. orbicularis*	*M. scutellata*
M. constricta	*M. littoralis*	*M. papillosa*	*M. truncatula*
M. coronata	*M. lupulina*	*M. polymorpha*	*M. turbinata*
M. disciiformis	*M. marina*	*M. praecox*	*M. varia*
M. elegans			

One forth of *Medicago sativa* collection are landraces, collected mainly from the higher altitudes. In coastal parts the landraces are very rare due to the presence of high yielding varieties such as Peru, Mesasirsa and Moapa.

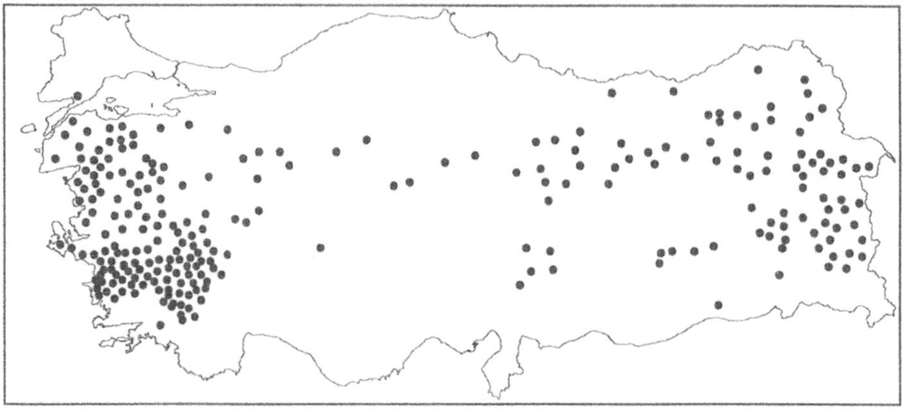

Figure 9. Medicago sativa collections

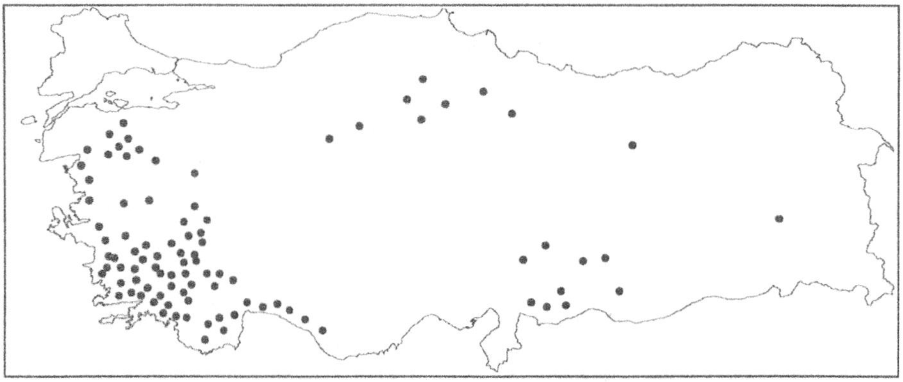

Figure 10. Medicago minima collections

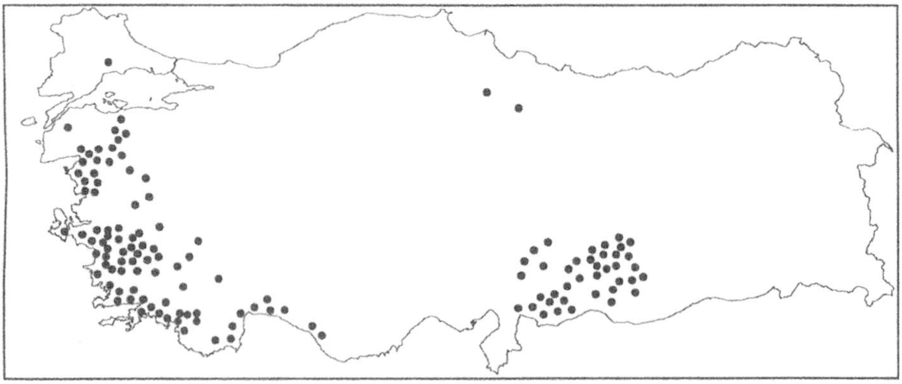

Figure 11. Medicago polymorpha collections

M. minima occurs in a wide range of environments (Fig. 10), particularly in the west and south-west. It is found on heavy or sandy soils, roadsides, low altitude forests, cultivated or fallow fields, pasture and meadows. *M. polymorpha* and M. *rigidula* are found in similar habitats but occur mainly in the west, east and south-east, at sea level to 2000 m (Figs. 11 and 12).

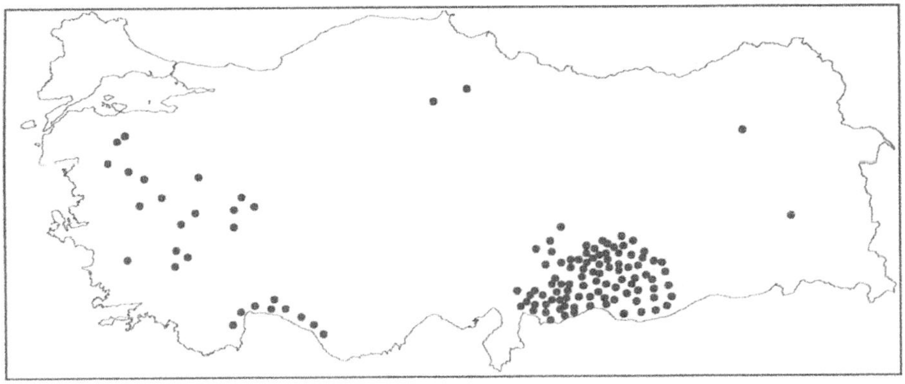

Figure 12. Medicago rigidula collections

3.3 *Trifolium* collections

Davis (1970) reported that there were 94 *Trifolium* species in Turkey and 11 species were endemics. Of these endemics, only three species are held in AARI collection, *T. aintabense*, *T. pannonicum* and *T. apertum* (Table 4).

Table 4. Trifolium collections

T. aintabense	*T. dubium*	*T. meneghianum*	*T. repens*
T. alpestris	*T. echinatum*	*T. michelianum*	*T. resupinatum*
T. angustifolium	*T. erosphaerum*	*T. multistriatum*	*T. retusum*
T. apertum	*T. fragiferum*	*T. nigrescens*	*T. scabrum*
T. argutum	*T. glanduliferum*	*T. pallidum*	*T. scutatum*
T. arvense	*T. globosum*	*T. pannonicum*	*T. spumosum*
T. batmanicum	*T. glomeratum*	*T. patens*	*T. squarrosum*
T. boissieri	*T. grandiflorum*	*T. pauciflorum*	*T. stellatum*
T. campestre	*T. haussknetchi*	*T. phleoides*	*T. subterraneum*
T. cherleri	*T. hirtum*	*T. physodes*	*T. sylvaticum*
T. clusii	*T. incarnatum*	*T. pilulare*	*T. tomentosum*
T. clypeataum	*T. lappaceum*	*T. pratense*	*T. vesiculosum*
T. dasyurum	*T. leucanthum*	*T. purpureum*	*T. xerocephalum*

Figs. 13 and 14 show the distributions of some *Trifolium* species. Two widespread perennial species , *T. repens* and *T. pratense*, are cultivated as fodder crops. The wild forms are found in pastures, meadows, field borders and roadsides.

T. resupinatum is a newly cultivated annual species grown in Aegean region. It is widely distributed as a weed especially in central, west and south Anatolia, from sea level up to 1500 m. It occurs in field borders, roadsides, waste and grassy places.

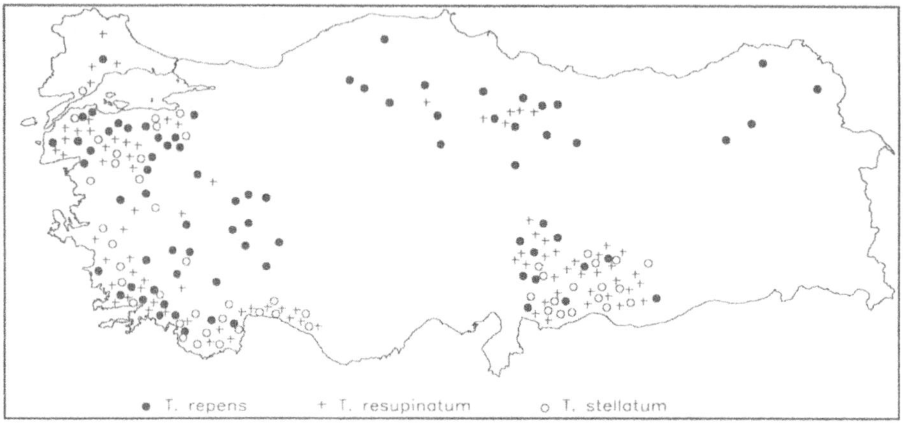

Figure 13. Collections of *Trifolium repens, T. resupinatum* and *T. stellatum*

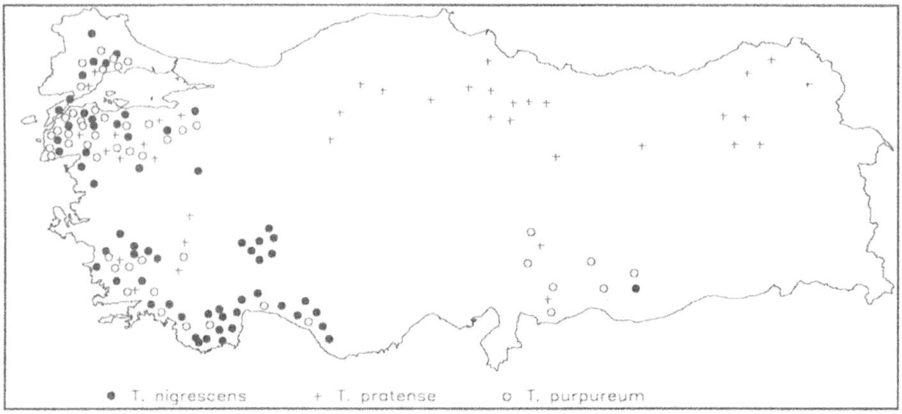

Figure 14. Collections of *Trifolium nigrescens, T. pratense* and *T. purpureum*

3.4 Lathyrus collections

There are some endemic species at local or regional level such as *L. cilicicus, L. elongatus, L. undulatus* out of about 60 species (Davis 1970). AARI *Lathyrus* collections is given in Table 5. Many species including some endemics have not been encountered during recent expeditions. This reflects to what extend the genetic erosion is high.

Table 5. Lathyrus collections

L. annuus	*L. digitatus*	*L. marmoratus*	*L. saxatilis*
L. aphaca	*L. gorgoni*	*L. nissolia*	*L. setifolius*
L. belinensis	*L. graminoides*	*L. ochrus*	*L. sphaericus*
L. blepharicarpus	*L. hierosolymitanus*	*L. pseudoaphaca*	*L. stenophyllus*
L. chrysanthus	*L. hirsutus*	*L. pseudocicera*	*L. tuberosus*
L. cicera	*L. inconspicuus*	*L. rotundifolius*	*L. undulatus*
L. cilicicus	*L. laxiflorus*	*L. sativus*	

L. aphaca is a widespread species (Fig. 15) found in fields, mainly on wheat, roadsides, field borders, pastures, meadows and *Pinus brutia* forests. It occurs from sea level to 2000 m. *L. cicera* is another widely distributed species represented by a great number of accessions in the AARI collections (Fig. 16), particularly in the western through to southern coastal areas.

Lathyrus sativus (Fig. 17) used to be widely grown as a fodder crop, mainly in central and east Anatolia, but is now almost impossible to find. The accessions in the genebank were collected before 1980. This species also exists as a weed in cultivated fields at altitudes of 20 to 1100 m.

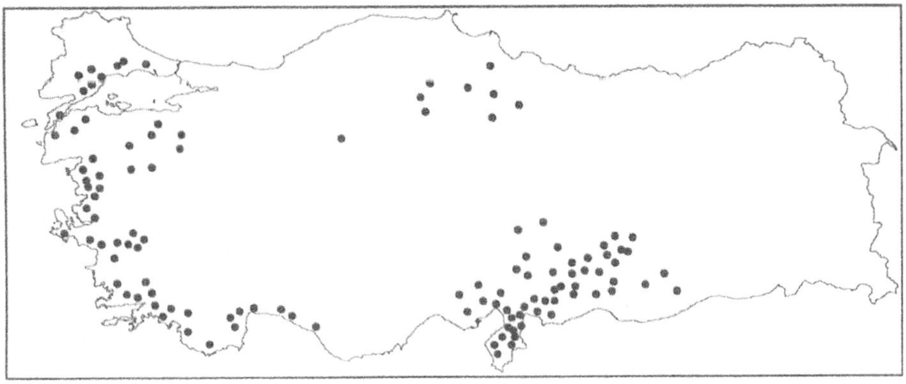

Figure 15. Lathyrus aphaca collections

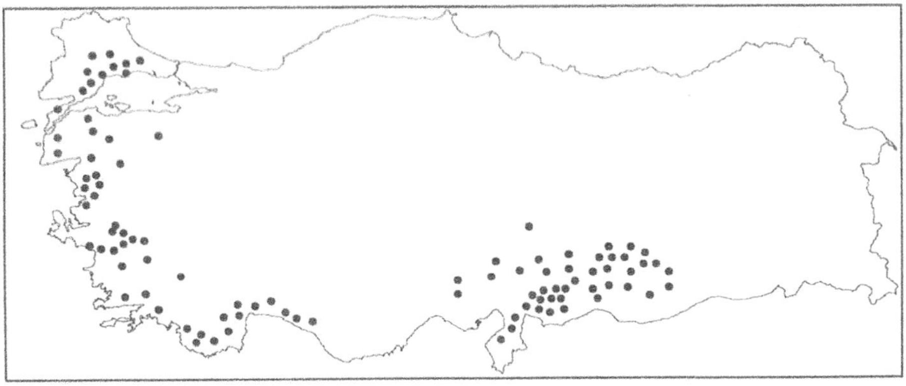

Figure 16. Lathyru cicera collections

4. CONCLUSION

The loss of plant genetic resources and genetic erosion is a serious problem in the world. AARI is responsible the plant genetic resources of Turkey and has been trying to reduce genetic erosion by collecting the plants and managing *ex*- and *in situ* conservation. There are many forage genera and species native to Turkey, and the Forage Crops Department of the Institute has been collecting and conserving these resources for many years. Although many seed samples have been collected and are now held in the genebank, there are still some areas that have not been visited. In a large country like Turkey, it is difficult to cover all areas, some of which may consist of species not found during previous expeditions. Therefore, there is still a need for further collecting missions to cover those areas.

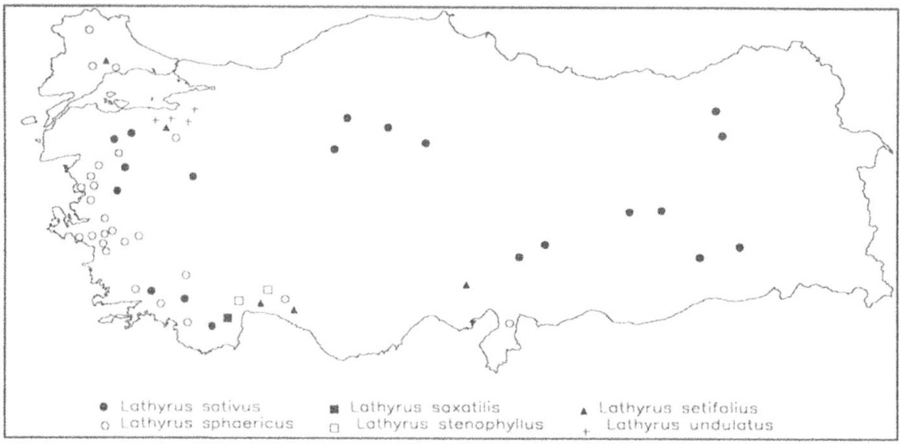

Figure 17. Lathyrus collections

REFERENCES

Anonymous. (1979-1996). Forage crops projects reports. (AARI: Izmir, Turkey).
Dokuzoguz, M. (1990). Bitki genetik kaynaklari. *In*: 'Turkiye'nin Biyolojik Zenginlikleri' (Ed.) Aykut Kence, p.25-40. Turkiye Cevre Sorunlari Vakfi. (Onder Matbaasý: Ankara).
Davis, P.H. (1970). 'Flora of Turkey and East Aegean Islands. Vol. 3.' (Edinburgh University Press: Edinburg).
Harlan, J.R. (1971). Agricultural Origins: Centre and non-centres. *Science*. **174**, 468-474.
Maxted, N. (1995). 'An ecogeographical study of *Vicia* subgenus *Vicia*.' (IPGRI: Rome, Italy).
Small, E. (1982). *Medicago* collecting in Turkey. *Plant Genetic Resources Newsletter* **49**, 11-12.
Vavilov, N.L. (1926). Studies on the origin of cultivated plants. *Bull. Appl. Bot.* **26**, 1-248.

CHAPTER 15

Forage and pasture legume genetic resources at ICARDA

M. Bounejmate, G. Gintzburger, S. Christiansen, L.D. Robertson, A. Abd El Moneim and J. Konopka
International Center for Agricultural Research in Dry Areas, P.O. Box 5466, Aleppo, Syria

Abstract The total holdings of forage and pasture legumes at ICARDA exceed 40,000 accessions. The largest collection is the genus *Medicago* (9085 accessions) followed by *Vicia* (5503 accessions), *Pisum* (3754 accessions), *Trifolium* (3605 accessions), *Lathyrus* (3161 accessions), *Astragalus* (1062 accessions), and *Onobrychis* (1000 accessions). A significant proportion of this material is indigenous germplasm collected by ICARDA in West Asia and North Africa. Of these, annual *Medicago* species (4249 accessions)make up the greatest proportion, followed by *Trifolium* (3157 accessions) and *Vicia* species (2109 accessions). Associated rhizobia were also occasionally collected. However, more effort is required to preserve the genetic variability available in the region.

Several comprehensive distribution studies have identified promising annual *Medicago* species. Information is also accumulating on *Vicia* and *Lathyrus* species, confirming their potential for the West Asia and North Africa region. Annual *Trifolium* species are also widely distributed in the region. Distribution studies have shown that a number of morphological and agronomic attributes of these legumes are related to soil and climatic parameters at the collection site.

The entire germplasm of the following species has been characterised: *Vicia sativa* subsp. *sativa* (1466 accessions), *V. narbonensis* (165 accessions), *V. ervilia* (245 accessions), *Medicago rigidula* (933 accessions) and *M. radiata* (135 accessions). Large variations were recorded for the descriptors used and accessions have been selected for further testing in the breeding programs at ICARDA. ICARDA's breeding programs have made significant progress in selecting species and lines adapted to the West Asia and North Africa region. Supplying national programs with adequate diversity and improved ICARDA material has resulted in the selection and release of several forage and pasture cultivars.

Key words Collecting, Conservation, Evaluation, ICARDA, West Asia, North Africa, Forage and Pasture Legumes.

1. INTRODUCTION

In West Asia and North Africa (WANA), rain-fed crops and livestock production have been integrated for centuries in an efficient and sustainable use of natural resources. Crops were usually followed by one or more years of fallow. Grazing pressure on native pastures and steppeland was adequate to allow the maintenance of species diversity and a high turnover of biomass. Population increases, however, have altered the existing equilibrium. To meet the increasing demands for food and feed, the cereal/fallow rotation has changed to continuous cereal cropping and rain-fed cereal cropping has extended into marginal land. The area devoted to grazing is becoming smaller with more animals grazing fewer hectares. This has generated serious problems including soil erosion, declining soil fertility and inadequate feed supply (Jones 1991, Gintzburger *et al.* 1998).

The International Center for Agricultural Research in the Dry Areas (ICARDA has devoted substantial efforts to help overcome problems of natural-resource degradation and inadequate feed supply in WANA. Given the potential agronomic and economic value of forage and pasture legumes in WANA (Al-Fakhry 1974, Leeuwrik 1976, Carter 1978, Saxena *et al.* 1993), ICARDA has collaborated closely with the National Agricultural Research

S. J. Bennett and P. S. Cocks (eds.), Genetic Resources of Mediterranean Pasture and Forage Legumes, 163–170.

Systems (NARS) to study the opportunities for including more legumes in farming systems. The present paper highlights ICARDA's achievements in forage and pasture legume genetic resource conservation and utilisation.

2. COLLECTIONS

2.1 Strategy and objectives

Germplasm was first gathered by importing accessions from major genebanks world-wide. However, ICARDA's forage and pasture legume breeders recognised the importance of the use of indigenous germplasm to develop locally adapted cultivars. Preliminary screening in Syria of thousands of annual *Medicago* (medics) accessions acquired from various genebanks throughout the world clearly showed that introduced material was not adapted to local conditions (Abd El Moneim and Cocks 1986, Cocks and Ehrman 1987).

From 1977 to 1983, collecting focused on medics. Based on Carter's (1978) report to ICARDA, pasture and forage research first concentrated on medics with a view of to replacing the traditional cereal/fallow rotations with a cereal/medic rotation. In 1984, collecting was extended to other annual legumes to improve the productivity of degraded marginal lands. Since 1990, the emphasis has been on *Vicieae* species. Results from long-term rotation trials showed that vetch is a better choice than medics as a feed resource grown in rotation on cultivated land.

Collecting procedures have also improved. All surveys conducted since 1984 have used the technique suggested by Marshall and Brown (1983) to maximise genetic diversity. Geographical features, rainfall, temperature, soil characteristics and grazing pressure have also been documented. This information has proved to be very important and may eliminate the need to conduct adaptability experiments (Cocks and Abd El Moneim 1991).

2.2 Achievements

ICARDA's current collection of forage and pasture legumes exceeds 40 000 accessions. The largest collection is that of the genus *Medicago* (9085 accessions), followed by *Vicia* (5503 accessions), *Pisum* (3754 accessions), *Trifolium* (3605 accessions), *Lathyrus* (3161 accessions), *Astragalus* (1062 accessions), and *Onobrychis* (1000 accessions). Of particular significance is the proportion of accessions collected by ICARDA in WANA. ICARDA, with the participation of national institutions, has been involved in more than 40 plant collection missions leading to the safeguarding of nearly 15,000 accessions (Table 1). Medics were the most frequently collected (4249 accessions), followed by *Trifolium* (3157 accessions) and *Vicia* species (2109 accessions). In a few instances, rhizobia were also collected: *Rhizobium leguminosarum* (91), *R. meliloti* (633) and *R. trifolii* (225). Rhizobial collections are maintained in a lyophilised form.

Collecting efforts have varied among countries. Significant diversity has been collected in Syria (5386), Morocco (2310), Jordan (2017), and Tunisia (1832). In addition, comprehensive collections have been made in WANA and other areas of the Mediterranean basin without ICARDA's involvement. These include Libya (Gintzburger 1980, Gintzburger *et al.* 1983), Morocco (Francis 1987, Feniche 1993), Algeria (Gintzburger 1984, Abdelguerfi *et al.* 1988), France (Prospéri *et al.* 1989), Iraq and Syria (Francis 1980), and Sardinia (Piano *et al.* 1982).

Table 1. Main forage legumes collected by ICARDA 1977-1995

Country	Genus and (samples) collected	Total
Algeria	*Anthyllis* (11), *Astragalus* (100), *Coronilla* (41), *Hippocrepis* (43), *Lathyrus* (33), *Lotus* (5), *Medicago* (395), *Melilotus* (43), *Onobrychis* (15), *Ononis* (1), *Pisum* (14), *Scorpiurus* (68), *Trigonella* (47), *Trifolium* (273), *Vicia* (208)	Plants: 1297
Bangladesh	*Lathyrus* (62), *Pisum* (10), *Vicia* (2)	Plants: 74
Cyprus	*Lathyrus* (5), *Vicia* (9), *Rhizobium meliloti* (6)	Plants: 14 Rhizobia: 6
Egypt	*Astragalus* (8), *Coronilla* (5), *Hippocrepis* (13), *Hymenocarpus* (1), *Lathyrus* (1), *Medicago* (49), *Melilotus* (1), *Onobrychis* (1), *Pisum* (1), *Scorpiurus* (7), *Trigonella* (10), *Trifolium* (17), *Vicia* (13), *Rhizobium leguminosarum* (1)	Plants: 127 Rhizobia: 1
Iran	*Astragalus* (46), *Coronilla* (2), *Hippocrepis* (6), *Hymenocarpus* (3), *Lathyrus* (5), *Lotus* (1), *Medicago* (144), *Onobrychis* (9), *Scorpiurus* (1), *Trifolium* (67), *Trigonella* (43), *Vicia* (12)	Plants: 339
Iraq	*Astragalus* (3), *Coronilla* (2), *Hippocrepis* (3), *Hymenocarpus* (4), *Lathyrus* (10), *Medicago* (70), *Pisum* (1), *Scorpiurus* (2), *Trifolium* (68), *Trigonella* (4), *Vicia* (13)	Plants: 180
Jordan	*Anthyllis* (1), *Astragalus* (205), *Bisserula* (6), *Coronilla* (39), *Hippocrepis* (78), *Hymenocarpus* (68), *Lathyrus* (32), *Lotus* (6), *Medicago* (641), *Melilotus* (2), *Onobrychis* (93), *Ononis* (3), *Pisum* (22), *Scorpiurus* (46), *Trifolium* (547)), *Trigonella* (128), *Vicia* (100), *Rhizobium leguminosarum* (65), *Rhizobium meliloti* (94), *Rhizobium trifolii* (56)	Plants: 2017 Rhizobia: 215
Lebanon	*Medicago* (20), *Onobrychis* (1), *Trifolium* (1), *Vicia* (11), *Rhizobium leguminosarum* (1), *Rhizobium meliloti* (30), *Rhizobium trifolii* (31)	Plants: 33 Rhizobia: 62
Morocco	*Anthyllis* (20), *Astragalus* (94), *Bisserula* (14), *Coronilla* (26), *Hedysarum* (6), *Hippocrepis* (45), *Lathyrus* (146), *Lotus* (47), *Lupinus* (18), *Medicago* (729), *Melilotus* (25), *Onobrychis* (12), *Ornithopus* (22), *Pisum* (20), *Scorpiurus* (174), *Tetragonolobus* (3), *Trifolium* (383), *Trigonella* (26), *Vicia* (500), *Rhizobium meliloti* (151), *Rhizobium trifolii* (83)	Plants: 2310 Rhizobia: 234
Pakistan	*Astragalus* (1), *Lathyrus* (20), *Melilotus* (5), *Pisum* (10), *Trigonella* (6), *Trifolium* (1), *Vicia* (16), *Rhizobium leguminosarum* (1)	Plants: 59 Rhizobia: 1
Syria	*Astragalus* (264), *Bisserula* (15), *Coronilla* (55), *Hedysarum* (1), *Hippocrepis* (62), *Hymenocarpus* (94), *Lathyrus* (477), *Lotus* (1), *Lupinus* (2), *Medicago* (1720), *Melilotus* (11), *Onobrychis* (150), *Pisum* (75), *Scorpiurus* (44), *Trifolium* (1280), *Trigonella* (192), *Vicia* (943), *Rhizobium leguminosarum* (23), *Rhizobium meliloti* (221), *Rhizobium trifolii* (55)	Plants: 5386 Rhizobia: 299
Tunisia	*Anthyllis* (18), *Astragalus* (103), *Bisserula* (1), *Coronilla* (34), *Hedysarum* (107), *Hippocrepis* (57), *Hymenocarpus* (11), *Lathyrus* (41), *Lotus* (77), *Lupinus* (3), *Medicago* (671), *Melilotus* (25), *Onobrychis* (1), *Ornithopus* (3), *Pisum* (5), *Scorpiurus* (141), *Tetragonolobus* (14), *Trifolium* (352), *Trigonella* (38), *Vicia* (130), *Rhizobium meliloti* (14)	Plants: 1832 Rhizobia: 14
Turkey	*Anthyllis* (1), *Astragalus* (29), *Coronilla* (2), *Hippocrepis* (6), *Hymenocarpus* (11), *Lathyrus* (39), *Lupinus* (2), *Medicago* (284), *Melilotus* (2), *Onobrychis* (5), *Pisum* (3), *Scorpiurus* (8), *Trifolium* (395), *Trigonella* (29), *Vicia* (96), *Rhizobium meliloti* (117)	Plants: 912
Total	***Anthyllis* (51), *Astragalus* (853), *Bisserula* (36), *Coronilla* (206), *Hedysarum* (114), *Hippocrepis* (313), *Hymenocarpus* (192), *Lathyrus* (871), *Lotus* (137), *Lupinus* (25), *Medicago* (4723), *Melilotus* (114), *Onobrychis* (287), *Ononis* (4), *Ornithopus* (25), *Pisum* (161), *Scorpiurus* (491), *Tetragonolobus* (17), *Trifolium* (3261), *Trigonella* (646), *Vicia* (2053), *Rhizobium leguminosarum* (91), *Rhizobium meliloti* (633), *Rhizobium trifolii* (225)**	**Plants: 14580** **Rhizobia: 949**

2.3 Prospects

In view of the extensive holdings of forage and pasture legume germplasm at ICARDA, increased emphasis will be placed on characterisation, evaluation, documentation and utilisation. However, WANA's rich flora is far from adequately sampled and preserved. More effort is needed to preserve the genetic variability available in the region. Gaps have been identified and priorities have been set in consultation with National Working Groups (Table 2). From an ecological viewpoint, emphasis will be on collecting the germplasm of native pastures and rangelands in low-rainfall areas which have been neglected. These areas are also affected by high rates of genetic erosion (Robertson and Bounejmate 1998). Therefore, they arehigh priority for germplasm collection.

Priorities have also been set at the sub-regional level. At the North African Region Co-ordination Meeting held 9-13 October 1993 in Algeria, priorities for germplasm collection, characterisation and evaluation in North Africa were set (ICARDA 1993). The participants unanimously endorsed the greater practical significance of prioritising species that are potentially exploitable by farmers. Their list (Table 3) includes nine major groups of pasture and forage species, including grasses and legumes, based on their need for research attention and the need for collection in each North African country.

Table 2. Priority species for ICARDA's region

High priority	Second priority	Low priority
Annual *Medicago* species, *Vicia sativa* ssp. *amphicarpa*, V. *sativa*, V. *narbonensis*, V. *ervilia*, V. *hybrida*, V. *villosa*, *Lathyrus sativus*, L. *ochrus*, L. *cicera*, annual *Trifolium* species	*Astragalus*, *Coronilla*, *Hedysarum*, *Hippocrepis*, *Hymenocarpus*, *Onobrychis*, *Scorpiurus*, *Trigonella*	*Melilotus*, *Pisum*, *Lotus*, *Ononis*

Table 3. Priorities for forage and pasture plant characterisation and evaluation, base collection site suggested for each group, and requirements for further collection in North Africa

Rank	Forage group	Base collection	Collection Needed
1	*Vicia* and *Lathyrus*	Morocco	Algeria, Libya
2a	*Medicago aculeata*	Algeria	-
2b	Dry-adapted *Medicago*	Libya	-
3	*Hedysarum*	Tunisia	Algeria, Libya, Morocco, Tunisia
4	*Avena*	Tunisia	Algeria, Libya, Morocco, Tunisia
5a	*Trifolium subterraneum*	Morocco	Algeria, Tunisia
5b	Other *Trifolium*	Algeria	-
6	*Medicago sativa*	Morocco/Tunisia	Algeria, Libya, Tunisia
7	*Lotus*	Morocco	Algeria, Libya, Tunisia
8	*Scorpiurus*	Algeria	-
9	*Lolium multiflorum*	Tunisia	Algeria, Libya, Morocco, Tunisia

Note: Other *Medicago, Lupinus, Pisum, Astragalus, Atriplex, Biserrula, Hippocrepis, Onobrychis, Dactylis* and *Festuca* species and *Trifolium alexandrinum* were also discussed but not ranked.

3. DISTRIBUTION STUDIES

During the 1970's, several attempts were made to introduce forage and pasture legumes into the farming systems of WANA. Despite considerable optimism and effort, failures have been common (Bakhtri 1983). Insufficient information on the ecology and distribution of these species, or a lack of recognition of their importance, were major causes of such failures.

Based on these past experiences, a new approach to evaluating forage and pasture legumes in WANA was developed at ICARDA (Cocks and Abd El Moneim 1991). In addition to emphasising the need for a better understanding of farmers reactions when introducing forage and pasture legumes to existing farming systems, this approach recognised the importance of understanding the adaptation and distribution of species and therefore ecological characterisation increased during collections.

Several comprehensive distribution studies have been undertaken. These include annual *Medicago* species (Ehrman and Cocks 1990, Bounejmate *et al.* 1992a 1992b, Zoghlami *et al.* 1996), annual legumes (Cocks and Ehrman 1987, Beale *et al.* 1991), annual *Trifolium* species (Beale *et al.* 1993, Zoghlami *et al.* 1995), and *Vicia* and *Lathyrus* (Francis *et al.* 1994, Robertson and Abd El Moneim 1996). Distribution studies have helped to establish the potential agricultural range for the successful use of these species indicate the species deserving greater attention.

In addition to indicating species deserving more attention, the distribution studies also yielded information which will assist in developing strategies for selecting adapted cultivars. They have shown that a number of morphological and agronomic attributes of legumes are related to soil and climatic parameters at the collection site. In Morocco, nodulation in the presence of nitrate, response to phosphorus, and response to temperature and frost tolerance were related respectively to soil nitrogen content, soil phosphorus content and winter temperature at the collection site (Bounejmate 1992). Similar results are reported by Cocks and Ehrman (1987) and Ehrman and Cocks (1990) in Syria.

4. GERMPLASM CHARACTERISATION, EVALUATION AND UTILISATION

4.1 Germplasm characterisation

So far, a total of 4026 accessions from four genera have been characterised. This includes the entire collection of *Lathyrus* (30 species, 1082 accessions), *Vicia sativa* subsp. *sativa* (1466 accessions), *V. narbonensis* (165 accessions), *V. ervilia* (245 accessions), *Medicago rigidula* (933 accessions) and *M. radiata* (135 accessions) germplasm held at ICARDA. For this evaluation, 21, 23 and 14 descriptors were observed for *Lathyrus*, *Vicia* and *Medicago* species, respectively (ICARDA 1993).

Large variations were observed for the descriptors used. This variation could be exploited to extend the agricultural range of these species in WANA. Several accessions of *Lathyrus*, *Medicago* and *Vicia* species were earlier flowering than the controls, indicating a potential for use in lower-rainfall areas. Also, accessions of *Vicia* varied in their seed size, a trait important during the selection of improved cultivars, as larger seeds are preferred for animal feed and achieve the highest price.

4.2 Forage legumes

Breeding programs for *Vicia* and *Lathyrus* started in the mid-1980s and have made significant progress in selecting species and lines adapted to the WANA region for seed and straw, hay or grazing (Robertson and Abd El Moneim 1996). Host-plant screening in *Vicia* and *Lathyrus* species for resistance to powdery mildew (*Erysiphe pisi* Syn.), botrytis blight (*Botrytis cinerea* Pers. ex Fr.) and ascochyta blight (*Ascochyta pisi* Lib.) have recently started. Several sources of resistance have been found but reconfirmation is still required. A screening program for root-knot nematode (*Meloidogyne artiellia* Frankland) was initiated in 1985 and several accessions with resistance have been identified. Lines resistant to *Orobanche crenata* in *Vicia narbonensis*, *V. sativa* subsp. *sativa*, *V. villosa* subsp. *dasycarpa* and *Lathyrus ochrus* have also been identified. Screening for cold tolerance has started with *Vicia* and *Lathyrus*. One accession of *Lathyrus ochrus* from Portugal showed a high level of cold tolerance. Five non-shattering genotypes of *Vicia sativa* subsp. *sativa* with earliness and erect plant type were selected (Abd El Moneim 1993). In *Lathyrus*, lines have been selected with low levels of anti-nutritional compounds (β-*N*-Oxalyl-L-α, β-Diaminopropionic Acid (β -ODAP). This neurotoxin causes lathyrism, a nervous disorder resulting in incurable paralysis of the lower limbs of people and domestic animals.

There have been significant improvements in the yields of *Lathyrus* species (Roberston and Abd El Moneim 1996). *Lathyrus sativus* lines have been selected which yield over one t/ha of seed in a low-rainfall environment. Dwarf chickling (*Lathyrus cicera*) has potential in low temperature, high altitude environments because of its cold tolerance. Lines have also been selected with promise for low- and medium-rainfall environments. Ochrus chickling (*Lathyrus ochrus*) is an important species in the WANA region because of its high level of resistance to *Orobanche crenata*. Large areas of WANA are infested with this parasitic weed and production of most legume species, including *Lathyrus* species, in these areas is difficult to impossible. This species is susceptible to cold, although one accession has been found with moderate levels of cold tolerance. In moderate winter environments selections from this species are significantly more productive than *L. sativus* or *L. cicera*.

4.3 Pasture legumes

As WANA has an immense resource of native pasture legumes (Mathison 1983), pasturelegume improvement at ICARDA relies on these resources. Two main selection criteria are used.

The first is adaptation to the environment of native species. This approach has been successful in identifying species and ecotypes adapted to the cold environment of West Asia. Native species including *Medicago rigidula*, *M. rotata* and *M. noeana* have proved to be better adapted to local conditions than introduced cultivars (Abd El Moneim and Cocks1986, Cocks and Ehrman 1987).

The ability to persist is the second important factor in the success of pasture legumes. Genotypes are selected from mixtures of 100 or more grown in a system similar to that which will be used by Syrian farmers. Sowing of a mixture of genotypes, and monitoring the change in proportion of each genotype under grazing, has proved a reliable guide potential cultivar success.

4.4 On-station and on-farm evaluation

Preliminary on-station evaluation of selected forage and pasture legume species and cultivars indicated suitable species of *Medicago*, *Pisum* and *Vicia* which are well adapted to the climatic conditions of north Syria (Osman and Nersoyan 1985). These species were used in 1985/86 to establish a long-term rotation trial of pasture and forage legumes compared to other common alternatives. After the first six years sites of medic and vetch species showed a significant accumulation of total nitrogen and organic matter in the soil. However, vetch was found to be better than medic as a feed resource grown in rotations on cultivated land, while medics performed better in the improvement of uncultivated land. Analysis also showed that only the heaviest medic grazing pressures were economically competitive with other treatments.

On-farm forage legume trials conducted for seven cropping seasons in north-west Syria, annual rainfall 270mm, confirmed the potential of *Vicia* and *Lathyrus* to increase protein production of the subsequent barley crop (ICARDA 1994). The trials compared fallow or barley with common vetch (*Vicia sativa*) or chickling (*Lathyrus sativus*) in rotation with barley. Yield of feed from both phases of the rotation was greatest from the barley/vetch and barley/chickling rotations. Total outputs of crude protein were twice as high for the rotations including legumes compared with the average output of barley/fallow and for barley/barley. The barley/vetch rotation gave the highest output of metabolisable energy.

5. SEED PRODUCTION

Seed production of forage and pasture legumes is insufficiently developed in WANA. Although several species have been recommended for the sustainable use of the natural resources in the region, their use is impeded by a lack of seed. Institutional, technical and economic constraints prevent the passage of promising, locally–adapted material from the evaluation stage to on-farm testing or broad-scale use.

To tackle these problems, and to alleviate some technical constraints, ICARDA has invested in the development of a participatory research approach to encourage production of forage and pasture species. A project was developed to demonstrate that pasture and forage and pasture legumes are valuable to farmers, in turn increasing demand that will, in turn, attract commercial seed producers (Christiansen 1993). ICARDA has also developed a functioning methodology to produce and harvest sufficient quantities of forage and pasture legume seed using machinery manufactured locally. The creation of small seed processing units will allow farmer co-operatives to produce their own locally adapted pasture–seed mixtures outside the formal seed industry channels. On-farm testing of machinery for seed production of pasture and forage legumes is currently in progress in several WANA countries including Syria, Morocco, and Lebanon.

ACKNOWLEDGEMENTS

Dr. Jan Valkoun made valuable comments on this paper, his contribution is very much appreciated.

REFERENCES

Abd El Moneim, A. 1993. Selection for non-shattering common vetch (*Vicia sativa* L.). *Plant Breeding* **110,** 168-171.

Abd El Moneim, A. M. and P.S. Cocks 1986. Adaptation of *Medicago rigidula* to a cereal pasture rotation in north-west Syria. *J. Agric. Science (Cambridge)* **107,** 179-186.

Abdelguerfi, A., J.Y. Chapot and A.P. Conesa 1988. Contribution à l'étude de la répartition des luzernes annuelles spontanées en Algérie selon certains facteurs du milieu. *Fourrages* **113,** 89-106.

Al-Fakhry, A.K. 1974. Improvement of dryland farming in northern Iraq. In: Recent Stages and Prospects of Agricultural Mechanisation in Developing Countries. Leipzig, GDR 1974; 237-242.

Bakhtri, M.N. 1983. La rotation céréale-luzerne annuelle en Afrique du Nord et au Proche Orient. AGPC:MISC/79, 35p. FAO, Rome.

Beale, P.E., M. Bounejmate, A. Lahlou, D.B. Marx and S. Christiansen 1993. Distribution of annual *Trifolium* species in Morocco. *Aust. J. Agric. Res.* **44,** 1303-10.

Beale, P.E., A. Lahlou and M. Bounejmate 1991. Distribution of wild annual legumes species in Morocco in relationship with soil and climatic factors. *Aust. J. Agric. Res.* **42,** 1217-30.

Bounejmate, M. 1992. Soil and climatic factors affecting the natural distribution of annual *Medicago* species in Morocco. PhD thesis, University of Western Australia.

Bounejmate, M., P.E. Beale and A.D. Robson 1992a. Annual *Medicago* species in Morocco. I. Species and their abundance. *Aust. J. Agric. Res.* **43,** 739-49.

Bounejmate, M., A.D. Robson and P.E. Beale 1992b. Annual *Medicago* species in Morocco. II. Distribution in relation to soil and climate. *Aust. J. Agric. Res.* **43,** 751-63.

Carter, E. D. 1978. A review of the existing and potential role of legumes in farming systems of the Near East and North African Region. A report prepared for ICARDA. Waite Agricultural Research Institute, University of Adelaide, South Australia.

Christiansen, S. 1993. A medic sweeper and a thresher help farmers produce pasture seed. World Agriculture 1994: The international review of agribusiness in developing markets. pp. 111-114.

Cocks, P.S. and A. Abd El Moneim 1991. A systems approach to the evaluation of pasture and forage legumes in west Asia. In: Legume Genetic Resources for Semi-Arid Temperate Environments A. Smith and L.D. Robertson, ed.), pp. 281-296. ICARDA, Aleppo, Syria.

Cocks, P.S. and T.A.M. Ehrman 1987. The geographic origin frost tolerance in Syrian pasture legumes. J. Appl. Ecol. **24**, 678-83.
Ehrman, T.A.M. and P.S. Cocks 1990. Ecogeography of annual legumes in Syria: Distribution patterns. *J. Appl. Ecol.* **27**, 578-91.
Feniche, F.A. 1993. Distribution naturelle et caractérisation agromorphologique des populations locales de *Medicago aculeata* au Maroc. Mémoire de 3ème cycle de l'IAVHII, Rabat.
Francis, C.M. 1980. Ecology and distribution of *Medicago* and other legume species in the upper Jezira region of Iraq and Syria. Consultant Report No. 2. Western Australian Overseas Projects Authority. Perth, Western Australia.
Francis, C.M. 1987. Morocco, a plant collection tour. A report compiled for the Grain Research Committee of Western Australia, Western Australian Department of Agriculture, Perth.
Francis, C.M., M. Bounejmate and L.D. Robertson 1994. Observations on the distribution and ecology of *Vicia* and *Lathyrus* species in northern Morocco. *Al Awamia* **84**, 17-27.
Gintzburger, G. 1980. Forage legume in Libya. *Plant Genetic Resources Newsletter* **47**, 7-11.
Gintzburger, G. 1984. Listing of the Algerian collection (June 1982): Annual medics and other legumes. Western Australian Department of Agriculture, Perth.
Gintzburger, G., T.L. Nordblom and A.E. Osman 1998. Agro-pastoral systems, feed calendars and feed resources in the Arab countries of the Mediterranean region. In: Integrated crop-systems in the dry areas of West Asia and North Africa. Proceedings of the workshop held 6-8 November 1995, Amman, Jordan (in press).
Gintzburger, G., A. Sbeta, H. Rojbani, M. Fituri and C.M. Francis 1983. Annual *Medicago* spp. collected from the Syrte Region of Libya. *Plant Genetic Resources Newsletter* **56**, 23-29. FAO-IBPGR, Rome
ICARDA 1993a. Proceedings of North African Region Co-ordination Meeting, 9-13 October 1993, Sidi Fredj, Algeria. ICARDA, Tunis Office, Tunisia.
ICARDA 1994. Pasture, Forage and Livestock Program Report for 1994. ICARDA, Aleppo, Syria.
Jones, M. 1991. Agricultural sustainability Research at ICARDA. A Presentation at Centers Week 1991, Washington, D. C. ICARDA, Aleppo, Syria.
Leeuwrik, D. M. 1976. The relevance of cereal-pasture legume rotation in the Middle East and the North African region. pp. 266-291. Proceedings of the third Regional Wheat Workshop, Tunis 1975. Cimmyt: Mexico.
Marshall, D.R. and A.H.D. Brown 1983. Theory of forage plant collection. In: Genetic Resources of Forage Plants J.G. McIvor and R.A. Bray, Ed.), pp. 135-48. CSIRO, Melbourne.
Mathison, M.J. 1983. Mediterranean and temperate forage legumes. In: Genetic Resources of Forage Plants J.G. McIvor and R. A. Bray, Ed.), pp. 63-81. CSIRO, Melbourne.
Osman, A. and N. Nersoyan 1985. Annual legumes for integrating rainfed crop and livestock production. In: Proceedings of the XV International Grassland Congress, Kyoto, Japan, p. 123-125.
Piano, E., M. Sardara, and S. Pusceddu 1982. Observations on the distribution and ecology of subterranean clover and other annual legumes in Sardinia. *Revista Di Agronomy* **18**, 273.
Prospéri, J.M., V. Gensollen, I. Olivieri and P. Mansat 1989. Observations sur la répartition et l'écologie des luzernes annuelles et de trèfle souterrain en corse. pp. 295-96. In: Proceedings of the XVI International Grassland Congress, Nice, France.
Robertson, L.D.and A. Abd El Moneim 1996. *Lathyrus* germplasm collection, conservation and utilisation for crop improvement at the International Center for Agricultural Research in the Dry Areas (ICARDA). Paper presented at the *Lathyrus* Genetic Resources Conservation Conference, India, December 1995.
Robertson, L.D. and M. Bounejmate 1997. Genetic erosion of pasture and forage legumes in the Mediterranean. In: Proceedings of CLIMA Workshop on Genetic Resources of Pasture and Forage Legumes in the Mediterranean 14-18 October 1996, York, Australia.
Saxena, M.C., A. Abd El Moneim and M. Ratinam 1993. Vetches (*Vicia* spp.) and chicklings (*Lathyrus* spp.) in the farming systems in West Asia and North Africa and improvement of these crops at ICARDA. In: J.R. Garlinge and M.W. Perry, (ed.), Potential for *Vicia* and *Lathyrus* species as new grain and fodder legumes for southern Australia, pp. 2-9. CLIMA, Nedlands, Western Australia.
Zoghlami, A., H. Hassen, L. Robertson and B. Reid 1995. Observations sur la répartition et l'écologie des espèces de *Trifolium* et de *Scorpiurus* au nord de la Tunisie. *Al Awamia* **90**, 115-120.
Zoghlami, A., H. Hassen, H. Seklani, L. Robertson and A.K. Salkini, 1996. Distribution des luzernes annuelles en Tunisie centrale en fonction des facteurs édaphiques et climatiques. *Fourrages* **145**, 5-16.

CHAPTER 16

From free access to fair trade: Political dimensions of collecting *in situ* plant genetic resources for food and agriculture (PGRFA)

Heather Paull
Centre for Development Studies, Flinders University of South Australia, Adelaide, SA.

Abstract Researchers requiring access to plant genetic resources for food and agriculture are currently facing a new regime of exchange. Whilst the precise nature of the regime remains unclear at this stage, once finalised, it will provide a clear defining of rights and obligations. In general, collectors can expect continued freedom of access if the intents of 'fair access' and 'fair exchange' are upheld.

Keywords PGRFA, Access, Exchange, Policies, Rights, Obligations

1. INTRODUCTION

This paper provides some preliminary comments on the political context in which the collection of farmer varieties, landraces, crop wild relatives and other wild species takes place, highlighting the extreme complexity of the international plant genetic resource (PGR)-related arena and recent climate of conflict in perspective and opinion over plant genetic resource issues. Recognising a shift in balance of power between the North (established market economies /OECD nations) and the South ('the rest'/G77 nations), at least in the realm of environmental issues, it indicates that we now face a new regime of exchange, the precise nature of which remains unclear, with negotiations between competing inter- and intra-national interests still in progress. Two main points are made. First, whilst North-South polarisation still exists, and whilst the North maintains a significant advantage in international negotiations per se, given its financial, high-technological and trade strength, it is now recognised that it is to the North's own detriment to negotiate in PGR-related flora without some degree of respect for the South's position. Second, is the observation that whether the new exchange regime is formulated on the basis of social justice or 'enlightened' self-interest, it will have to have a clear defining of rights and obligations. Transfers will need to be carried out on the basis of fair and equitable sharing of benefits by users and donors of germplasm and funds and technology, however that 'fairness' and equity be finally defined between the interested parties.

The paper provides an historical context to the current PGRFA debate, the responses, emerging policies and some of the international principles that have been developed in consequence - i.e. the rules and standards co-ordinating international behaviour upon which future agreements on PGR access are likely to be based. Finally, a few examples of current access-related policies are used to illustrate some emerging trends, giving a flavour of the bases on which collection and exchange are likely to take place in the near future.

2. BACKGROUND: ACCESS

Today, genetic resources can be considered either state property, common / community property (for example, as the right of local farming or indigenous communities), as individual 'intellectual' property or open / free access material. As this paper will show, such state control is now clearly backed by customary international law, through instruments such as the Convention on Biological Diversity (CBD) (negotiated under the auspices of the United Nations

S. J. Bennett and P. S. Cocks (eds.), Genetic Resources of Mediterranean Pasture and Forage Legumes, 171–181.

Environment Programme, UNEP) and the International Undertaking on Plant Genetic Resources (Undertaking) (of the UN's Food and Agricultural Organisation, the FAO), providing nations with the right to deny or impose conditions on access to biological (and thus, genetic) resources within their territory. At the same time, however, reflecting a growing challenge to the assumed 'natural' state of national control, is the international recognition and confirmation of varying degrees of community rights. Such rights are reflected in the notion of Farmers' Rights (in the International Undertaking), or indigenous people's rights over lands and resources in, for example, the 1989 International Labour Convention No. 169 (tenurial right), the Conventional on Biological Diversity (also refered to, herein, as the CBD or Convention), and the UN's Conference on Environment and Development's (UNCED's) Agenda 21. Individual rights, such as Intellectual Property Rights (IPRs), are also becoming increasingly relevant. IPRs confer limited proprietary rights in bioresources as patents (over plants, animals, micro-organisms, DNA, genes), Plant Breeder's Rights (over varieties) or Trade Secrets (Sehgal and van Rompey, ND). Whilst primarily a Northern (and controversial) phenomena (being based in western private property and natural rights systems), these rights appear set to become more prevalent and global in the future in response to the General Agreement on Tariffs and Trade (GATT) Agreement on Trade-Related Aspects of Intellectual Property (TRIPs) which requires all GATT (now World Trade Organisation - WTO) members to provide *sui generis* schemes to protect plants (GATT, ND). Such IPR harmonisation between nation states is meant to provide the right environment (commercial inducements) to support transfer of genetic resources (Commonwealth of Australia 1994).

This mixed scenario, however, is a relatively recent phenomena. For most of the 20th century, at least, exchange of genetic resources across international boundaries for formal sector research and breeding has largely occurred on the basis of 'free access'. Common, state or individual property rights have figured little in influencing the process of exchange. Most plant breeders today have been used to obtaining material from colleagues on an informal *quid pro quo* basis with formal contracts being a fairly recent innovation. Countries have, on the whole, permitted collecting missions with few conditions applied other than requiring participation of scientists of the host nation in the collection and the provision of samples and related informed for storage at a local facility. (IPGRI 1996) Also, the possibility that local communities (and indigenous peoples) have rights which require their consultation and consent prior to access has largely been ignored.

Breeders and researchers would be well familiar with the general ease of obtaining samples from accessions held in national collections as well as those held by the international network of research centres under the Consultative Group on International Agricultural Research (CGIAR). The basic operating maxim of the CGIAR Centres, for example, has been open access (for *bona fide* research workers), as explicitly stated in their PGR Policy of 1989 (CGIAR 1989). Such free flow policies were clearly facilitated by the general lack of interference from state regulatory authorities in exchange and distribution of major food crops other than phyto-sanitary controls or quarantine regulations (Barton and Siebeck 1994).

Over the years, however, open access has come under question with the distribution, collection, ownership and utilisation of agronomically useful PGR becoming volatile international issues, with one often finding the words 'appropriation' or 'exploitation' used in the literature, especially that relating to the North-South PGR relationship, inferring that all was not as equitable in exchange as first thought. What, then, has caused collection to become more overtly 'political' in recent years with conflicts rising over access?

3. POLITICISING PGRFA: THE MAJOR FACTORS INVOLVED

The first, perhaps obvious point, is that PGRFA are such an issue because of their essential nature, providing the foundation and stability to farming systems world wide and representing insurance against unknown future needs and conditions (Cooper *et al.* 1994). Second, as the term genetic *resource* implies, PGRFA are also valuable (and hence politically important) in terms of their contribution to a nation's economy. A strong agricultural sector can mean export revenue or, at least, reduced need to spend revenue on imports. Just as important is the fact that, today, this resource can also be turned for a profit in the light of the new biotechnologies and Intellectual Property Rights (IPR) - a point of particular relevance to the PGRFA debate and one which will be returned to. Suffice to say at this point, is that access to, and control over, resources per se are both national concerns (political sovereignty, after all, has always relied on control of natural resources) and commercial concerns (for example, of companies, or public research institutions required to self-fund their activities), the potential for profit from allowing access to, and from the actual exploitation of, genetic resources, drawing both public and private interest.

Third, whilst PGRFA are valuable, their distribution is uneven, *in situ* PGRFA being found in the centres of domestication and diversity which, with few exceptions, are in the nations of the South. Despite the large degree of inter-dependency between nations on germplasm originating in other regions for all major agricultural crops (FAO 1996a; IPGRI 1996) and their reliance on access to additional diversity from original sources to maintain their industrial agricultural systems, the North-South flavour of the uneven *in situ* distribution remains quite apparent. Industrialised regions, for example, have a dependency rate on non-native crops of over 95% (Australia and North America's being around 100%) (Cooper *et al.* 1994), whilst the 'developing' world has a lower, albeit significant, dependency level (FAO 1996a). In terms of access, then, the PGRFA-rich South has a degree of leverage with the North, particularly in international bioresource-related fora (UNEP and FAO) and in bilateral relationships. Balancing this, however, is the high-technological and financial power of the North and the apparent dominancy of trade-related issues in international relations, including those relating to IPR in bioresources. The influences of the latter clearly extent past the GATT and its heir, the World Trade Organisation, imposing upon the explicit bioresource / PGRFA-related negotiations such as those associated with the Convention and International Undertaking (WTO 1995).

The fourth exacerbating factor is the threat to in-field diversity and the resulting states of increased risk and uncertainty, a threat now well recognised (FAO 1996a). It was the main impetus for the systematic collections of the 1960s and the acceleration, since the 1970s, of collecting missions and the construction and expansion of the genebank network as co-ordinated by the, then, International Board for Plant Genetic Resources (IBPGR). Despite these efforts, however, genetic erosion remains a serious concern (FAO 1996a) gene banks being insufficient by themselves as mechanisms for maintaining diversity. *In situ* initiatives, especially in 'developing' nations are now also seen as essential, not only in protected areas (for wild species) or through management of rangelands and forests, but on-farm as well (UNEP 1992; FAO 1996a).

In terms of managing land races and traditional crop varieties, it is now apparent that *in situ* agrobiodiversity is the result of informal innovations knowledge and practices of farmers and local or indigenous communities to whom a great debt is owed. Typically, the providers today are those farming families without strong links with the modern agricultural sector who have limited financial resources and live on marginal lands. The question has become one of how to promote this *in situ* conservation - for example, how to encourage farmer community innovators and assist them in using their knowledge and skills to select and build their innovative capacity (FAO 1996a). At issue has been whether this promotion will be undercut (constraining the farmer's use of protected seed) by the adoption of a Plant Breeder's Right (PBR) type of *sui generis* system of IPR protection for varieties in developing nations (in the light of TRIPs) (for example, Crucible Group 1994). Of concern too, are the potentially negative effects of implementing the distinct, uniform and stable (D.U.S.) criteria of PBR in 'developing' nations, as these are considered to be at odds with the intra-varietal variation required by small scale and resource poor farmers (Cromwell *et al.* 1992) and hence the maintenance of *in situ* biodiversity.

Both the maintenance of a living, evolving 'bank' of *in situ* resources and having access to these resources are required to maintain a viable industrial agricultural sector. Industrial agriculture is therefore not only dependent on the PGRFA producer's capacity to continue their valuable work (and thus incentives to promote this), but on their continuing goodwill (or acquiescence to national authorities) in permitting access to their fields for germplasm (and related cultural information) collection. As has become evident, such goodwill and national openness can no longer be taken for granted without a revision in thinking and approach by those seeking to collect and utilise the resource.

4. EQUITY

This brings the discussion to perhaps the key issue around which the whole access debate turns - equity - the concern that the regime of access, utilisation and ownership has been inequitable and how this can be rectified. Three interconnected factors are at work here: i) perceptions, particularly, but certainly not only emanating from the South, of an uneven evaluation of PGRFA (and hence of knowledge); ii) perceptions that the one-way flow from farmer to formal sector has not been in the interest of farmers whilst the formal sector has, in effect, had a 'free ride' on the back of these farmers; and finally iii) perceptions that the North has benefited from, and continues to benefit from, the flow more than the South.

Concern over the uneven evaluation of PGR (in the broad sense of the term) has much to do with the availability of new biotechnologies which add market value to 'useful' PGR (including PGRFA), the development of the aforementioned incentive mechanisms (Intellectual Property Rights - IPRs) for formal sector innovations and the lack of similar incentives and reward mechanisms for the community innovation sector - an uneven evaluation now considered unjust (Shiva 1993). The question has been how to redress this inequity, especially given that IPRs are now well entrenched in the North and, as indicated, are set to extend into developing nations under TRIPs pressure. One of the main problems, it appears, is that current IPR mechanisms recognise only the western industrial model of innovation and are not suited to recognising or rewarding the knowledge and innovations of farming communities and indigenous peoples (Shiva 1994).

Related to this has been the growing realisation that the one-way flow of PGRFA from the farmer-provider to the genebank, researcher and, perhaps, private sector works against the interests of the farming community, lowering their innovation ceiling (lack of access to a wider gene pool certainly not enhancing their capacity to maintain diversity). A further concern is the perception that the formal sector has had a 'free ride' on the back of PGRFA-producer communities - especially for those who have benefited little in terms of appropriate varieties from formal sector research.

There has also been a national element to this with many in the South perceiving the flow through the genebank system to be of primary benefit to the already advantaged North (even taking the work of the Consultative Group (CGIAR) centres into account), subsidising Northern private industry into the bargain. The flow to the North has made undeniably rich contributions to their agricultural sectors. For example, the contribution of CIMMYT germplasm to Australian wheat production has been at the level of an estimated US $82 million (Brennan 1986) up to $122 million per annum (GRAIN 1992). Critically, some materials released by the CGIAR centres to the North have, after only minor adaptations, been registered by private industry for a Plant Breeder's Right (PBR) (Commonwealth of Australia 1984). Also, the concentration of germplasm in the North, with at least half of all collected crop seed originating in the South now held in Northern gene banks (GRAIN 1992; RAFI, ND), was seen as exploitative of the South's local genetic wealth and a potential political weapon, to be used against 'recalcitrant' Southern governments (through denial of access). An historical element to this has also played its part with the flows of germplasm during the Colombian and colonial era perceived to be one of the root causes of the current economic (and hence political) weakness of many 'developing' nations - part of their colonial legacy and central to their unequal starting position in the global market place (Kloppenburg 1988). This legacy has been especially galling, no doubt, in the context of the current net transfer of resources from South to North in the wake of the debt crisis.

The outflow of genetic material from the CGIAR centres to private and public breeders in the North raised questions regarding the legal status of their international collections and, in the context of its professed 'trust' status, for whom it was being held (Siebeck and Barton 1992) (a concern heightened by perceptions that CGIAR centres were controlled by Northern boards and funders.) The question was: how strong a standing would the South have in the management, development and control over agricultural germplasm held in these research centres? At issue was not only who owns the *ex situ* collections held in the CGIAR centres but who owns the accessions collected by the CGIAR than are stored in Northern institutions? Complicating this scenario further has been the increased privatisation of research in the North, an increasingly vulnerable world economy and the increasingly restrictive nature of Plant Breeder's Rights (PBR) with the International Union for the Protection of New Varieties of Plants (UPOV) agreement of 1991, in which the breeder's right was strengthened vis-à-vis the rights of farmers and between breeders (UPOV 1991).

5. THE RESPONSE: LEVELLING THE PLAYING FIELD

It is perhaps understandable that the PGRFA-rich nation states, farmers and local and indigenous communities, have sought some means to level the playing field. The result has been a range of specific reform proposals and an on-going debate contested at the international (FAO, UNEP/CBD and GATT/WTO), regional (e.g. Andean Pact) and national levels, concerning rights over, and the conditions of access to, PGRFA and biodiversity per se. To understand this debate, and current policies impacting on collection from the field, some understanding of the concepts developed within the FAO and Convention on Biological Diversity negotiations, in particular, is essential.

6. EVOLVING SYSTEMS, POLICIES AND CONCEPTS IMPACTING PGRFA COLLECTION

The backbone to international negotiations are the aforementioned International Undertaking (Undertaking) and Convention on Biological Diversity (Convention or CBD), the first relating to PGRFA, the second to biodiversity per se, including PGRFA. The Undertaking is a voluntary agreement and part of, and the legal framework to, the FAO's Global System. First adopted in 1983 with annexes added in 1989 and 1991, the Undertaking is intended to promote international conservation efforts, with the South to take charge of conservation of genetic material in their territories and the North to meet the cost of conservation (FAO 1991). As it is now being renegotiated (in the light of the Convention), it tries to develop a broad multilateral approach to the conservation, use and sharing of benefits of PGRFA, albeit, one without excluding bilateral deals (GRAIN 1995). The Global System is monitored by the Commission on Plant Genetic Resources (CPGR) (now Commission on Genetic Resources for Food and Agriculture (CGRFA)), which negotiated a Code of Conduct for Plant Germplasm Collecting and Transfer in 1993 (see below), which provides guide lines for collection and transfer of PGR. The CPGR also established the Network of *Ex Situ* Collections Under the Auspices of the FAO which now includes the collections of the CGIAR network of centres since their agreement of 1994 (FAO and CGIAR 1994), thus clarifying the 'trust' status of those collections (see below).

The Convention on Biological Diversity (UNEP 1992), in force since 29 December 1993, is an internationally binding agreement covering genetic resources which gears their management towards bilateral agreements. It applies to *in situ* genetic resources and those *ex situ* resources acquired in accordance with the Convention (a clear difference it has with the Undertaking, which covers those PGRFA collected before the entry into force of the CBD). It has a three-fold objective (Art.1): the conservation of biodiversity, sustainable use of its components and the fair and equitable sharing of benefits arising out of genetic resource use. It seeks to realise its objectives by ensuring appropriate access to genetic resources (Art.15), the appropriate transfer of relevant technologies (Art. 16 and 19 especially), and by appropriate funding (Art. 20 and 21). In effect, it seeks a regulated system of access, which enables originators of genetic diversity (in particular, developing nations) to share in "...a fair and equitable way, the results of research and development and the benefits arising from the commercial and other utilisation" of the genetic resource they provide (Art. 15.7).

The key concepts encapsulated in both these agreements are as follows. First, PGRFA are the 'common concern' of humankind, not the 'common heritage' and, second, access is subject to the overriding claims of states, the above mentioned 'national sovereignty' claim (CBD, Art.15.1; Undertaking, Resolution 3/91, Annex III) and 'intellectual' ownership (Annex III of the Undertaking, with respect to Breeder's Rights, and Art. 16.2 of the CBD, with respect to Intellectual Property Rights per se). The third provision is optional - that is, where access to genetic resources is granted, it is to be on the basis of prior informed consent (PIC) of the source country "unless otherwise determined by that Party"(CBD, Art. 15.5). Further, access is to be based on mutual benefit (Undertaking, Annex III), the contractual establishment of 'mutually agreed terms' (CBD, Art. 15.4), such as the right to be involved in research and development based on the resource supplied (CBD, Art. 15.6) and the right to share in research results and benefits arising from the resource's use (including commercialisation) in a fair and equitable manner (CBD, Art. 15.7). Finally, both agreements establish the need to recognise, and support, farming community and indigenous people's knowledge, innovations and practices (Undertaking, Resolution FAO 5/89 and 3/91; CBD, (Art. 8(j) and 10(c)). Together, these key concepts emphasise the fact that whilst access may continue to be facilitated, it will no longer be totally free.

A few points need to be made. The Convention goes further than the Undertaking with respect to Intellectual Property Rights (IPRs). The Undertaking only recognises Plant Breeder's Rights (PBRs) whilst the Convention, which refers to IPRs per se, appears to validate the intent of the TRIPs agreement, although its wording is ambiguous and open to wide interpretation. Whilst it requires that access to technology for 'developing' nations is to be under "fair and most favourable terms", at the same time, it is to be "on terms... consistent with the adequate and effective protection" of IPR (Art. 16.2), wording which appears to put it in line with the TRIPs document. The Convention does indicate, however, that states are to make sure that they do not impose restrictions running counter to the Convention's objectives (Art. 15.2 and 16.5).

With national sovereignty, clearly the way is open for bilateral contractual deals. With conditionality (by source states) now validated, states can legislate the terms by which non-nationals must pay for access. Given expectations of a mutually agreed basis to the exchange, however, access seekers can also negotiate over the terms of supply although, again, the Convention does urge states and individuals to impose as few restrictions as possible (Art. 15.2). In state terms, this means that access could rely on having something to trade, clearly a problem for some of the smaller or resource-poor nations which may have little with which to bargain. Most obviously, 'developed' nations and companies are at an advantage in negotiations, having funds and technologies or educational services to trade in exchange for access (GRAIN 1995). The much feted Costa Rican Merck-InBio contract (Reid *et al.* 1993)) and the agreement between Astra Pharmaceuticals and Griffith University and its unit, the Queensland Pharmaceutical Research Institute, are but two examples of the sorts of bilateral deals facilitated by the Convention (albeit for the pharmaceutical industry). (QPRI 1996).

In comparison to Intellectual Property Rights (IPRs) and National Sovereignty, community or indigenous people's rights are far less secure to date. Whilst the Farmers' Rights (FR) (FAO Resolution C5/89 and C3/91) is, in part, an equity response to the uneven evaluation of PGRFA, in practice it has been cast in terms justifying an effort to establish a mechanism to support on-farm conservation and plant breeding. The right rests with the international community "as trustee" (Undertaking, Annex II) rather than the producers and is to be recognised and rewarded through financing the FAO's Global Plan of Action (FAO 1993b). It remains to be formalised or funded, however, and is subject to many interpretations, including the controversial suggestion, outlined below, that it can be strengthened towards ensuring control of, and access to agricultural biodiversity by local communities, i.e. to the legal status of an IPR-like, *sui generis,* collective /community right (Crucible Group 1994; GRAIN 1995; WTO 1996). Still, even in its current state, it has engendered expectations that farming communities deserve something in return for their knowledge and innovative capacity. Negotiations regarding Farmers' Rights were stalled at the FAO's International Technical Conference at Leipzig (June 1996), postponing their discussion until the negotiations over the IU revision debate in December.

Farmers' Rights are not referred to in the Convention's Articles, although they are acknowledged as an outstanding issue in the text. However, the knowledge, innovations and practices of local and indigenous communities are recognised, as is the need to encourage the equitable sharing of the benefits derived from using such knowledge, innovations and practices (Art. 8(j) and 10(c)), although such rights remain subject to national legislation. The lack of a clear defining of 'traditional' or 'indigenous' knowledge however, has left it open to debate over how it can be protected and rewarded. As alluded to above, one defining has led to the notion of a *sui generis* local community rights system based on Heritage, Territoriality and Communality which some claim could be incorporated into both the Undertaking and the Convention. This system, it is suggested, would not be incompatible with national sovereignty and would devolve (or retain) part of the control over biodiversity to local farming and indigenous communities (GRAIN 1995). Countering this, is the suggestion that the Convention was not meant to create a legal instrument to protect indigenous innovation or knowledge. Further, it is claimed, 'traditional' knowledge is not, legally speaking, Intellectual Property, and can not be treated as such. In the context of the Convention, traditional and indigenous knowledge could be recognised and rewarded, though, through the benefit sharing, mutually agreed approach of voluntary, contractual arrangements (WTO 1996).

Clearly, if communal rights , as put forward by the Group of 77, could be defined as a *sui generis* rights (as required by TRIPs), giving collective ownership by farming communities over their genetic resources, then this would have clear implications for collection (and with whom to share benefits), with a strong role for local and indigenous communities in the setting of the terms and conditions of access. Currently, proposals for *sui generis* systems to protect biodiverse resources and innovation are being (or have been) worked on at regional (Andean Pact) and national levels (e.g. India and Philippines) (GRAIN 1995).

Community rights relating to biodiversity remain at the heart of current negotiations, to be discussed in the Conference of the Parties III (COP III to the Convention on Biological Diversity) in November 1996 (i.e. prior to the revision of the Undertaking). It does seem unlikely, however, that the OECD nations would countenance any reinterpretation of IPRs (to include collective rights) outside the World Trade Organsiation (TRIPs agreement) arena. Hence, OECD/WTO pressure is likely to mean that an international agreement pertaining to collective community rights is only a remote possibility at this stage. One thing remains clear, though. Until the issue of rights over biological resources are resolved, it is doubtful that the access issue (and hence, the sharing of benefits from their exploitation) will be either.

7. REMAINING ISSUES

As the above discussion indicates, within the context of these evolving concepts, policies and systems, there are several main issue areas remaining for intense debate. In essence, a sort of contest is being waged among those protecting profits (IPR), national sovereignty, community rights and biodiversity. First, is whether sovereignty lies with the national state (national sovereignty) or local communities (common / farmer's rights). Connected with this is the question of who grants access and with whom should the benefits be shared. At this stage, the Undertaking and Convention still need to be reconciled in some way. Whilst both talk about sharing benefits, for example, the question remains how this is to be achieved, through a multilateral framework (as per the Undertaking and Farmers' Rights) or through the bilateral deals facilitated by the 'mutually agreed terms' between contracting parties of the Convention. There IPGRI integrated model approach - a Multilateral System for Exchange (MUSE) - is also up for debate (IPGRI 1996). The proposed system is based on the general principles within the context of the Convention which would operate to standard rules in relation to: i) conditions of membership, ii) terms of access to genetic resources, iii) mechanisms for sharing benefits among participants and iv) relationships with non members of the system (e.g. release under Material Transfer Agreements). It is promoted as a means for recognising the advantages of the multilateral approaches to exchange, whilst at the same time providing for full - including bilateral - benefit sharing.

To date, however, no operational system has been established. Indeed, the whole question of how the international community will handle a whole range of interrelated agricultural biodiversity issues, such as the rights of control and access, the sharing of benefits, conservation, trade and finance, remains an open one. Significant divisions of opinion are evident on North-South lines over funding; measures to ensure access to, and equitable sharing of, benefits derived from genetic resources; and Farmers' Rights. The issue of Convention support for Intellectual Property Rights remains a point of contention with many 'developing' nations, with the Convention looked upon these nations as a means for creating space to counter the WTO push for harmonisation. However, with many industrialised nations (including Australia), and some developing ones too, defending the IPR system, it is likely that the debate over IPRs will be an on-going and lively one for some time yet. Besides the North-South division of opinion, there remains a degree of 'turf' wars, not only between the UN agencies involved (UNEP and FAO) over who should be doing what (GRAIN 1996), but, as alluded to earlier, between these and the World Trade Organisation. Clearly, who has ultimate control over international management and access decisions relating to PGRFA remains a moot point.

8. CURRENT POLICES RELATING TO ACCESS

In closing, some reference needs to be made to current polices relating to access. One model for regulating movement of genetic resources already exists, the previously mentioned, International Code of Conduct for Plant Germplasm Collecting and Transfer (FAO 1993a). It is in line with the needs of individual nations and with the Convention (Art. 5) and has been suggested (by the Sub-regional Meeting for West and Central Africa) as a good model on which to base national legislation (FAO 1996b). The code describes the shared responsibilities of collectors, donors, sponsors, curators and users of germplasm, and emphasises the need for "co-operation and a sense of reciprocity" among donors, curators and users of PGR (Art.1). In effect, it : "provides a set of general principles which governments may wish to use in developing their national regulations, or formulating bilateral agreements on the collection of germplasm" (Art. 3.6); seeks to "promote mechanisms to facilitate compensation of local communities and farmers" (giving practical expression to the principles of Farmers' Rights); and to "avoid situations whereby benefits currently derived from plant genetic resources by these local communities and farmers are undermined by the transfer or use by others of the resources" (Art. 1.8) (FAO 1993a).

Amongst other things the Code suggests i) permit systems for collectors (Art. 6); ii) the respect of local customs, values and property rights and the need to inform the local communities and farmers concerned about the purpose of the mission, how and where they could request and obtain samples of the collected germplasm and leave duplicates should these be requested (Art.10); iii) that curators should use, e.g. Material Transfer Agreements (MTAs), to promote the objectives of the code (Art. 13); iv) that "without prejudice to the concept of Farmers Rights", users should consider providing some form of compensation for the benefits derived from the use of germplasm including access to new, improved varieties, on mutually agreed terms; support for research of relevance to conservation and utilisation of PGR; training at the institutional and farmer levels; transfer of appropriate technology; and support for programmes to evaluate and enhance local land races and other indigenous germplasm (Art. 14).

At the moment, national or institutional policies relating to access are in a state of change. In general, the legal status of existing *Ex Situ* Collections remains undefined (given that the provisions of the Convention do not relate to those PGR outside the country of origin and acquired prior to its signing - Art. 15.3 of the Convention), with the issue to be solved within the context of the FAO Global System. For these, the responsibility for distribution of material lies, in practice at least, with each curator and rarely with a higher authority (FAO 1996b). One important option explored in this regard has been to establish agreements between the FAO and owners of gene banks and for the latter to voluntarily place their collections under the auspices of the FAO and under the umbrella of the Undertaking. Thirty two nations had indicated such a willingness by 1994 (IPGRI 1996), a move also undertaken by the CGIAR Centres, as noted above. Transfer of material from these CGIAR centres is now covered by an interim Material Transfer Agreement (CGIAR 1995).

Some countries have already begun to put mechanisms in place to enforce 'prior informed consent' and access on mutually agreed terms; whilst others are waiting for the outcome of current discussion on these issues by the CGRFA and the Conference of Parties to the Convention on Biological Diversity. The Philippines is one of the few nations which, through its Presidential Executive Order No, 247 of 1995, has national regulation governing access to PGR. Amongst other things, the implementing rules and regulations require a system of mandatory Research Agreements; payment of negotiated royalties on commercial use; in the case of endemic species, access to the relevant technology without royalty and that prior informed consent be obtained from concerned local and indigenous communities (Republic of the Philippines 1996).

In the Mediterranean, for example, Greece is seeking to prevent any 'illegal' collection within its borders, an activity which have been quite frequent in the past (Stravropoulos and Samaras *pers comm.*). As a member of the EU, it is largely obliged to follow the common European approach which favours multilateral agreements avoiding preferential agreements or exclusions although this does not preclude bilateral agreements with states with whom Greece has particularly good relations. In the absence of the new Undertaking and of a concrete EU policy, Greece follows a modified version (i.e. recognising national sovereignty) of the "common heritage of mankind'/ free distribution to *bonafide* users arrangement of the past. Access requires formal permission and is granted on two premises: i) of "fair and equitable sharing of benefits" and ii) that "free access doesn't necessarily mean at no charge". Collection and exchange of PGR is currently regulated by a Presidential Decree: No 80/1990 and whilst no formal arrangements for germplasm access and collecting have yet been made post-Convention, it is evident that changes are on the way and benefit sharing is going to be part of the reality of collecting in Greece. In this regard, suggestions go beyond the royalty notions, to the more concrete, perhaps capacity building in the form of training of local scientists (to PhD) or other forms of advanced research in the 'developed' nation; perhaps donations of modern laboratory equipment, funding expeditions and evaluation work with mixed scientific teams etc.

Another Mediterranean nation, Cyprus, has recently prepared legislation for the protection of Nature (biodiversity) in accord with the EU relevant regulations and directives although this is still in the final draft stage. It includes *inter alia* legislation for the protection of the National Herbarium and the national genebank and incorporates regulation for the national implementation of those commitments undertaken by the ratification of the relevant conventions (such as the Convention and Undertaking). Until the ratification of this legislation, Cyprus refers to these international agreements and to the FAO Code of Conduct, including farmers rights on genetic resource issues. (Della *pers comm.*)

At the regional level, for example, the African Ministerial Conference on the Environment, AMCEN, adopted a common position regarding the Convention, recommending that African nations impose a temporary ban on transfer from African nations of bio-resources not covered by existing Conventions and where PIC is not in effect pending elaboration of definitive legislation on the subject. The Andean Pact nations have discussed a proposal (called the Cartagena Agreement) for a common approach to conservation, protection and access to genetic resources. Also, Asian scientists have prepared two declarations on the issues, one calling for ethical approaches to collection, the other for the regulation of movement of PGR in the region (FAO 1996b).

9. CONCLUSION

Evidence suggests that national sovereignty, the ideals of benefit sharing, and of Farmers' Rights, are being taken seriously by the PGRFA donor nations. In general, collectors can expect continued freedom of access if the intents of 'fair access' and 'fair exchange' are upheld. The key to this is, not only should fair and equitable access be entrenched in noble words and legislation, but that it should be the working maxim of all stakeholders in PGRFA. Whilst practitioners may all regret the possibility of red tape, of bureaucracy getting in the way of good research, if a longer-term perspective is taken, then it may be that formalising fairness in exchange (rather than relying on "gentleman's agreements") will be the only method that will make relationships between institutions, nations and communities smoother and long-lasting. It appears to be the key to future exchange and international co-operation in striving for food security.

It is worth noting that it is not just the so-called 'developing' nations who are keen to regulate access to their genetic resources. For example, Australia, is working towards control over access to its genetic resources in the light of the Convention on Biological Diversity and to ensuring that "an appropriate return for any permitted access" accrues to that nation (Commonwealth of Australia 1996) (National Strategy for the Conservation of Australia's Biological Diversity, Objective 2.8). Plant science researchers in Australia, then, would be hard pressed to seek less regulatory mechanisms from their overseas colleagues. Clearly regulation is a world wide trend which is likely to be the norm of the future.

In terms of access for the industrial North, perhaps the following quote from the Curator and Deputy Curator of the Greek Gene Bank sums the potentials of the new PGRFA exchange context : "with modest expectation, good will and earnest scientific motivation, some logical long standing reliable agreements could be achieved" (Stavropoulos and Samaras pers comm.).

REFERENCES

Barton, J. and Siebeck, W. (1994). 'Material Transfer Agreements in Genetic Resources Exchange.' Issues in Genetic Resources, No. 1. May 1994. IPGRI, Rome.

Brennan, J. (1986). 'Impact of Wheat Varieties from CIMMYT on Australian Wheat Production.' Agricultural Economics Bulletin No. 5. Department of Agriculture, NSW, September.

Commonwealth of Australia (1984). 'Plant Variety Rights.' Parliament, Senate Standing Committee on National Resources, May.

Commonwealth of Australia (1994). 'Access to Australia's Biological Resources: A Discussion Paper.' Office of Chief Scientist, Department of the Prime Minister and Cabinet. A Paper prepared for the Co-ordination Committee on Science and Technology, March 1994, AGPS, Canberra.

Commonwealth of Australia (1996). 'National Strategy for the Conservation of Australia's Biological Diversity.' Department of Environment, Sport and Territories, (URL: http: www.erin.gov.au/portfolio/esd/biodiv/strategy/chap2.html).

Consultative Group on International Agricultural Research (CGIAR) (1989). 'CGIAR Policy on Plant Genetic Resources.' (IBPGR: Rome).

Consultative Group on International Agricultural Research (CGIAR) (1995). Material Transfer Agreement, Interim Agreement produced by the Inter-Centre Working Group on Genetic Resources of the CGIAR at its meeting of Jan 15-19 1995 (ICWC MTA), CGIAR.

Cooper, D., Engels, J., and Frison, E. (1994). 'A Multilateral system for plant genetic resources: imperatives, achievements and challenges.' Issues in Genetic Resources No. 2. May 1994. IBPGR Occasional Series.

Cromwell, E., Friss-Hansen, E. and Turner, M. (1992) 'The Seed Sector in Developing Countries: A Framework for Performance Analysis.' Working Paper 65. (Overseas Development Institute: London).

Crucible Group (1994). 'People, Plants and Patents: The Impact of Intellectual Property on Biodiversity, Conservation, Trade and Rural Society.' (International Development Research Centre: Ottawa).

Food and Agricultural Organisation of the United Nations (FAO) (1991). FAO Global System for the Conservation and Utilisation of Plant Genetic Resources: Progress Report and Matters of Decision by the Conference, 26th Session, Rome, 9-28th. November 1991.

Food and Agricultural Organisation of the United Nations (FAO) (1993a). Draft International Code of Conduct for Plant Germplasm Collecting and Transfer, FAO Council, Hundred and Third Session, Rome 14-25 June 1993, CL 103/16-Sup.1.

Food and Agricultural Organisation of the United Nations Council (FAO) (1993b). Report of the 5th Session of the Commission on Plant Genetic Resources, Rome 19-23 April 1993. Hundred and Third Session of the FAO Council, Rome. June 1993. CL 103/16.

Food and Agricultural Organisation of the United Nations Council (FAO) (1996a). State of the World's Plant Genetic Resources (Full Background Documentation). Draft Report for the FAO International Technical Conference on Plant Genetic Resources. Work in Progress.

General Agreement on Tariffs and Trade (GATT). Agreement on Trade-Related Aspects of Intellectual Property Rights, Including Trade in Counterfeit Goods. Annex III of the Draft Final Act of the Uruguay Round. MTN.TNC/W/FA, GATT, Geneva, pp. 57-90.

Genetic Resources Action International (GRAIN) (1992). Biodiversity at Rio: Conservation or Access? *Seedling* **9**, 2-5.

Genetic Resources Action International (GRAIN) (1995). Towards a Biodiversity Community Rights Regime. *Seedling* **12**, 2-14.

Genetic Resources Action International (GRAIN) (1996). From Leipzig to Buenos Aires. *Seedling* **13**, 17-21.

International Plant Genetic Resources Institute (IPGRI), (1996), 'Access to plant genetic resources and equitable sharing of benefits: A contribution to the debate on systems for the exchange of germplasm', Issues in Genetic Resources, No. 4.

International Union for the Protection of New Varieties of Plants (UPOV), (1991), Diplomatic Conference for the Revision of the International Convention for the Protection of New Varieties of Plants, Geneva, March 4-9. Final Draft. DC/91/138. UPOV, Geneva.

Kloppenburg, J.R. (1988) 'First The Seed: The Political Economy of Plant Biotechnology 1492-2000'. (Cambridge University Press: Cambridge).

Queensland Pharmaceutical Research Institute (QPRI) (1996). Natural Product Drug Discovery. (http://www.qpri.gu.edu.au/joint/htm).

Reid, W., Laird. S., Meyer, C., Gamez, R., Sittenfeld, A., Janzen, D., Gollin, M., and Juma, C. (1993). 'Biodiversity Prospecting: Using Genetic Resources for Sustainable Development.' (World Resources Institute: USA).

Republic of the Philippines, Department of Environment and Natural Resources (1996). Implementing Rules and Regulations on the Prospecting of Biological and Genetic Resources. Department Administrative Order No. 96-20.

Rural Advancement Foundation International (RAFI) Conserving Indigenous Knowledge: Integrating Two Systems of Innovation. Study Commissioned by the UNDP.

Sehgal, S.M. and van Rompaey, J. (no date). Intellectual Property Rights and the Global Seed Industry.' (Mimeograph, Plant Genetic Systems, N.V.: Gent, Belgium).

Siebeck, W. and Barton, J. (1992). 'The Implications of Applying the Legal Concept of Trust to Germplasm Collections at CGIAR Research Centres' *Diversity* **8**, 29-35.

Shiva, V. (1993) 'Intellectual Property Rights May Block Biodiversity Conservation and Affect Local Community Rights', Mimeograph, Third World Network Briefing Paper for CSD, June 30.

Shiva, V.(1994) 'The Need for *Sui Generis* Rights' *Seedling* **12**, 11-15.

United Nations Environment Programme (UNEP) (1992). Convention on Biological Diversity. Mimeograph. 5 June.

World Trade Organisation, Committee on Trade and Environment (1995). Environment and TRIPs, WT/CTE/W/8, 8 June (http://www.wto.org/Trade+Env/tetrips.htm)

World Trade Organisation, Committee on Trade and Environment (1996). Trade and Environment Bulletin 10, http://www.wto.org/wto/Trade+Env/te010.htm.

CHAPTER 17

A generic conspectus of the forage legumes of the Mediterranean basin

Katherine Woodgate[1], Nigel Maxted[1] and Sarita Jane Bennett[2]
[1] *School of Biological Sciences, University of Birmingham, Edgbaston, Birmingham, B15 2TT, U.K.*
[2] *Co-operative Research Centre for Legumes in Mediterranean Agriculture (CLIMA), University of Western Australia, Nedlands, Western Australia 6907, AUSTRALIA*

Abstract An ecogeographic / taxonomic conspectus and identification aids are presented for 29 temperate forage legume genera commonly found in the Mediterranean basin. The conspectus for each genus contains a summary of the current accepted classification, taxonomic history and affinities, suggestions of useful keys, ecological and distributional information and a summary of uses. The conspectus is complemented by novel descriptions, a bracketed key, a tabular key and a generic character set that can be used with an interactive identification system.

1. INTRODUCTION

The genera included in the conspectus are all temperate forage legume genera and therefore belong to the family Leguminosae, sub-family Papilionoideae. The Leguminosae, as a whole, are morphologically diverse and contain approximately 650 genera. The family is usually divided into three sub-families, of which the largest is the Papilionoideae containing 400-500 genera (Polhill, Raven and Stirton 1981). The Papilionoideae are distributed through the temperate, sub-tropical and tropical regions. The leaves are usually pinnate; flowers are irregular with lateral petals enclosed by the standard in the bud; there are usually 10 stamens, commonly diadelphous but sometimes monadelphous or free. Traditionally the Papilionoideae have been divided into 10 or 11 tribes based on habit, vegetative and floral characters, although more recently the number of tribes has been increased to 32 (Polhill and Raven 1981).

Legumes provide an important genetic resource throughout the world, only ranking second to the Gramineae in economic importance to man (Heywood 1978). They are uses include: human consumption, fodder, forage, green manure and land management. However, genetic erosion of the entire legume gene pool is currently occurring in the Mediterranean Basin, due to changes in land-use, replacement of crop landraces with improved or higher-yielding varieties and increasing grazing pressure. This erosion presents a serious threat to their efficient and effective exploitation. Both effective *in situ* and *ex situ* conservation strategies are required if the genetic resources of this family are to remain available for utilisation. The underlying framework for the success of such complementary strategies is taxonomic. There is a dual requirement for a better taxonomic understanding of the taxa being conserved and for better, more user-friendly identification systems, which can be used by those lacking a formal taxonomic training. Unfortunately, most of the traditional keys currently available are difficult to use because of the technical terms they employ, their reliance on obscure characters that are difficult to record and their general lack of flexibility for the user (Maxted and Crust 1995).

The use of computers to assist in plant identification has been an important development in recent years. Through the use of systems, such as DELTA (DEscriptive Language for Taxonomy) and associated programs (Dallwitz 1980), data can be used to generate either printed keys or computer-based interactive keys. Inter-active identification programmes utilise the principle of multi-access keys. In other words, unlike traditional dichotomous keys

S. J. Bennett and P. S. Cocks (eds.), Genetic Resources of Mediterranean Pasture and Forage Legumes, 182–226.

that force the user to commence identification in a pre-ordained sequence, multi-access keys allow entry through any characters in the data set, with no preordained sequence of questions needed to achieve an accurate identification. Interactive keys can often cope with a degree of user-error or atypical material and still obtain an accurate identification. These keys are therefore thought to be more user-friendly than conventional identification aids. The addition of new taxa, amending the data set or production of specialist keys is relatively simple.

The focus of this taxonomic study is generic. It is important to develop identification systems at all levels of the taxonomic hierarchy, so keys are required to families and genera, as well as species and subspecies. This project will complement other keys based at the species level where data are currently being developed (eg. *Vicia*, *Cicer*, *Trifolium*, *Lens* and *Medicago*) by Maxted and various collaborators. The 29 genera included in this study are all common or economically useful temperate forage and food legumes of the Mediterranean Basin; they are listed in Table 1. The ecogeographic / taxonomic conspectus is summarised in Appendix 1. The identification aids are presented in the form of a dichotomous key in Appendix 2 and multi-access key in Appendix 3.

Table 1. Genera included in study.

Tribe	Genera in Tribe	Genera included in study	
Galegeae-Astragaleae	20	Astragalus	
Hedysareae	7	Hedysarum	Onobrychis
		Ebenus	
Loteae	6	Lotus	Cytisopsis
		Dorycnium	Hymenocarpus
		Anthyllis	Tetragonolobus
Coronilleae	6	Securigera	Scorpiurus
		Coronilla	Ornithopus
		Hippocrepis	Hammatolobium
Vicieae	5	Vicia	Lathyrus
		Lens	Pisum
		Vavilovia	
Cicereae	1	Cicer	
Trifolieae	7	Ononis	Trifolium
		Melilotus	Trigonella
		Medicago	
Crotalarieae	16	Lotononis	
Genisteae - Lupininae	1	Lupinus	

2. METHODS AND RESULTS

Characters and their states were chosen from various literature sources including Floras, monographs and identification keys (Schischkin and Bobrov 1948; Davis 1970; Hutchinson 1964; Lesins and Lesins 1979; Maxted 1993; Meikle 1977; Polhill and Raven 1981; Polunin 1969; Van der Maesen 1972; Zohary and Heller 1984). In total 62 characters thought to be useful for distinguishing temperate forage legume genera were used, principally morphological but also taxonomic and ecological. The character and character states are listed in Appendix 4 and a glossary of some of the technical terms used is provided in Appendix 5. Care was taken to ensure that all the states of a character were logically related, mutually exclusive and given as few states as possible without loss of information (mostly six or fewer). All genera had their characteristics recorded for the 62 characters, based primarily on information from literature sources.

The data set was recorded in DELTA format (Dallwitz 1980). The data-set was manipulated using the program CONFOR (Dallwitz 1980). The natural language description is included in the conspectus (Appendix 1). In the ecogeographic and taxonomic conspectus the 29 temperate forage legume genera are arranged alphabetically. For each genus details of current classification, taxonomic history, taxon description, taxonomic affinities, useful keys, habitat, distribution and uses are provided. The information included is largely based on an extensive literature review of Floras, monographs, ecogeographic surveys and more general texts, but the primary sources were Allen and Allen 1981; Davis 1970; Duke 1981; Kupicha 1983; Lock 1989; Maxted 1995; Smartt 1990; Smartt and Simmond 1995 and Zohary and Hopf 1994.

The data set recorded in DELTA format was used to generate the bracketed and tabular keys given in Appendices 2 and 3 respectively. The DELTA data was also further incorporated into the interactive identification system, EXPLORE (Crust and Maxted, unpublished). The program starts with a complete, logically ordered character matrix set which can be re-ordered as required. The characters can be ordered by separation values, a figure calculated by the computer which indicates the strength of a character in separating the taxa within the data set. Characters with the highest values are potentially the best characters to use for identification. The characters can also be ordered by character weight, which is an estimate of the reliability or usefulness of a character provided subjectively by the person preparing the taxon data set. The programme differentiates certain taxa from the unknown specimen based on different character states and provides a list of remaining taxa. This is repeated until an identification is achieved, followed by matching and checking with a full description. The threshold level can be altered to allow for a number of errors to be made in the identification process and yet still obtain an accurate identification. A copy of the programme EXPLORE and various forage legume data sets can be obtained by contacting the third author.

3. DISCUSSION

One of the prerequisites of producing a DELTA data set is the availability of sufficient comparative data to allow the recording of each character for each taxon. In the present generic level study, the descriptive data was largely obtained from Floras and other literature sources which are notorious for their inconsistent use of characters. For example, the presence of a terminal tendril or mucro on the leaf in the *Vicieae* is diagnostic for the five genera included, but this character is seldom recorded for other genera. The difficulty in obtaining comparative character data made the production of the generic character set time consuming. Another problem encountered was that both within and between literature sources a large number of descriptive synonyms are used to describe characters and character states. For example, eighteen terms were found to describe leaflet shape, while in practice six states were considered sufficient to describe all the variation between the genera without loss of information.

It was also found that working at a generic level meant that many of the common characters used in species distinction become 'variable' within genera. This makes it even more difficult to produce acceptable keys which are not dependent on obscure characters (Partridge *et al.* 1988; 1993). This can, at least partially, be overcome by using comments such as 'rarely', 'usually', 'often' and 'occasionally' in the natural language descriptions and by coding for more than one state per character. However, this problem does reinforce the point that when an identification is made the specimen should always be compared to a description of the genus to confirm the identification.

The bracketed key produced is a traditional single-access key and bears all the normal problems associated with such a key. The user is forced to start at question one and work through alternative diagnostic descriptions until the identification is made. These key offer little or no flexibility. If a character used in the key is not present on the specimen being identified the identification process is halted. Unfortunately, some of the leads in the key are based on more obscure characters, such as seed shape. Ideally one would like to add confirmatory characters but this is not always possible, especially when dealing with generic characteristics. This problem is circumvented in the presented tabular key and the interactive computer key (EXPLORE). Both of these are multi-access keys where the user has a choice of character sequences to obtain an identification. Of these two, the tabular key is more exacting as it requires recurrent reference to the character list. Perhaps the simplest identification can be obtained using the interactive identification system, where the user can make changes to sequences, ranking, listing remaining taxa, weighting of characters, giving taxon descriptions and so on. This system can be operated with minimal knowledge of computers or detailed plant taxonomy by a wide range of users.

Identification, as Maxted (1996) observes, involves two steps; first, the decision as to which taxon the specimen represents and, second, the decision as to what is the 'accepted' name for that taxon. This study aims to help the user at both steps, by using the identification aids the user will be able to determine the genus of a specimen and consultation of the ecogeographic and taxonomic conspectus will provide the accepted name. The conspectus entry for each genus contains a description which can be used for matching with the specimen and confirmation of its identification. The conspectus is intended as general introduction to the genus for conservationists and leads to further literature and information.

ACKNOWLEDGEMENTS

The authors wish to thank Rick Crust for his advice, guidance to DELTA and for associating the generic character set with EXPLORE.

REFERENCES

Allen, O.N. and Allen, E.K. (1981). 'The Leguminosae. A Source Book of Characteristics, Uses and Nodulation'. (Macmillan Publishers Ltd: London)

Anon. (1992). 'Earth Summit '92. The United Nations Conference on Environment and Development'. (The Regency Press: London).

Askevold, I. F. and O'Brien, C.W. (1994). DELTA, an invaluable computer program for generation of taxonomic monographs. *Annals of the Entomological Society of America* **87**(1), 1-16.

Auricht, G., Bisby, F., Ehrman, T. and Maxted, N. (1988) 'Wild Forage Legume Collection In Turkey. Phase II - South-East Turkey'. (IBPGR/Vicieae Project: Rome).

Barneby, R.C. (1964) Atlas of North American Astragalus. *Mem. N.Y. Bot. Gard.* **13**, 1-1188.

Coles, S., Maxted, N. and Van der Maesen, L.J.G. (1998). Identification Aids for *Cicer* (Leguminosae, *Cicereae*) taxa. *Edinburgh Journal of Botany*. (In press).

Chrtkova-Zertova, A., Van der Maesen, J. G. and Rechinger, K. H. (1979). 'Flora Iranica. Papilionaceae I - *Vicieae*'. (Akademische Druck-u. Verlagsanstalt: Graz).

Cullen, J. (1970). Anthyllis L. in 'Flora of Turkey and East Aegean Islands' (Ed. P.H. Davis). pp. 3: 534. (Edinburgh University Press: Edinburgh).

Dallwitz, M.J. (1980). A General System for Coding Taxonomic Descriptions. *Taxon* **29**, 41-46.

Dallwitz, M.J., Paine, T.A. and Zurcher, E. J. (1996). 'Users guide to the DELTA System. A general System for Processing Taxonomic Descriptions, Edition 4.04 (January)'. (CSIRO Division of Entomology: Canberra, Australia).

Davis, P. H. (1970). 'The Flora of Turkey and East Aegean Islands, Vol.3'. (Edinburgh University Press: Edinburgh).

Davis, P.H. and Heywood, V.H. (1973). 'Principles of Angiosperm Taxonomy'. (Krieger: New York).

DELTA Home page: http://www.keil.ukans.edu/delta

Duke, J.A. (1981). 'Handbook of legumes of world economic importance'. (Plenum Press: New York).

Fitter, R., Fitter, A., and Blamey, M. (1985). 'The wild flowers of Britain and northern Europe'. (Willian Collins and Co.: Glasgow).

Global Environment Facility. (1995). *In situ* Conservation of Genetic Diversity in Turkey. Newsletter 1.

GRIN (1997). Http://www.ars-grin.gov/

Heywood, V. H. (1978). 'Flowereing Plants of the World'. (Oxford University Press: Oxford).

Hickey, M. and King, C. (1988). '100 Families of Flowering Plants'. (Cambridge University Press: Cambridge).

Hutchinson, J. (1964). 'The Genera of Flowering Plants, Vol. 1'. (Clarendon Press: Oxford).

IBPGR (1985). Report from a Working Group on Forages for Mediterranean and Adjacent Semi-arid/Arid Type Areas. International Board for Plant Genetic Resources, Rome.
Jackson, M.T. and Yungas, A.G.(1984) Variation in the Grass Pea (*Lathyrus sativus* L.) and wild species. *Euphytica* **33**, 549-559.
Kearney, J. and Smartt, J. (1995) The Grasspea. In 'Evolution of Crop Plants' (Eds. J. Smartt and N.W. Simmonds). (Longman Scientific and Technical: London).
Kupicha, F.K. (1976). The Infragenic Structure of *Vicia. Notes from the Royal Botanic Garden, Edinburgh.*, **34**, 287-326.
Kupicha, F.K. (1977). The delimitation of the tribe *Vicieae* and the relationships of *Cicer* L. *Botanical Journal of the Linnean Society* **74**, 131-162.
Kupicha, F.K. (1983) The infragenic structure of *Lathyrus. Notes from the Royal Botanic Garden Edinburgh*, **41**, 209-244.
Lamont, E-J. (1996). An Ecogeographic Survey of the *Trifolium* species of Turkey. Unpublished eighteen month report. University of Birmingham and IGER.
Lesins, K. A. and Lesins, I. (1979). 'Genus *Medicago* (Leguminosae). A Taxogenetic Study'. (Dr. W. Junk. bv Publishers:The Hague).
Lock, J. M. (1989). 'Legumes of Africa. A Check-list'. (Royal Botanic Gardens: Kew).
Lock, J. M. and Simpson, K. (1991). 'Legumes of West Asia. A Check-list'. (Royal Botanic Gardens: Kew).
Lock, J.M. and Heald, J. (1994). 'Legumes of Indo-China. A Check-list'. (Royal Botanic Gardens: Kew).
Loi, A., Cocks, P.S., Howieson, J.G. and Carr, S.J. (1997). Morphological characterisation of Mediterranean populations of *Biserrula pelecinus* L. *Plant Breeding* **116**, 171-176.
Maxted, N. (1993). A phenetic investigation of *Vicia* L. subgenus *Vicia* (Leguminosae, *Vicieae*). *Botanical Journal Of The Linnean Society*, **111**, 155-182.
Maxted, N. (1995). 'An Ecogeographical Study of *Vicia* subgenus *Vicia'*. (IPGRI: Rome).
Maxted, N. (1996). Taxonomy and Systematics. In 'Conservation Biology' (Ed. I.F. Spellerberg). pp. 91-104. (Longman Group Ltd: Harlow).
Maxted, N. and Crust, R. (1995). Aids to Taxonomic Identification. In 'Collecting Plant Genetic Diversity' (Eds. L. Guarino, V. Ramanatha Rao and R. Reid). (CAB International: Wallingford, UK).
Meikle, R. D. (1977). 'The Flora of Cyprus, Vol. 1'. (Bentham-Moxon Trust, Royal Botanic Gardens: Kew).
Mouterde, P. (1966). 'Nouvelle Flore du liban et de la Syrie. Vol. 2'. (Librairie Orientale: Beirut, Lebanon).
Partridge, T.R., Dallwitz, M.J. and Watson, L. (1988). A Primer for the DELTA System on MS-DOS and VMS. Division of Entomology Report No. 38. CSIRO, Australia.
Partridge, T.R., Dallwitz, M.J. and Watson, L. (1993). A Primer for the DELTA. Division of Entomology, CSIRO, Australia.
Podlech, D. (1983) Zur taxonomie und Nomeclatur der Tragacanthoiden Aatragali. Mitt. *Bot. Munchen* **19**, 1-23.
Polhill, R.M. and Raven, P.H. (1981). 'Advances in Legume Systematics, Part One'. (Royal Botanic Gardens: Kew).
Polhill, R.M., Raven, P.H. and Stirton, C.H. (1981). Evolution and Systematics of the Leguminosae. In 'Advances in Legume Systematics' (Eds. R.M. Polhill and P.H. Raven). (Royal Botanic Gardens: Kew).
Polunin, O. (1969). 'Flowers of Europe. A Field Guide'. (Oxford University Press: Oxford).
Polunin, O. (1987). 'Flowers of Greece and the Balkans. A Field Guide'. (Oxford University Press: Oxford).
Press, B. (1993). 'Field guide to the wild flowers of Britain and Europe.' (New Holland Publishers Ltd.: London).
Rechinger, K. H. (1984). 'Flora Iranica. Papilionaceae II'. (Akademische Druck-u. Verlagsanstalt: Graz).
Schischkin, B.K. and Bobrov, E.G. (1948). 'Flora of the USSR Vol. XIII'. (Izdatel'stvo Akademii Nauk SSSR: Moscow).
Smartt, J. (1990). 'Grain Legumes. Evolution and Genetic Resources'. (Cambridge University Press: Cambridge).
Smartt, J. and Simmonds, N.W. (1995). 'Evolution of Crop Plants. 2nd Ed.'. (Longman Scientific and Technical: Harlow).
Stace, C.A. (1989). 'Plant Taxonomy and Biosystematics'. (Cambridge University Press: Cambridge).
Toothill, E. (1984). 'The Penguin Dictionary of Botany'. (Penguin Books: London).
Townsend, C. C. (1974). 'Flora of Iraq. Vol.3'. (Ministry of Agriculture and Agrarian Reform of the Republic of Iraq: Bagdad).
Tutin, T.G., Heywood, V.H., Burges, N.A., Moore, D.M., Valentine, D.H., Walters, S.M., and Webb, D.A. (1968). 'Flora Europaea'. (Cambridge University Press: Cambridge).
Van der Maesen, L. J. G. (1972). '*Cicer* L., a Monograph of the Genus with Special Reference to the Chickpea, its Ecology and Cultivation'. Mededelingen Landbouwhogeschool: Wageningen 72-10).
Yakovlev, G. P., Sytin, A. K. and Roskov, Yu. R. (1996). 'Legumes of Northern Eurasia. A Check-list'. (Royal Botanic Gardens: Kew'.
Zhang, Z., Wen, Q. and Liu, C. (1990). Hepatoprotective effects of Astragalus root. *J. Ethnopharmacol.* **30**, 145-149.
Zohary, M. and Heller, D. (1984). 'The Genus *Trifolium'*. (The Israel Academy of Sciences and Humanities: Jerusalem).
Zohary, D. and Hopf, M. (1994). 'Domestication of Plants in the Old World. 2nd Ed.'. (Clarendon Press: Oxford).

APPENDIX 1. ECOGEOGRAPHIC / TAXONOMIC CONSPECTUS FOR TEMPERATE FORAGE LEGUME GENERA.

Anthyllis L.

Classification (from Cullen 1970).
Family Leguminosae
Subfamily Papilionoideae
Tribe Loteae
Genus *Anthyllis* L.

Taxonomic History

The current accepted classification includes approximately 25 species, but monographic work is required.

Taxon Description

Anthyllis L. Sp. Pl. 719 (1753); Gen. Pl. ed.5:355 (1754)

Type species: *A. vulneraria* L.

Synonyms: *Barba-Jovis* (Tourn.) ex. Adans (1753); *Triopodion* Medik. (1787); *Zenopogon* Link (1831); *Pogonitis* Reichb. (1837); *Fakeloba* Raf. (1838); *Aspalathoides* K.Koch (1853); *Acanthyllis* Pomel (1874).

Annual, biennial or perennial herbs or shrubs. Height range 30-90 cm. Prostrate. Surface sparsely pubescent. Spines absent. Leaves imparipinnate, rarely trifoliate or simple. 1-15 leaflets per leaf. Leaflet shape elliptic, oblong, obovate or cuneate; margins entire; lamina pilose; glands absent. Stipules small or absent, fused with petiole. Flowers rarely solitary or arranged in heads; terminal. 1 (-many) flowers per inflorescence. Sessile or pedunculate. Corolla white, purple, yellow or red (may be part-coloured). Standard ovate, claw long; wing ovate or obtuse; keel blunt or pointed; wings longer than keel. Calyx tubular or campanulate, teeth sub-equal or bilabiate. Bracts small or absent, setaceous; bracteoles present or absent. Stamens monadelphous or diadelphous. Anthers monomorphic. Stigma terminal. Style glabrous. Ovary stipitate or rarely sessile, few to many ovules. Fruit oval or arched to linear; joints present or absent; indehiscent; enclosed in calyx (calyx often inflated). Usually 1-3 seeds per fruit.

20-25 species (Polhill 1981).

Useful Taxonomic Treatments

Cullen, J. (1970). 'Flora of Turkey and East Aegean Islands. Vol. 3' (Ed. P.H. Davis). P.534 (Edinburgh University Press: Edinburgh).

Polhill, R.M. (1981) 'Loteae, Vol. 1'. (Eds. R.M. Polhill and P.H. Raven). pp. 371-374 (Royal Botanic Gardens: Kew).

Polunin, O. (1969). 'Flowers of Europe. A field Guide'. pp.210-211. (Oxford University Press: Oxford).

Ecology

The species are common in meadows and pastures, rocky ground, scrub and woodland.

Distribution

Distributed around the Mediterranean, extending into Europe, western Asia and North Africa.

Uses

Members are used as ground-cover and fodder plants. *A. vulneraria* (Sand Clover or Kidney Vetch) has been cultivated in Europe since about 1858 as a forage plant and is especially good for goats and sheep (Duke 1981). It was used in the past to heal wounds (Allen and Allen 1981). *A. cytisoides* is used for honey production (GRIN 1997). The flowers of some species are ornamental.

Astragalus L.

Classification (from Yakovlev *et al.* 1996).
Family Leguminosae
Subfamily Papilionoideae
Tribe Galegeae
Genus *Astragalus* L.
Sections *Acanthophace, Acmothrix, Aegacantha, Albertregelia, Alexandri, Alopecioides, Alopecuroidei, Ammodendron, Ammodytes, Ammotrophus, Ankylotus, Annulares, Astenolobium,*

Brachycarpus, Buceras, Bulimioides, Bungeastrum, Campanella, Campylotrichon, Cartilaginella, Cenantrum, Chaetodon, Chlorosphaerus, Chomutoviana, Christiania, Christianopsis, Chronopus, Coluteocarpus, Corethrum, Craccina, Cysticalyx, Cystium, Cystodes, Cytisodes, Dipelta, Drepanodes, Ermophysa, Eremophysis, Erioceras, Erionotus, Eriopodus, Eudmus, Extremiorientalis, Galegiformis, Glycyphyllus, Gontscharoviella, Grammocalyx, Halicacabus, Harpilobus, Helmia, Hemiphaca, Hemiphragmium, Heterodontus, Hololeuce, Holophyllum, Hymenostegis, Hypoglottidei, Hypsophilus, Irinaea, Komaroviella, Laguropsis, Leptophysa, Leucophysa, Lithoon, Lithophilus, Macrocarpon, Macrocystis, Macrocystodes, Macropodium, Macrosemium, Macrotrichoides, Malacothrix, Megalocystis, Melilotopsis, Mesocarpon, Myobroma, Nuculiella,Onobrychiopsis, Onobrychium, Ophiocarpus, Ornithopodium, Orobella, Oroboidei, Oxyglottis, Paracystium, Pelta, Pedina, Pendulina, Phacodes, Phyllolobium, Picrophace, Platyglottis, Polares, Popovianthe, Proselius, Pseuderioceras, Pseudoammotrophus, Robusti, Scheremeteviana, Sesamei, Sesbanella, Sewerzowia, Sphaerocystis, Stereothrix, Stipitella, Tamias, Tanythrix, Theiochrus, Thlaspidium, Trachycercis, Tricholobus, Tropidolobus, Uliginosi, Xenophaca, Xiphidium

Taxonomic History

The name *Tragacantha* was first used as the generic rank by Miller in 1759. It has subsequently been reduced to subgeneric status by Kuntze in 1891 who combined all currently accepted species of *Astragalus* into one taxon. This genus is the largest of the Leguminosae and is currently grouped into more than 100 subdivisions. It is generally agreed among legume taxonomists that subgeneric and generic divisions are inadequate, Podlech (1983) has begun this work of clarification by separating the genus *Astracantha*. Much taxonomic work was carried out by Barneby (1964), based on the flora or North America but there are still considerable problems in the delimitation of taxa both at the section and species level.

Taxon Description

Astragalus L. Sp. Pl. 755 (1753); Gen. Pl. 335 (1754).

Type species: *A. christianus* L.

Synonyms: *Phaca* L., *Stella* Medic.; *Alopecias* Stev.; *Myobroma* Stev.; *Ankylotus* Stev.; *Tragacantha* Miller.

Annual, biennial or perennial herbs or shrubs. Height range 2-100 cm. Erect, ascending, decumbent or prostrate. Surface glabrous to pilose or villous. Spines present or absent. Leaves paripinnate, imparipinnate, occasionally trifoliate and rarely simple. 1-80 leaflets per leaf. Leaflet shape linear or filiform to obovate or cuneate (very variable); lamina pubescent; glands absent. Stipules conspicuous, scarious or foliaceous, fused or free from petiole; stipels absent. Flowers rarely solitary or arranged in racemes or spikes; pseudo-terminal or axillary. 1 - many flowers per inflorescence. Pedunculate or sessile. Corolla white, pink, purple, blue or yellow. Claw long; wing oblong; keel up curved and often blunt at apex; wings and keel usually shorter than standard. Calyx tubular to campanulate, teeth subequal or bilabiate. Bracts small and/or inconspicuous; bracteoles present or absent. Stamens diadelphous. Anthers monomorphic. Stigma terminal. Style linear or up curved, usually glabrous. Ovary sessile or stipitate, few to numerous ovules. Fruit oval to linear (very variable), turgid or inflated; septa present; dehiscent or indehiscent; exserted from or enclosed in calyx. Seed reniform, often compressed. Several - many seeds per fruit.

About 2000 species (Polhill 1981).

Taxonomic Affinities

Oxytropis is closely allied to *Astragalus* while Asian members of *Caragana* are similar to the woody species.

Useful Taxonomic Treatments

Chamberlain, D. F. and Matthews, V. A. (1970). Astragalus L. 'Flora of Turkey and East Aegean Islands. Vol. 3'. (Ed. P.H. Davis). pp.49-51. (Edinburgh University Press: Edinburgh).

Mouterde, P. (1966). 'Nouvelle Flore du Liban et de la Syrie. Vol. 2'. pp.315-327. (Librairie Orientale: Beirut, Lebanon).

Polhill, R.M. (1981) 'Galegeae, Vol. 1'. (Eds. R.M. Polhill and P.H. Raven). pp. 357-363 (Royal Botanic Gardens: Kew).

Townsend, C. C. (1974). 'Flora of Iraq. Vol.3'. pp. 239-249. (Ministry of Agriculture and Agrarian Reform of the Republic of Iraq: Bagdad).

Zohary, M., (1972). 'Flora Palaestina, Vol. 2'. Pp. 55-57. (Israel Academy of Sciences and Humanities, Jerusalem: Israel).

Ecology

Most species of *Astragalus* prefer sunny areas with light, porous soil. Many are drought resistant (Allen and

Allen 1981). Typical habitats include steppe, desert, fields, roadsides, stony areas and woods.

Distribution

Found mainly in temperate Northern hemisphere, representatives are cosmopolitan outside the tropics and Australia. The distribution of the genus extends into the Arctic zone and the mountains of South America and Asia. The centre of diversity is in the arid continental area of Middle Asia and the plateau of Iran (Yakovlev *et al.* 1996).

Uses

Comparatively few members of this large genus are important agriculturally. *A. cicer* (cicer milkvetch) is used for pasture, although some caution must be used when selecting them for forage due to poisoning from either selenium accumulation or locoism (toxic poisoning which commonly causes hallucinations, delirium, defective vision and poor hair quality (Allen and Allen 1981). It was originally used in Europe before being introduced into the USA in 1923 and Canada in 1931. Several cultivars have since been released (Smartt and Simmonds 1995). A number of other species such as *A. bisulacatus, A. emoryanus, A. lentiginosus, A. mollissimus* and *A. nothoxys* can cause vertebrate poisoning due to accumulation of selenium. *A. arenarius* and *A. glycyphyllos* are used for forage in Poland, *A. hypoglottis* is favoured in the European Alps, *A. hamosus* is used in the Mediterranean, while *A. chinensis, A. cicer, A. falcatus* and *A. sinicus* are used both as forage and as green manure in Taiwan and Japan (Allen and Allen 1981). *A. arrectus, A. cicer, A. dauricus, A. mortoni* and *A. rubyi* are useful in soil erosion control. Other members are used as selenium or uranium indicator plants. *A. alopecurioides, A. aristatus, A. echinus* and *A. boeticus* have ornamental values.

Members of the *Astragalus* genus (eg *A. adscendens, A. cerasocrenus, A. echidnaeformis, A. gossypinus, A. gummifer, A. leiocladus, A. parnassi* and *A. vernus*) are important sources of Tragacanth gum in Syria, Turkey, Iraq, Iran, South-east Europe and Central Asia. The gum, collected by tapping the rootstock, is used in the dressing of yarn, the sizing of textiles and paper, in wall-paper paste and in mixtures for water-proofing and stiffening felt materials. Finer grade gum is used in confectionary and food-stuffs such as ice-cream, syrups, jellies, mayonnaise etc. It is also important medicinally, used for example, to firm pill mixtures, in glycerine, shaving and dental creams, jellies, lotions and nail polishes. *A. gummifer* is used as a remedy for coughs and diarrhea, and even as an aphrodisiac (Duke 1981). It is also used as a lubricant for surgical instruments (Allen and Allen 1981).

Astragalus species, and in particular *A. membranaceus*, are used in Chinese herbal medicine to replenish vital energy and accelerate wound healing. *Astragalus* also enhances the immune system and has antiviral properties (Zhang *et al.* 1990).

Cicer L.

Classification (from Van der Maesen 1984)

Family Leguminosae

Subfamily Papilionoideae

Tribe Cicereae

Genus *Cicer* L.

Subgenus *Pseudononis*

Sections *Monocicer, Chamaecicer*

Subgenus *Viciastrum*

Sections *Polycicer, Acanthocicer*

Taxonomic History

The genus *Cicer* was described by Linnaeus, based on *C. arietinum*, using the Tournefort name. A revision by Jaubert and Spach (1842) distinguished four sections, later adapted by Alefeld (1859) and Boissier (1872). Popov (1929) and Linczevski (1948) split the genus into two subgenera, *Pseudononis* and *Viciastrum*, and this position was adopted by Van der Maesen (1972). Traditionally *Cicer* was placed in the tribe Vicieae, but recently *Cicer* has been removed and now constitutes a monogeneric tribe, the Cicereae (Kupicha 1977).

Taxon Description

Cicer L. Sp. Pl. 738 (1753).

Type species: *C. arietinum* L.

Synonyms: *Nochotta* S.G. Gmel. (1774); *Spiroceras* Jaub. and Spach (1842).

Annual or perennial herbs. Height range 10-80cm. Erect, ascending, decumbent or prostrate. Surface pubescent. Spines present or absent. Leaves paripinnate, imparipinnate or trifoliate, with terminal tendril, mucro or leaflet. 3-36 leaflets per leaf. Leaflet shape ovate, cordate, elliptic, oblong, obovate or cuneate; margins dentate; lamina pubescent; glands present or absent. Stipules large and conspicuous or small and/or inconspicuous, foliaceous, free from petiole; stipels absent. Flowers solitary or arranged in racemes; axillary. 1-5 flowers per inflorescence. Pedunculate. Corolla white, pink, blue or purple (veined). Standard mostly obovate, claw broad; wing obliquely obovate; keel pointed or blunt. Calyx teeth subequal or bilabiate. Bracts small and/or inconspicuous; bracteoles absent. Stamens diadelphous. Anthers monomorphic. Stigma terminal. Style filiform or incurved, glabrous or pubescent (near ovary) on abaxial surface. Ovary sessile, few or numerous ovules. Fruit oval or linear, turgid or inflated; surface pubescent. Pod length 9-30mm. Seeds sub-globose or ovoid; 1-10 per fruit.

About 42 species (Coles *et al.* 1998).

Taxonomic Affinities

Although formerly included in the Vicieae, *Cicer* are closely related to the Trifolieae, especially *Ononis*, and differ only significantly in their possession of tendrils, stipules being connate from the petiole and diadelphous stamens.

Useful Taxonomic Treatments

Chrtkova-Zertova, A., Van der Maesen, J. G. and Rechinger, K. H. (1979). 'Flora Iranica. Papilionaceae I - Vicieae: 2-3'. (Akademische Druck-u. Verlagsanstalt: Graz).

Coles, S., Maxted, N. and Van der Maesen, L.J.G. (In press). Identification Aids for *Cicer* (Leguminosae, *Cicereae*) taxa. *Edinburgh Journal of Botany*.

Davis, P. H. (1970). 'Flora of Turkey and the East Aegean Island. Vol. 3'. pp. 267-268. (Edinburgh University Press: Edinburgh).

Van der Maesen, L. J. G. (1972). '*Cicer* L., A Monograph of the Genus with Special Reference to the Chickpea (*Cicer arietinum* L.), its Ecology and Cultivation'. pp. 20-23. (H. Veenman and Zonen N.V: Wageningen).

Van der Maesen, L. J. G. (1984). Taxonomy, distribution and evolution of the chickpea and its wild relatives. In 'Genetic Resources and their Exploitation - Chickpea, Faba Beans and Lentils' (Eds. J.R. Witcoumbe and W. Erskine), pp. 95-104. (Martinus Nijhoff / Dr. W. Junk Publishers: The Hague).

Ecology

Plants tend to be drought-resistant and do not tolerate heavy rains, sandy or wet soils or hot climates. They occur from sea level to 5600m, although some annual species do not grow above 1000m. Typical habitats include rocky ground, scrub, deciduous woodland, coniferous woodland, fallow, meadow or pasture.

Distribution

Wild species are distributed in Central and Western Asia. Endemics occur in Morocco, Ethiopia, Greece and Crete. *C. arietinum* is widely cultivated from South West Asia to India, throughout the Middle East and Mediterranean. It is also common in Ethiopia, Tanzania and as a fairly recent introduction in South America and Australia. Vavilov (1926) indicated several centres of diversity in South west Asia and the Mediterranean. Ethiopia was considered a secondary centre.

Uses

Only *C. arietinum* (the Chickpea) is cultivated for its edible seeds (14-30% protein - Duke 1981). A sandy-loam and a cool climate is needed to give high yields. The peas are eaten fresh, dried, boiled or roasted while the green pods and young shoots can be eaten as a vegetable. It is the third most important pulse crop in the world, and the area under cultivation is approximately 9-10 Mha (Smartt and Simmonds 1995). Although sometimes used as a forage crop, high quantities of oxalic acid lower its value as a forage (Duke 1981). Some of the non-spiny species (*C. echinospermum, C. incisum*) may be eaten by cattle. *C. echinospermum* and *C. reticulatum* are both potential gene sources for cultivar improvement in chickpea (GRIN 1997).

Coronilla L.

Classification (from Lassen 1989).
Family Leguminosae
Subfamily Papilionoideae
Tribe Coronilleae
Genus *Coronilla* L.

Taxonomic History

The genus has recently had many species excluded and transferred to *Securigera*. Lassen (1989) accepted *C. valentina* as the type species, because *C. varia*, the former type, was transferred to *Securigera*.

Taxon Description

Coronilla L. Sp. Pl. 742 (1753).

Type species: *C. valentina* L.

Synonyms: *Emerus* Tourn. ex Mill. (1752); *Scorpius* Medik. (1787); *Aviunculus* Fourr. (1868); *Callistephana* Fourr. (1868).

Annual or perennial herbs or shrubs. Height range 6-250cm. Ascending, decumbent or prostrate. Surface usually glabrous or sometimes pubescent. Spines absent. Leaves imparipinnate or upper trifoliate and lower simple. (3-)5-7(-9) leaflets per leaf. Leaflet obovate or cuneate; margins entire; lamina glabrous, pubescent or glaucous; glands absent. Stipules large and conspicuous or small and/or inconspicuous, scarious or foliaceous, fused with petiole; stipels absent. Flowers arranged in umbels; axillary. (2-)4-20 flowers per inflorescence. Long peduncle. Corolla yellow or rarely purple and white. Standard sub-orbicular, claw long; wing obliquely obovate or oblong; keel beaked or pointed. Calyx campanulate, teeth subequal. Bracts small and/or inconspicuous, scarious; bracteoles absent. Stamens diadelphous. Anthers monomorphic. Stigma terminal. Style subulate or incurved, glabrous. Ovary sessile, numerous ovules. Fruit arched to linear, sometimes slightly compressed; surface smooth and glabrous to pubescent; joints present (may be reduced); dehiscent or rarely indehiscent; exserted from calyx. Pod length 1.5-110mm. Seeds transversely oblong.

About 20 species (Polhill 1981).

Taxonomic Affinities

Closely related to *Securigera* as until recently the *Securigera* species were considered part of *Coronilla.* It is more distantly related to *Lotus* (Polhill and Raven 1981).

Useful Taxonomic Treatments

Chamberlain, D.F. (1970). 'Flora of Turkey and East Aegean Islands. Vol.3'. (Ed. P.H. Davis). pp. 538. (Edinburgh University Press: Edinburgh).

Lassen, P. (1989). A new delimitation of the genera *Coronilla, Hippocrepis* and *Securigera. Willdenowia* 19, 49-62.

Meikle, R. D. (1977). 'The Flora of Cyprus. Vol. 1'. pp. 511. (Bentham-Moxon Trust, Royal Botanic Gardens: Kew).

Mouterde, P. (1966). 'Nouvelle Flore du Liban et de la Syrie. Vol. 2'. Pp. 380. (Librairie Orientale: Beirut, Lebanon).

Polhill, R.M. (1981) 'Coronilleae, Vol. 1'. (Eds. R.M. Polhill and P.H. Raven). pp. 375 (Royal Botanic Gardens: Kew).

Polunin, O. (1969). 'Flowers of Europe. A field Guide'. pp. 211-212. (Oxford University Press: Oxford).

Ecology

Typically found on waste areas, roadsides, grasslands, rocky ground and scrub. They are best adapted to well-drained, fertile soils with a pH of 6 or more.

Distribution

Found in central and southern Europe, the Mediterranean, West Asia and North east Africa.

Uses

The major forage species of this genus, crown vetch (*S. varia*), has recently been transferred to *Securigera. C. emerus, C. scorpoides* and *C. valentina* are used as ornamentals (GRIN 1997).

Cytisopsis Jaub. and Spach

Classification (from Chamberlain 1970).

Family Leguminosae

Subfamily Papilionoideae

Tribe Loteae

Genus *Cytisopsis* Jaub. and Spach

Taxonomic History

Has sometimes been included within *Anthyllis.*

Taxon Description

Cytisopsis Jaub. and Spach. Illustr.1:154,t.84 (1842).

Type species: *C. dorycniifolia* Jaub. and Spach.

Annual shrubs. Height range 10-30cm. Often prostrate. Surface pilose or villous. Spines absent. Leaves simple or digitate. 5-7 leaflets per leaf. Leaflet shape oblong; margins entire; lamina pilose; glands absent. Stipules absent. Flowers solitary or arranged in umbels; axillary. 1-3 flowers per inflorescence. Sessile. Corolla yellow. Standard ovate, claw long; wing obtuse; keel beaked and slightly blunt or curved; wings and standard longer than keel. Calyx tubular, teeth bilabiate. Bracts conspicuous; bracteoles present. Stamens diadelphous. Anthers monomorphic. Style slightly incurved, glabrous. Ovary sessile, numerous ovules. Fruit circular, sub-terete; surface pubescent; thin joints present; dehiscent; exserted from calyx. Pod length 15-20mm. Seeds sub-globose, many.

Two species (Polhill 1981).

Useful Taxonomic Treatments

Chamberlain, D.F. (1970). 'Flora of Turkey and East Aegean Islands. Vol.3'. (Ed. P.H. Davis). pp. 511-512 (Edinburgh University Press: Edinburgh).

Polhill, R.M. (1981) 'Loteae, Vol. 1'. (Eds. R.M. Polhill and P.H. Raven). pp. 371-374 (Royal Botanic Gardens: Kew).

Ecology

Members of this genus are typically found in pine forests, macchie, rocky ground and on coastal slopes.

Distribution

One species occurs in the eastern Mediterranean and the other in North Africa.

Uses

No reference can be found to species of this genus being commercially used.

Dorycnium Mill.

Classification (from Demiriz 1970).

Family Leguminosae

Subfamily Papilionoideae

Tribe Loteae

Genus *Dorycnium* Miller

Sections *Bonjeania, Dorycnium*

Taxonomic History

Historically, the genus has been included in *Lotus* by several authors.

Taxon Description

Dorycnium Mill. Gard. Dict. Abridg. ed.4 (1754).

Type species: *D. suffructicosa* Vill.

Synonyms: *Bonjeanea* Reichb. (1832); *Bonjeania* Reichb. (1832); *Dorychnium* Brongn. (1843); *Ortholotus* Fourr. (1868).

Perennial herbs or shrubs. Height range 15-50(-100)cm. Often erect. Surface glabrous to pilose or villous. Spines absent. Leaves imparipinnate. (4-)5(-9) leaflets per leaf. Leaflet shape oblong, obovate or cuneate; margins entire; lamina glabrous to pilose; glands absent. Lower leaflets often stipuloid. Stipules small or absent. Flowers arranged in heads or umbels; axillary or pseudo-terminal. 4-40 flowers per inflorescence. Pedunculate or sessile. Corolla white, pink or purple (keel often dark). Standard ovate to oblong, claw short; wing falcate to obovate; keel curved or blunt. Calyx campanulate, teeth subequal to slightly bilabiate. Bracts often small and/or inconspicuous or rarely conspicuous, sometimes foliaceous; bracteoles absent. Stamens diadelphous. Anthers monomorphic. Stigma terminal. Style incurved, glabrous. Ovary sessile, numerous ovules. Fruit circular, turgid or inflated to terete; joints often present; dehiscent. Pod length 3-14 mm. Seeds compressed to globose; 1-8 per fruit.

12 species (Meikle 1977).

Useful Taxonomic Treatments

Demiriz, H. (1970). 'Flora of Turkey and East Aegean Islands. Vol. 3'. (Ed. P.H. Davis) pp. 512-518. (Edinburgh University Press: Edinburgh).

Meikle, R. D. (1977). 'The Flora of Cyprus. Vol. 1'. pp. 475. (Bentham-Moxon Trust, Royal Botanic Gardens: Kew).

Mouterde, P. (1966). 'Nouvelle Flore du Liban et de la Syrie. Vol. 2'. pp. 303. (Librairie Orientale: Beirut, Lebanon).

Polunin, O. (1969). 'Flowers of Europe. A field Guide'. pp. 208-209. (Oxford University Press:, Oxford).

Ecology

Species are usually found in areas with a low fertility, especially on light, sandy soils. Typical habitats include rocky ground, steppes, macchie, deciduous woodland, coniferous woodland, fallow, meadow or pasture and seashore.

Distribution

Common in the Mediterranean, southern Europe and western Asia.

Uses

D. hirsutum (hairy Canary Clover) is sometimes used as an ornamental, forage crop and for control of soil erosion (Allen and Allen 1981). *D. pentaphyllum* is used in honey production (GRIN 1997).

Ebenus L.

Classification (from Huber-Morath 1970).
Family Leguminosae
Subfamily Papilionoideae
Tribe Hedysareae
Genus *Ebenus* L.

Taxon Description

Ebenus L. Sp. Pl. 764 (1753).

Type species: *E. cretica* L.

Synonyms: *Ebeniduium* Jaub. and Spach (1843).

Perennial herbs or shrubs. Height range 5-60 cm. Surface pubescent, pilose or villous. Spines present or absent. Leaves imparipinnate or trifoliate (may be folded). 3-30 leaflets per leaf. Leaflet shape linear, filiform, lanceolate or elliptic; margins entire; lamina pubescent; glands absent. Stipules small and/or inconspicuous, often scarious, free from petiole; stipels absent. Flowers arranged in spikes; axillary. Numerous flowers per inflorescence. Long peduncle. Corolla purple or yellow. Standard obovate or obcordate, claw short and wide; keel oblique and blunt; wings short, keel equal to standard or shorter. Calyx subulate or campanulate, teeth subequal. Bracts conspicuous, foliaceous; very small bracteoles. Stamens diadelphous. Anthers monomorphic. Stigma terminal. Style filiform, glabrous. Ovary sessile, few ovules. Fruit circular to linear, compressed; surface smooth and glabrous or pubescent; joints present; indehiscent; enclosed in calyx. 1-2(-3) seeds per fruit.

About 20 species (Polhill 1981).

Useful Taxonomic Treatments

Huber-Morath, A. (1970). Ebenus L. In 'Flora of Turkey and East Aegean Islands. Vol. 3'. (Ed. P.H. Davis). pp. 590-591. (Edinburgh University Press: Edinburgh).

Polhill, R.M. (1981) 'Hedysareae, Vol. 1'. (Eds. R.M. Polhill and P.H. Raven). pp. 367-370 (Royal Botanic Gardens: Kew).

Taxonomic Affinities

This genus is very closely related to *Onobrychis*.

Ecology

Ebenus species are found on cliffs and rocks up to 2700m. Typical habitats include rocky ground, steppe, scrub and coniferous woodland.

Distribution

The genus has its centre of diversity in Anatolia, Turkey. Only six species are known outside Turkey and these are found in surrounding countries and India.

Uses

No reference can be found to species of this genus being commercially used.

Hammatolobium Fenzl

Classification (from Chamberlain 1970).
Family Leguminosae
Subfamily Papilionoideae
Tribe Coronilleae
Genus *Hammatolobium* Fenzl.

Taxon Description

Hammatolobium Fenzl Ill. Pl. Syr., t.1 (1843).

Type species: *H. lotoides* Fenzl.

Synonyms: *Hamatolobium* Fenzl (1842); *Ludovicia* Coss. (1856).

Perennial herbs or shrubs. Height range 10-20cm. Prostrate. Surface pubescent to pilose or villous. Spines absent. Leaves sub-digitate, imparipinnate or trifoliate. 3-5 leaflets per leaf. Leaflet shape ovate or cordate; margins entire; lamina pubescent; glands absent. Lower leaflets may be stipuloid. Stipules small and/or inconspicuous, scarious, free from petiole; stipels absent. Flowers solitary arranged in umbels; axillary. Often 2 flowers per inflorescence. Pedunculate. Corolla yellow or purple. Standard sub-orbicular, claw long; wing obliquely obovate to oblong; keel beaked or pointed; wings longer than keel. Calyx teeth subequal. Bracts conspicuous, foliaceous. Stamens monadelphous. Anthers monomorphic. Stigma terminal. Style sub-spatulate and incurved at apex, glabrous. Ovary sessile. Fruit linear, terete or compressed; surface pubescent; joints present; indehiscent. Pod length 10-25mm. Seeds arcuate; 2-6 per fruit.

2 species (Polhill 1981).

Useful Taxonomic Treatments

Chamberlain, D.F. (1970). *Hammatolobium* Fenzl. 'Flora of Turkey and East Aegean Islands. Vol.3'. (Ed. P.H. Davis). pp. 547. (Edinburgh University Press: Edinburgh).

Polhill, R.M. (1981) 'Coronilleae, Vol. 1'. (Eds. R.M. Polhill and P.H. Raven). pp. 375 (Royal Botanic Gardens: Kew).

Taxonomic Affinities

The general taxonomic view is that *Hammatolobium* species seem to be closer to *Anthyllis* and *Lotus* than *Coronilla.*

Ecology

Habitats include rocky ground and deciduous woodland.

Distribution

The two species are morphologically dissimilar and found at either end of the Mediterranean.

Uses

No reference can be found to species of this genus being commercially used.

Hedysarum L.

Classification (from Rechinger 1984).

Family Leguminosae

Subfamily Papilionoideae

Tribe Hedysareae

Genus *Hedysarum* L.

Sections *Obscura, Multicaulia, Subacaulia, Crinifera*

Taxonomic History

The genus has sometimes classified in the tribe Desmodieae (Lock 1994).

Taxon Description

Hedysarum L. Sp. Pl. 745 (1753).

Type species: *H. coronarium* L.

Synonyms: *Hedisarum* Neck. (1768); *Sulla* Medik. (1787); *Echinolobium* Desv. (1813).

Annual or perennial herbs (or rarely shrubs). Height range 5-70cm. Erect, ascending, decumbent or prostrate. Surface pubescent. Spines absent. Leaves imparipinnate. 8-24 leaflets per leaf. Leaflet shape linear, filiform, ovate, cordate, elliptic or oblong; margins entire; lamina often pitted or glabrous; glands absent. Stipules large and conspicuous to small and/or inconspicuous, scarious, fused or free from petiole; stipels absent. Flowers arranged in heads or racemes; axillary. 3-20 flowers per inflorescence. Pedunculate. Corolla white, purple, red or yellow (rarely). Standard obovate or obcordate, claw very short or almost absent; wing oblong; keel oblique, blunt or curved (rarely slightly); wings shorter than keel and standard. Calyx tubular or campanulate, teeth subequal. Bracts small and/or inconspicuous, scarious or setaceous; bracteoles present. Stamens diadelphous. Anthers monomorphic. Stigma terminal. Style filiform, glabrous. Ovary sub- sessile, numerous ovules (4-5). Fruit arched (not strongly), compressed; surface smooth and glabrous or rugose; joints present; indehiscent; exserted from calyx. Pod length 5-16 mm. Seeds reniform and compressed; 3-many per fruit.

About 100 species (Meikle 1977).

Taxonomic Affinities

Hedysarum is closely allied to *Onobrychis*. The principal differences are shown below:

Character	*Hedysarum*	*Onobrychis*
Fruit	Mostly two or more segments	One segment
Wings	Large, only slightly shorter than keel	Small, much smaller than keel
Ovary	Linear, two to many ovules, constricted between	Oblong or sub-spherical, one or two ovules, not constricted between

Useful Taxonomic Treatments

Hedge I. C. (1970). *Hedysarum* L. 'Flora of Turkey and East Aegean Islands, Vol. 3'. (Ed. Davis, P. H.) pp. 549-550. (Edinburgh University Press: Edinburgh).

Mouterde, P. (1966). 'Nouvelle Flore du Liban et de la Syrie. Vol. 2'. pp. 383-384. (Librairie Orientale: Beirut, Lebanon).
Rechinger, K. H. (1984). 'Flora Iranica. Papilionaceae II'. pp. 367-369. (Akademische Druck-u. Verlagsanstalt: Graz).
Townsend, C. C. (1974). 'Flora of Iraq. Vol.3'. pp. 467. (Ministry of Agriculture and Agrarian Reform of the Republic of Iraq: Bagdad).

Ecology

The genus is well adapted to calcareous soils with rocks or gravel. Typical habitats include steppe, deciduous woodland, fallow, meadow or pasture, cultivated land, and scrub. Some species are found up to 3500m.

Distribution

Well distributed in temperate Europe, the Mediterranean region of North Africa, Asia, Siberia and North America. The genus is most diverse in the eastern Mediterranean and West Asia.

Uses

The foliage of all species is palatable but the plants are not naturally plentiful enough to be of great importance. The native European *H. coronarium* L. (Sulla sweetvetch, Spanish or Italian sainfoin) is cultivated as fodder, forage and green manure in the Mediterranean. If fed green or as hay it has a comparable nutritional value to red clover (Duke 1981). This is also a successful crop for soil improvement. *H. hedysaroides* is cultivated as a fodder crop, especially in the Alps. *H. splendens* was introduced into Columbia, South America as a forage. *H. arcticum* and *H. fruticosum* are used for fodder, and *H. arcticum* is also grown as a vegetable (GRIN 1997). The roots of *H. alpinum* are collected in northern America and eaten either raw, roasted or boiled (Allen and Allen 1981). A few species are cultivated as ornamentals.

Hippocrepis L.

Classification (from Chamberlain 1970).
Family Leguminosae
Subfamily Papilionoideae
Tribe Coronilleae
Genus *Hippocrepis* L.

Taxonomic History

The genus was revised by Lassen 1989. Although the genus consists mostly of herbs, recent changes in its definition have lead to the inclusion of some shrubby species (Lock and Simpson 1991).

Taxon Description

Hippocrepis L. Sp. Pl. 744 (1753).

Type species: *H. unisiliquosa* L.

Synonyms: *Hipocrepis* Neck. (1768); *Ferrumequinum* (ex Tourn.) Medic. (1787); *Hippocris* Raf. (1814).

Annual or perennial herbs or rarely shrubs. Height range 5-50cm. Sub-erect, prostrate, sprawling or creeping. Surface glabrous to pilose or villous. Spines present or absent. Leaves imparipinnate. 6-16 leaflets per leaf. Leaflet shape oblong to obovate or cuneate (sometimes narrowly); margins entire; lamina pubescent; glands absent. Stipules small or absent, scarious, fused with petiole; stipels absent. Flowers solitary or arranged in umbels; axillary. 1-6 flowers per inflorescence. Pedunculate (or in axil). Corolla yellow. Standard sub-orbicular, claw long; wing falcate to obovate or oblong; keel beaked to pointed. Calyx campanulate, teeth subequal to bilabiate. Bracts small and/or inconspicuous, scarious; bracteoles absent. Stamens diadelphous. Anthers monomorphic. Stigma terminal. Style subulate and incurved (at apex), glabrous. Ovary sessile, numerous ovules. Fruit arched to linear, rarely sub-terete or compressed; surface smooth and glabrous, pubescent or spiny (barbellate); U-shaped joints present; indehiscent. Pod length 15-45mm. Seeds arcuate; 6-13 per fruit.

21 species (Polhill 1981).

Taxonomic Affinities

Closely allied to *Ornithopus* and *Coronilla*

Useful Taxonomic Treatments

Chamberlain, D. F. (1970). *Hippocrepis* L. 'Flora of Turkey and East Aegean Islands. Vol. 3'. (Ed. Davis, P. H). pp. 545 (Edinburgh University Press: Edinburgh).
Lassen, P. (1989). A new delimitation of the genera *Coronilla*, *Hippocrepis* and *Securigera*. *Willdenowia* 19, 49-62.
Meikle, R. D. (1977). 'The Flora of Cyprus. Vol. 1'. pp. 519. (Bentham-Moxon Trust, Royal Botanic Gardens: Kew).

Polhill, R.M. (1981) 'Coronilleae, Vol. 1'. (Eds. R.M. Polhill and P.H. Raven). pp. 375 (Royal Botanic Gardens: Kew).
Rechinger, K. H. (1984). 'Flora Iranica. Papilionaceae II'. pp. 353-354. (Akademische Druck-u. Verlagsanstalt: Graz).
Townsend, C. C. (1974). 'Flora of Iraq. Vol.3'. pp. 460. (Ministry of Agriculture and Agrarian Reform of the Republic of Iraq: Bagdad).

Ecology

Members are common in dry, warm regions with calcareous, rocky soils. Typical habitats include scrub, fallow, meadow or pasture, and cultivated land.

Distribution

The distribution extends from the Mediterranean centre of diversity North into Europe and East to West Asia.

Uses

Useful forage plants. A number of species are sometimes grown in rock gardens (e.g. *H. comosa,* and *H. emerus*).

Hymenocarpus Savi

Classification (from Chamberlain 1970).
Family Leguminosae
Subfamily Papilionoideae
Tribe Loteae
Genus *Hymenocarpus* Savi

Taxonomic History

The genus was originally viewed as two species, *H. circinnatus* and *H. nummularis*. The distinction being based on differences in the wing of the pod. However, more recent authors (Rechinger 1984) have reduced the taxa to subspecies.

Taxon Description

Hymenocarpus Savi Fl. Pisana 2:205 (1798).

Type species: *H. circinatus* L.

Synonyms: *Circinnus* Medik. (1787); *Circinus* Medik. (1789).

Annual herbs. Height range 10-35cm. Prostrate. Surface sparsely pubescent. Spines absent. Leaves simple or imparipinnate, often with terminal leaflet enlarged. 4-8 leaflets per leaf. Leaflet shape oblong to obovate or cuneate; margins entire; lamina sparsely pubescent; glands absent. Lower leaflets not stipuloid. Stipules small and/or inconspicuous, scarious, fused or free from petiole. Flowers arranged in umbels; axillary. 2-6 flowers per inflorescence. Pedunculate. Corolla yellow to red. Standard sub-orbicular, claw short; wing obovate; keel beaked; wings longer than keel, shorter than standard. Calyx campanulate, teeth subequal. Bracts conspicuous, mainly foliaceous; bracteoles absent. Stamens diadelphous. Anthers monomorphic. Stigma terminal. Style incurved, glabrous. Ovary sessile, few ovules. Fruit sub-oval or circular to arched, compressed; surface smooth and glabrous to pubescent (or reticulate-veined); joints often present; indehiscent; exserted from calyx. Pod length 12-25 mm. Seeds reniform; 1-2 per fruit.

1 species (Polhill 1981).

Useful Taxonomic Treatments

Chamberlain, D. F. (1970). *Hymenocarpus* Savi. 'Flora of Turkey and East Aegean Islands. Vol. 3'. (Ed. Davis, P. H.) pp. 533 (Edinburgh University Press: Edinburgh).
Polhill, R.M. (1981) 'Loteae, Vol. 1'. (Eds. R.M. Polhill and P.H. Raven). pp. 371-374 (Royal Botanic Gardens: Kew).
Townsend, C. C. (1974). 'Flora of Iraq. Vol.3'. pp. 196. (Ministry of Agriculture and Agrarian Reform of the Republic of Iraq: Bagdad).

Ecology

Found on lower mountain slopes and steppe grasslands, on clay, sandy or silty soil.

Distribution

This monotypic genus is endemic to the Mediterranean region and West Asia.

Uses

A minor forage constituent of foothill and steppe pastures.

Lathyrus L.

Classification (from Kupicha 1983).
Family Leguminosae
Subfamily Papilionoideae
Tribe Vicieae
Genus *Lathyrus* L.
Sections *Orobus, Lathyrostylis, Pratensis, Neurolobus, Orobon, Orobastrum, Viciopsis, Linearicarpus, Lathyrus, Aphaca, Nissolia, Clymenum, Notolathyrus.*

Taxonomic History

The taxonomic history of genus *Lathyrus* is fairly limited with only five significant treatments (Kupicha 1983). Although there is no world-wide monograph of *Lathyrus*, the morphology and taxonomy has been studied most recently by Bassler (1966 1973 1981) and Kupicha (1983), leading to an improvement in infragenic classification.

Linneus recognised the two genera, *Lathyrus* and *Orobus*, but Godron (1849) united the two into *Lathyrus*, characterised by a dorsally compressed, pubescent style, in which he recognised six sections. They were given generic status again by Boissier (1872), but this has not been followed by subsequent authors. Bassler (1966) separated the oroboid taxa into a distinct *Orobus* subgenus within *Lathyrus*. Davis (1970) classified the genus into ten sections, including *Orobus*, although he warns that the classification is provisional. Since his work two other classifications have been proposed. Czefranova (1971) divides the genus into six subgenera although this has not been widely accepted. Kupicha (1983) maintains that although the genus is well-structured at the sectional level the correlated characters are not strong enough to give subgeneric recognition. Therefore, she divides Davis's section of *Orobastrum* into *Orobastrum, Viciopsis* and *Linearicarpus*; combines *Lathyrus* and *Cicercula* into a single *Lathyrus* section and adds section *Notolathyrus*.

Taxon Description

Lathyrus L. Sp. Pl. 729 (1753).

Type species: *L. sylvestris* L.

Synonyms: *Orobus* L. (1753); *Aphaca* Adans. (1763); *Clymenum* Tourn. ex Adans. (1763); *Nissolia* Tourn. ex Adans. (1763);*Ochrus* Tourn. ex Adans. (1763); *Lathirus* Neck. (1768); *Cicercula* Medik. (1787); *Athyrus* Neck.(1790); *Nissolia* Moench (1794); *Oxypogon* Raf. (1819); *Cicerella* DC. (1825); *Platystylis* Sweat (1828); *Spatulima* Raf. (1836); *Anurus* C. Presel (1837); *Astrophia* Nutt. ex Torrey and Gray (1838); *Graphiosa* Alef. (1861); *Lastila* Alef. (1861); *Navidura* Alef. (1861); *Lathyros* St. Lag. (1880).

Annual or perennial herbs. Height range 5-250cm. Often climbing. Surface glabrous or pubescent. Spines absent. Stems usually winged. Leaves sub-imparipinnate or paripinnate, with terminal tendril or mucro. 2-16 leaflets per leaf. Leaflet shape linear or filiform; margins entire; lamina glabrous or pubescent; glands absent. Stipules large and conspicuous, foliaceous, free from petiole. Flowers solitary or arranged in racemes; axillary. 1-25 flowers per inflorescence. Pedunculate. Corolla white, pink, purple, blue, yellow or red. Claw short and broad; wing falcate to obovate or oblong; keel blunt or pointed; wings longer than keel. Calyx tubular to campanulate, teeth subequal or bilabiate. Bracts small or absent; bracteoles absent. Stamens monadelphous or diadelphous. Anthers monomorphic. Stigma terminal. Style linear to spatulate or incurved, pubescent or glabrous (rarely), pubescence on adaxial surface. Ovary sub-stipitate or sessile, numerous ovules or rarely few. Fruit compressed; surface ridged or grooved (may be winged); joints absent; dehiscent. Pod length 15 mm (+). Seeds globose or rarely compressed; 2-many per fruit.

About 150 species (Kupicha 1983).

Taxonomic Affinities

Lathyrus species are very closely related to *Vicia* species but usually have fewer leaflets which are also larger, thicker and more prominently veined. Other distinguishing characters include winged stems, a pubescent, non-grooved style and wings almost or completely free from the keel. Species which may be easily confused are *L. aureus* (similar to *Vicia crocea*), and *L. niger* (similar to *V. cassubica*).

Useful Taxonomic Treatments

Ball, P.W. (1968). 'Flora Europaea. Vol. 2'. (eds. Tutin, T.G., Heywood, V.H., Burges, N.A., Moore, D.M., Valentine, D.H., Walters, S.M., and Webb, D.A.). pp. 136-138 (Cambridge University Press: Cambridge).

Campbell, C.G. (1997). Grass pea: *Lathyrus sativus* L. Promoting the conservation and use of underutilised and neglected crops. 18. Institute of Plant Genetics and Crop Plant Research, Gatersleben / International Plant Genetic Resources Institute, Rome, Italy.

Chrtkova-Zertova, A., Van der Maesen, J. G. and Rechinger, K. H. (1979). 'Flora Iranica. Papilionaceae I - Vicieae'. pp. 62-64. (Akademische Druck-u. Verlagsanstalt: Graz).

Davis, P. H. (1970). 'Flora of Turkey and the East Aegean Island. Vol.3'. pp. 328-333. (Edinburgh University Press: Edinburgh).
Kupicha, F.K. (1983). The Infragenic Structure of *Lathyrus*. *Notes from the Royal Botanic Garden Edinburgh*. 41, 209-244.
Mouterde, P. (1966). 'Nouvelle Flore du Liban et de la Syrie. Vol. 2'. pp. 416-418. (Librairie Orientale: Beirut, Lebanon).
Polunin, O. (1987). 'Flowers of Greece and the Balkans. Field Guide'. pp. 269-299. (Oxford University Press: Oxford).

Ecology

Lathyrus species are found in a wide range of habitats, including forests, macchie, rocky slopes, grassland and pastures, cultivated land (as a crop and weed), hedges, roadsides, water meadows, by streams or marshes. In general they are well adapted to rather dry areas but can tolerate water-logging, cool weather and soil with poor nutrients.

Distribution

This genus is widespread in the temperate regions of both the northern and southern hemispheres and extends into tropical East Africa, Australia and South America. The natural distribution has been obscured by cultivation. Its main centre of origin and therefore diversity is in the Mediterranean and South-West Asia.

Uses

Important world-wide, *Lathyrus* species have a wide range of uses as minor crops including human consumption, fodder, forage, land stabilisation and ornamental cultivation. Of particular importance is *L. sativus* (Grass or Indian Pea), cultivated since Neolithic time from South Europe to India, through North Africa and Ethiopia for human consumption as a pulse. Today it is particularly important in India, Pakistan, Bangladesh, Java and Ethiopia. Herbaceous members of the genus are palatable to livestock. In Europe *L. sativus, L. latifolius, L. sylvestris* and *L. annuus* are grown as fodder crops, while *L. cicera* and *L. ochrus* are cultivated in West Asia. A full list of cultivated species is given in Smartt and Simmonds (1995). Some species (e.g. *L. sylvestris, L. cicera, L. hirsutus, L. ochrus* and *L. sativus)* are used for soil cover and stabilisation, erosion control, rehabilitation of cut-over or burned-over land and as green manure. The flowers of many *Lathyrus* species are attractive and cultivated as ornamentals (e.g. *L. odoratus, L. latifolius* and *L. maritimus*).

L. sativus is drought resistant and was a major pulse for human or animal consumption in India, Southern Europe and North Africa. In recent years cultivation has been drastically reduced, due to the presence of b-*N*-oxalyl-L-a,b diaminopropionic acid (ODAP), a neurological poison, which causes lathyrism (Smartt 1990). Recent interest in *L. sativus* has increased with the discovery of low or negligible ODAP lines (Campbell 1997).

Lens Mill.

Classification (from Davis and Plitmann 1970).

Family Leguminosae

Subfamily Papilionoideae

Tribe Vicieae

Genus *Lens* Miller

Taxonomic History

Lens is closely related to *Vicia* and *Lathyrus*. It has historically been grouped with a subset of species from these genera in the genus *Ervum* L.

Taxon Description

Lens Mill. Gard. Dict. Abridg., ed.4 (1754).

Type species: *L. culinaris* Medik.

Synonyms: *Ervum* L., Sp. Pl., ed.1, 738 (1753) pro parte.

Annual herbs. Height range 8-40cm. Sub-erect, sprawling or creeping. Surface pubescent to pilose or villous. Spines absent. Stems not winged. Leaves paripinnate with terminal tendril or mucro. 4-18 leaflets per leaf. Leaflet shape linear to filiform, ovate, cordate, elliptic, oblong, obovate or cuneate; margins entire or emarginate; lamina sub-glabrous to pilose; glands absent. Stipules large and conspicuous, foliaceous, free from petiole; stipels absent. Flowers solitary or arranged in racemes; axillary. 1-3 flowers per inflorescence. Pedunculate. Corolla white, purple or blue. Standard sub-orbicular or obovate, claw short and broad; wing obliquely obovate; keel pointed or beaked; wings longer than keel. Calyx campanulate, teeth subequal. Bracts small and/or inconspicuous or absent; bracteoles absent. Stamens monadelphous or diadelphous. Anthers monomorphic. Style filiform or incurved, pubescent on adaxial surface. Ovary sub-sessile, usually few ovules. Fruit oval to linear, strongly compressed;

surface smooth and glabrous or pubescent; joints absent. Pod length 8-22mm. Seeds globose or compressed; 1-2(-3) per fruit.

6 species (Ladizinsky 1986).

Taxonomic Affinities

Lens holds an intermediate position between *Lathyrus* and *Vicia*.

Useful Taxonomic Treatments

Ball, P.W. (1968). 'Flora Europaea. Vol. 2'. pp. 136 (Eds. Tutin, T.G., Heywood, V.H., Burges, N.A., Moore, D.M., Valentine, D.H., Walters, S.M., and Webb, D.A.). (Cambridge University Press: Cambridge.

Davis, P. H. and Plitmann, U. (1970). *Lens* Miller. 'Flora of Turkey and East Aegean Islands. Vol. 3'. (Ed. P.H. Davis). pp. 325-328 (Edinburgh University Press: Edinburgh).

Ladizinsky, G. (1986). A new *Lens* from the Middle East. *Notes from the Royal Botanic Garden, Edinburgh* 43, 489-492.

Townsend, C. C. (1974). 'Flora of Iraq. Vol.3'. pp. 544. (Ministry of Agriculture and Agrarian Reform of the Republic of Iraq: Bagdad).

Zohary, M., (1972). 'Flora Palaestina, Vol. 2'. pp. 210. (Israel Academy of Sciences and Humanities: Jerusalem, Israel).

Ecology

Lens species grow well on poor soils and loams. Typical habitats include rocky ground, steppe, scrub, coniferous woodland, fallow, meadow or pasture, and cultivated land. *L. culinaris* (the crop lentil) can grow in marginal environments but will not tolerate humid, tropical climates.

Distribution

Distributed within the Mediterranean Basin and western Asia, although there is significant differences in the distribution of individual species (Smartt 1990). For example, *L. orientalis* has an eastern distribution from Turkey to the Central Asian republics; *L. nigricans* is found largely on the northern shores of the Mediterranean and Algeria, Morocco and the Canary Islands; while *L. ervoides* has a similar but more restricted distribution with outlying populations in Ethiopia and Uganda.

Uses

Of the six species only *L. culinaris* is cultivated, it is grown from Spain and Morocco to India and is eaten as dhal, soups or as flour. The seeds of lentils are very nutritious (about 25% protein) and have been used for human consumption for thousands of years, being one of the first plants to be brought under cultivation (Allen and Allen 1981). Most of the crop is consumed in the major areas of production. *Lens* is a characteristic companion of wheat and barley in rotation. It is also a good forage plant.

Lotononis Eckyl. and Zeyh

Classification (from Polhill 1981).

Family Leguminosae

Subfamily Papilionoideae

Tribe Crotalarieae

Genus *Lotononis* Eckyl. and Zeyh

Taxonomic History

Sometimes included in the monogeneric tribe Lotononideae (Davis 1970) or the Genisteae (Zohary 1972).

Taxon Description

Lotononis Eckyl. and Zeyh Enum. 176 (1836).

Type species: *L. prostrata* L.

Synonyms: *Amphinomia* DC. (1825); *Leobordea* Delile (1830); *Aulacinthus* E. Mey. (1835); *Capnitis* E. Mey. (1835); *Lipozygis* E. Mey. (1835); *Telina* E. Mey. (1835); *Krebsia* Eckl. and Zeyh. (1836); *Leptis* E. Mey. (1836); *Polylobium* Eckl. and Zeyh. (1836); *Leptidium* C. Presl (1844); *Leobardia* Pomel (1874).

Perennial herbs or shrubs. Height range 15-30cm. Surface pubescent. Spines absent. Leaves digitate or trifoliate. 1, 3 or rarely 5 leaflets per leaf. Glands absent. Stipules small or rarely absent, foliaceous, free from petiole; stipels absent. Flowers solitary or arranged in racemes or umbels; sub-axillary or terminal. 1-6 flowers per inflorescence. Corolla yellow or pale red. Standard orbicular to oblong; wing obliquely ovate or oblong; keel beaked. Calyx teeth bilabiate. Stamens monadelphous. Anthers dimorphic. Stigma terminal. Style incurved, glabrous. Ovary sessile, numerous ovules. Fruit linear, rarely turgid, inflated or compressed; surface pubescent; joints absent; dehiscent; slightly exserted from calyx. Pod length 6-10mm.

About 100 species (Polhill 1981).

Useful Taxonomic Treatments

Chamberline, D.F. (1970). Lotononis Eckyl. and Zeyh. 'Flora of Turkey and East Aegean Islands. Vol. 3'. (Ed. P.H. Davis). pp. 36 (Edinburgh University Press: Edinburgh).

Polhill, R.M. (1981) 'Crotolarieae, Vol.1'. (Eds. R.M. Polhill and P.H. Raven). pp. 399-402 (Royal Botanic Gardens: Kew).

Zohary, M., (1972). 'Flora Palaestina, Vol. 2'. pp. 38-39. (Israel Academy of Sciences and Humanities: Jerusalem, Israel).

Ecology

Typical habitats include rocky ground and coniferous woodland.

Distribution

The centre of diversity is Southern Africa, with extensions to the Mediterranean, Africa, India and Australia.

Uses

Members of *Lotononis* have value in soil-improvement and erosion-control projects. *L. bainesii* was introduced into Australia in 1952 and has since spread to other Pacific areas, especially Taiwan (Allen and Allen 1981).

Lotus L.

Classification (from Rechinger 1984).

Family Leguminosae

Subfamily Papilionoideae

Tribe Loteae

Genus *Lotus* L.

Sections *Lotus, Erythrolotus, Ononidium, Lotea*

Taxonomic History

The delimitation of individual species remains controversial, with authors disagreeing over the specific or sub-specific rank of the various taxa. (Lock 1991).

Taxon Description

Lotus L. Sp. Pl. 773 (1753).

Type species: *L. corniculatus* L.

Synonyms: *Lotea* Medik. (1787); *Krokeria* Moench (1794); *Lotulus* Raf. (1818); *Andaca* Raf. (1836); *Pedrosia* Lowe (1856); *Heinekenia* Webb ex Benth. and Hook. f. (1865); *Lotos* St.Lag. (1880).

Annual or perennial, herbs or more rarely semi-shrubs. Height 5-100cm. Erect, ascending, decumbent or prostrate. Surface glabrous to pilose or villous. Spines absent. Leaves imparipinnate or trifoliate. Usually 3 or 5 leaflets per leaf. Leaflet shape lanceolate, ovate, cordate, oblong, obovate or cuneate; margins entire; lamina pubescent; glands absent. Lower leaflets stipuloid. Stipules small or absent, tubercule-like, fused with petiole. Flowers solitary or arranged in umbels; axillary. 1-5(-many) flowers per inflorescence. Pedunculate. Corolla white, pink, yellow or red. Standard sub-obovate, orbicular or ovate to acuminate, claw with thickened, infolded margins; wing obovate; keel beaked and possibly curved; standard often longer than wings and keel. Calyx campanulate, teeth subequal or bilabiate. Bracts conspicuous, foliaceous; bracteoles absent. Stamens diadelphous. Anthers monomorphic. Stigma terminal or lateral. Style incurved, glabrous. Ovary sessile or shortly stipitate, numerous ovules. Fruit arched to linear, turgid, inflated or rarely compressed; surface ridged or grooved to rugose; joints present or rarely absent; dehiscent; exserted from calyx. Pod length 3.5-70 mm. Seeds sub-globose to arcuate; 1-many per fruit.

About 100 species (Polhill 1981).

Taxonomic Affinities

The tribes Loteae and Coronilleae are extremely similar, although recent surveys of pollen suggest that separation into two tribes is appropriate (Polhill 1981). *Lotus* and *Coronilla* not only show an affinity to one another but also to the astragaloid part of Galegeae, especially in Europe.

Useful Taxonomic Treatments

Heyn, C. C. (1970). *Lotus* L. 'Flora of Turkey and East Aegean Islands. Vol. 3'. (Ed. Davis, P. H). pp. 518-520 (Edinburgh University Press: Edinburgh).

Mouterde, P. (1966). 'Nouvelle Flore du Liban et de la Syrie. Vol. 2'. Ppp. 305-306. (Librairie Orientale: Beirut, Lebanon).

Polhill, R.M. (1981) 'Loteae, Vol. 1'. (Eds. R.M. Polhill and P.H. Raven). pp. 371-374 (Royal Botanic Gardens: Kew).

Rechinger, K. H. (1984). 'Flora Iranica. Papilionaceae II'. pp. 328-329. (Akademische Druck-u. Verlagsanstalt: Graz).
Zohary, M., (1972). 'Flora Palaestina, Vol. 2'. pp. 88-89. (Israel Academy of Sciences and Humanities: Jerusalem, Israel).

Ecology

Species of *Lotus* are adapted to diverse habitats, ranging from sea-level to over 3000m, and from very wet soils to xerophytic desert conditions (Allen and Allen 1981). Typical habitats include rocky ground, steppe, macchie, coniferous woodland, fallow, meadow or pasture, cultivated land, marsh or damp areas, and seashore.

Distribution

Common in north temperate regions, the genus extends to North and East Africa, Australia (two endemics) and extra-tropical South America (one species). Its two main centres of origin are the Mediterranean (with a southern spread to the Sahara and western to Asia) and western North America. The three agriculturally important species are of European origin.

Uses

There are three important perennial pasture species of *Lotus*. *L. corniculatus* (Birdsfoot trefoil) is used for hay, silage and pasture forage as it is winter-hardy, drought resistant, long-lived and is tolerant of less fertile soils. It is grown mainly in the north-eastern and western USA, eastern Canada, much of Europe, southern South America, India, Australia and New Zealand. *L. pedunculatus*, syn: *L. uliginosus* or *L. major*, (Big trefoil) is a non-bloat promoting forage, which is limited to areas with milder winters and acidic soils. Its is important in north-eastern North America, southern South America, southern Europe, New Zealand and Australia, especially on peat soils which are not suitable for white clover. *L. tenuis* (Narrowleaf trefoil) is grown for pasture and hay. It is tolerant of dry soils and salt and is widely found in Europe. *L. gaber* is also grown for fodder and *L. tetragonolobus* is grown locally for human consumption (GRIN. 1997).

Lupinus L.

Classification (from Rechinger 1984).
Family Leguminosae
Subfamily Papilionoideae
Tribe Genisteae
Genus *Lupinus* L.

Taxonomic History

The tribal placement of the genus is controversial, Bentham (1865) linked it with numerous other genera in the tribe Genisteae, then Bisby (1981) retained the genus in the Genisteae, but separates it into the monogeneric subtribe Lupininae, however, Yakovlev *et al.* (1996) raises the subtribe to tribal rank. Yakovlev *et al.* (1996) also comment that the genus is very internally variable and warrants further infra-generic separation.

Taxon Description

Lupinus L. Sp. Pl. 721 (1753).

Type species: *L. albus* L.

Annual or perennial, herbs or more rarely semi-shrubs. Erect or ascending. Surface pilose or villous. Spines absent. Leaves imparipinnate, palmate. Usually 5-15 leaflets per leaf. Leaflet shape linear or oblanceolate; margins entire; lamina pubescent; glands absent. Stipules small, foliaceous, adnate to base of petiole. Flowers arranged in racemes or spike; terminal. Corolla white, purple, blue or yellow. Standard orbicular or ovate. Wing falcate, obovate or oblong; keel beaked and curved; standard often longer than wings and keel. Calyx campanulate, teeth bilabiate. Bracts often caducous; bracteoles persistent, usually adanate to the base of the calyx. Stamens monodelphous. Anthers dimorphic. Stigma terminal. Style incurved, glabrous. Ovary sessile, 2- to many ovuled. Fruit linear, turgid, more or less laterally compressed; surface usually with silky hairs; septa present; dehiscent; exserted from calyx. Seeds globose; 1-many per fruit.

About 200 species (Bisby 1981).

Taxonomic Affinities

Lupinus is isolated within the Papilionoideae, but has been historically linked to the *Cytisus-Genista* complex in the Genisteae., from which it is distinguished on leaflet number, arrangement and size, and also on more obscure wing and keel characters (Bisby 1981).

Useful Taxonomic Treatments

Bisby, F.A. (1981). Genisteae (Adans.) Benth. In Advances in Legume Systematics Vol. 1. (Eds. R.M. Polhill and P.H. Raven). pp. 409-425. (Royal Botanic Gardens: Kew).
Chamberline, D.F. (1970). *Lupinus* L. 'Flora of Turkey and East Aegean Islands. Vol. 3'. (Ed. Davis, P. H). pp. 38-40. (Edinburgh University Press: Edinburgh).
Mouterde, P. (1966). 'Nouvelle Flore du Liban et de la Syrie. Vol. 2'. (Librairie Orientale: Beirut, Lebanon).
Rechinger, K. H. (1984). 'Flora Iranica. Papilionaceae II'. pp. (Akademische Druck-u. Verlagsanstalt: Graz).
Zohary, M., (1972). 'Flora Palaestina, Vol. 2'. pp. (Israel Academy of Sciences and Humanities: Jerusalem, Israel).

Ecology

Species of *Lupinus* are adapted to diverse habitats, ranging from sea-level to over 3000m, and from very wet soils to xerophytic desert conditions, but they are most commonly found on dry sites, stony slopes, and calcareous and sandy-loam soils (Allen and Allen 1981).

Distribution

The genus has two centres of diversity in the Andes and Rockies mountain chains of South and North America, and secondly in the Mediterranean Basin, extending into the tropical highlands of Africa. Species are widely naturalised around the Mediterranean, in South Africa, New Zealand, Australia and south-eastern United States (Bisby 1981).

Uses

Species of the genus are used for ornament, fodder, forage, green manure, soil improvement and more rarely human food (Allan and Allan 1981) Fresh seeds are poisonous, but prolonged treatment by boiling or steeping in water removes the bitter alkaloids In peru lupine flour is added to bread, biscuits and noodles (Duke 1981). *L. alba* has been cultivated in the Mediterranean for over 3,000 years. *L. luteus* and *L. angustifolius* are also cultivated. The wild *L. angustifolius* is poisonous, especially to sheep, but, alkaloid-free (sweet) cultivars are available and provide a useful protein supplement in animal feed (Duke 1981).

Medicago L.

Classification (from Lesins and Lesins 1979).

Family Leguminosae

Subfamily Papilionoideae

Tribe Trifolieae

Genus *Medicago* L.

Subgenera *Medicago, Orbicularia, Lupularia, Spirocarpos*

Taxonomic History

Urban (1873) produced a monograph of *Medicago*, defining the infrageneric ranks and their main characteristics. The taxonomic study carried out by Lesins and Lesins (1979) delimits four subgenera, although other authors accept sectional rank (Heyn and Davis 1970). Taxonomic treatments vary considerably in the number of species recognised (Lock and Simpson 1991).

Taxon Description

Medicago L. Sp. Pl. 778 (1753).

Type species: *M. sativa* L.

Synonyms: *Medica* Adans. (1763); *Medicula* Medik. (1787); *Trifillium* Medik. (1787); *Radiata* Medik. (1789); *Triphyllum* Medik. (1789); *Cochleata* Medik. (1789); *Diplprion* Viv. (1824); *Lupulina* Noulet (1837); *Lupularia* Opiz (1852); *Spirocarpus* Opiz (1852).

Annual, biennial or perennial herbs or rarely shrubs. Height 5-70cm. Erect, decumbent or prostrate. Surface glabrous, pilose or villous. Spines absent. Leaves trifoliate. Leaflet shape oblong, obovate or cuneate; margins dentate, lobed or rarely entire; lamina glabrous or pilose; glands absent or rarely present. Stipules fused with petiole. Flowers sub-solitary or arranged in heads or racemes; axillary. Few-many flowers per inflorescence. Pedunculate. Corolla yellow or purple. Standard obovate or oblong, claw free from staminal tube; wing oblong; keel blunt; wings longer than keel. Calyx tubular or campanulate, teeth subequal. Bracts small and/or inconspicuous, scarious; bracteoles absent. Stamens diadelphous. Anthers monomorphic. Stigma terminal or lateral. Style subulate, glabrous. Ovary sessile or shortly stipitate, numerous ovules. Fruit coiled or arched (rarely linear); surface spiny, smooth or pubescent; indehiscent or rarely dehiscent; exserted from calyx. Seeds reniform. 1-20+ per fruit.

About 50 species (Heyn 1981).

Taxonomic Affinities

Medicago is closely related to *Trigonella* and *Melilotus*. *Medicago* species share numerous characters with *Trigonella*, from which they are difficult to distinguish.

Useful Taxonomic Treatments

Heyn, C.C. (1981) 'Trifolieae, Vol. 1'. (Eds. R.M. Polhill and P.H. Raven). pp. 383-385 (Royal Botanic Gardens: Kew).

Heyn, C.C. and Davis, P.H. (1970). Medicago L. 'Flora of Turkey and East Aegean Islands. Vol. 3'. (Ed. P.H. Davis). pp. 483-486 (Edinburgh University Press: Edinburgh).

Lesins, K. A. and Lesins, I. (1979). 'Genus *Medicago* (Leguminosae). A Taxonomic Study'. pp. 62-66. (Dr. W. Junk. bv Publishers: Hague).

Mouterde, P. (1966). 'Nouvelle Flore du Liban et de la Syrie. Vol. 2', pp. 252-255. (Librairie Orientale: Beirut, Lebanon).

Zohary, M., (1972). 'Flora Palaestina, Vol. 2'. pp. 137-139. (Israel Academy of Sciences and Humanities: Jerusalem, Israel).

Ecology

Species prefer deep, well-drained, light soils with a pH range of 6.5 - 7.5. They are well adapted to many climatic conditions and tend to be drought-resistant. Typical habitats include roadsides, cultivated fields and grasslands.

Distribution

Found in Europe (except the north), central and west Asia, and north Africa to Ethiopia. A few species in tropical and South Africa. The main centre of diversity is around the Mediterranean. It has been widely introduced outside its natural distribution.

Uses

M. sativa subsp. *sativa* (lucerne or alfalfa) is one of the most valuable forage crops of warm temperate and cool sub-tropical regions, and has been for thousands of years. This is due to its high yield and quality, wide soil and climatic adaptation and general flexibility. It also recovers rapidly from defoliation, is resistant to environmental stress, is long-lived and good at fixing nitrogen. The total area used for lucerne cultivation was approximately 32 Mha in 1995 (Smartt and Simmonds 1995). Alfalfa sprouts are of increasing importance in many countries for human consumption. It is also used to prevent soil erosion and its fibre has been used in paper production (Duke 1981). Several other *Medicago* species are also cultivated: *M. arabica* (Spotted or Southern Burclover), *M. falcata* (Sickle medic), *M. lupulina*, *M. orbicularis* (Button clover, Large disc-medic), *M. polymorpha* (California or Toothed burclover), *M. scutellata* (Snail medic) and *M. truncatula* (Barrel medic) are all grown as forage or fodder crops. *M. arborea* is sometimes grown as an ornamental.

Melilotus Mill.

Classification (from Smith and Gorz 1965).

Family Leguminosae

Subfamily Papilionoideae

Tribe Trifolieae

Genus *Melilotus* Mill.

Subgenera Eumelilotus, Micromelilotus

Taxonomic History

The genus has been divided into two subgenera, Eumelilotus (typically biennial species) and Micromelilotus (typically annuals) by Smith and Gorz (1965).

Taxon Description

Melilotus Mill. Gard. Dict. Abridg., ed.4 (1754).

Type species: *M. officinalis* Willd.

Synonyms: *Sertula* O. Ktze. (1891); *Melilota* Medik. (1787); *Melilothus* Hornem. (1819); *Meliotus* Steud. (1841); *Brachylobus* Dulac (1867).

Annual, biennial or perennial herbs. Height 10-250cm. Erect or ascending. Surface glabrous. Spines absent. Leaves trifoliate. Leaflet shape oblong, obovate or cuneate; margins dentate; lamina glabrous, pubescent or pilose; glands absent. Stipules fused with petiole. Flowers arranged in racemes; axillary. 3-100 flowers per inflorescence. Pedunculate. Corolla white or yellow. Standard obovate or oblong, claw may be adnate to staminal tube; wing oblong; keel blunt; wings longer than keel. Calyx campanulate, teeth subequal. Bracts small or absent; bracteoles absent. Stamens monadelphous or diadelphous. Anthers monomorphic. Stigma terminal. Style filiform and incurved above, glabrous. Ovary sessile or stipitate, few ovules. Fruit oval to circular or linear, sometimes compressed; surface smooth, ridged to rugose or spiny; indehiscent; exserted from calyx. Pod length 1.5-7mm. Seeds globose to ovoid, sometimes compressed; 1-many per fruit.

About 20 species (Heyn 1981).

Taxonomic Affinities

Closely related to *Medicago* and *Trigonella.*

Useful Taxonomic Treatments

Chamberlain, D. F. (1970). *Melilotus* Mill. 'Flora of Turkey and East Aegean Islands. Vol. 3'. (Ed. P.H. Davis). pp. 448-452 (Edinburgh University Press: Edinburgh).

Heyn, C.C. (1981) 'Trifolieae, Vol. 1'. pp. 383-385 (Eds. R.M. Polhill and P.H. Raven). (Royal Botanic Gardens: Kew).

Mouterde, P. (1966). 'Nouvelle Flore du Liban et de la Syrie. Vol. 2'. pp. 265. (Librairie Orientale: Beirut, Lebanon).

Polunin, O. (1969). 'Flowers of Europe. A field Guide'. pp. 201-202. (Oxford University Press: Oxford).

Rechinger, K. H. (1984). 'Flora Iranica. Papilionaceae II'. pp. 199-200. (Akademische Druck-u. Verlagsanstalt: Graz).

Ecology

Melilotus species prefer temperate to warm climates, and are found from sea level to 2000m. They are moderately winter-hardy and drought-resistant. Alkaline or neutral soils are preferred. Typical habitats include rocky ground, cultivated land, and the seashore.

Distribution

About 11 annual species are common to Mediterranean Europe, Asia Minor and north Africa, the remainder being perennial species of Europe and adjacent Asia. Some have been widely introduced either accidentally or as fodder crops.

Uses

Melilotus species are valued for forage and green manure. They are also used in the improvement of alkaline soils and erosion control, especially in areas of North America where *Melilotus* species are not productive. *M. indica* (Indian sweetclover, Senji, Sourclover) has been used in India for a long time as a forage. *M. officinalis* (Yellow sweetclover) is sometimes grown as a fodder crop, although ingestion of large amounts can lead to poisoning of cattle (due to the formation of dicoumarol from coumarin, a compound present in all species of *Melilotus*), and *M. alba* is grown as a fodder crop known as Bokhara clover (Fitter *et al.* 1985). Gene transfer programmes have been carried out to reduce the effects of this disease (Allen and Allen 1981). Species are also used in the production of commercial honey (*M. alba* and *M. officinalis*), as an insect repellent and as a flavouring in cheese (*M. altissima* and *M. officinalis*).

Onobrychis Mill.

Classification (from Rechinger 1984).

Family Leguminosae

Subfamily Papilionoideae

Tribe Hedysareae

Genus *Onobrychis* Mill.

Subgenus *Onobrychis*

Sections *Dendrobrychis, Lophobrychis, Onobrychis, Laxiflorae*

Subgenus *Sisyrosema*

Sections *Anthyllium, Afghanicae, Heliobrychis, Hymenobrychis, Insignes*

Taxon Description

Onobrychis Mill. Gard. Dict. ed.4 (abr.ed.) q (1754).

Type species: *O. viciifolia* Scop.

Synonyms: *Onobruchus* Medik. (1787); *Eubrychis* DC. (1825); *Dendrobrychis* DC. (1825); *Eriocarpaea* Bertol. (1843); *Sartoria* Boiss. (1849).

Annual or perennial herbs (rarely shrubs). Height 4-50(-150)cm. Erect, ascending, decumbent or prostrate. Surface rarely glabrous, often pilose or villous. Spines rarely present. Leaves imparipinnate. 4-28 leaflets per leaf. Leaflet shape linear, filiform, lanceolate, ovate, cordate, elliptic or oblong; margins entire; lamina glabrous to pilose; glands absent. Stipules small and/or inconspicuous, scarious, fused or rarely free from petiole; stipels absent. Flowers arranged in racemes or spikes; axillary. 3-20 flowers per inflorescence. Pedunculate. Corolla white, pink, purple or yellow (often with dark venation). Standard obovate or obcordate, claw short; keel oblique and blunt; wings much shorter than keel and standard, keel equal or shorter than standard. Calyx campanulate, teeth subequal to bilabiate. Bracts small and/or inconspicuous, scarious; bracteoles present or absent. Stamens diadelphous. Anthers monomorphic. Stigma terminal. Style filiform or linear, glabrous. Ovary sessile or shortly stipitate, few ovules. Fruit arched to circular, sometimes compressed; surface smooth and glabrous to rugose or spiny; joints absent; indehiscent; exserted from calyx. Pod length 2.5-20mm. Seeds reniform to oblong; 1-2(-3) per fruit.

About 130 species (Meikle 1977).

Taxonomic Affinities

Closely allied to *Hedysarum* (see *Hedysarum* for further details). *Onobrychis* is believed to be phylogenetically derived from *Hedysarum*, based on morphological, distributional and chemical information.

Useful Taxonomic Treatments

Hedge I. C. (1970). *Onobrychis* Adans. 'Flora of Turkey and East Aegean Islands. Vol. 3': 561-564. Ed. Davis, P. H. Edinburgh University Press, Edinburgh.

Meikle, R. D. (1977). 'The Flora of Cyprus, Vol. 1'. pp. 523-526. (Royal Botanic Gardens: Kew).

Mouterde, P. (1966). 'Nouvelle Flore du Liban et de la Syrie. Vol. 2'. pp. 387-388. (Librairie Orientale: Beirut, Lebanon).

Rechinger, K. H. (1984). 'Flora Iranica. Papilionaceae II'. pp. 389-391. (Akademische Druck-u. Verlagsanstalt: Graz).

Townsend, C. C. (1974). 'Flora of Iraq. Vol.3'. pp. 472-474. (Ministry of Agriculture and Agrarian Reform of the Republic of Iraq: Bagdad).

Ecology

Typical habitats include banks and grassy places (robust perennials), stony dry places, steppe, woodland, cultivated land, seashore and roadsides (slender annuals). The plants prefer light, well-drained calcareous soils in areas with low precipitation. They tend to be resistant to frost.

Distribution

This genus is well represented in Europe, the mediterranean region, western and central Asia and Ethiopia. Some species are geographically localised while others are more widely distributed.

Uses

Various members of this genus have been used as fodder, forage and soil erosion control, particularly in dry, calcareous areas (Allen and Allen 1981). *O. viciifolia* (Sainfoin) has been widely cultivated as a fodder plant since the fifteenth century in Europe and North America and is a nutritious, highly-palatable perennial forage and the whole genus is also important in honey production (Allen and Allen 1981)

Ononis L.

Classification (from Polhill and Raven 1981).

Family Leguminosae

Subfamily Papilionoideae

Tribe Trifolieae

Genus *Ononis* L.

Taxonomic History

Ononis is frequently separated from the Trifolieae and classified in the tribe Ononideae (Hutchinson 1964; Davis 1970; Zohary 1972 and Townsend 1974). Various characters suggest the need for this separation, e.g. the plants are often viscid, the leaves are sometimes reduced to a single leaflet, the androecium is monadelphic and dimorphic. Other authors (Bentham, 1865; Taubert, 1891; Meikle 1977 and Pohill and Raven 1981) believe that the evolutionary relationships, expressed by morphological characters, are better expressed by uniting *Ononis* with the Trifolieae. Bronn (1822) even went so far as to propose the inclusion of *Ononis* in the Genisteae. Species delimitations are also often still controversial (Lock and Simpson 1991).

Taxon Description

Ononis L. Sp. Pl. 716 (1753).

Type species: *O. spinosa* L.

Synonyms: *Anonis* (Tourn. ex) Adans. (1763); *Bonaga* Medik. (1787); *Natrix* Moench (1794); *Bugranopsis* Pomel (1874).

Annual or perennial, herbs or rarely shrubs. Height 3-100cm. Erect, ascending, decumbent or prostrate. Surface pilose or rarely glabrous. Spines present or absent. Leaves trifoliate or sometimes imparipinnate, (rarely unifoliate by reduction). Usually 3 leaflets per leaf, rarely 1-10. Leaflet shape lanceolate, elliptic, oblong; margins dentate; lamina pilose; glands present. Stipules large and conspicuous, scarious or foliaceous, fused or free from petiole (at base); stipels absent. Flowers solitary or arranged in racemes; axillary or terminal. 1-3 flowers per inflorescence. Sessile or pedunculate. Corolla white, pink, purple or yellow (may be part-coloured). Standard sub-orbicular or obovate, claw short; wing obovate to oblong; keel pointed to beaked; keel shorter than standard. Calyx tubular or campanulate, teeth subequal or bilabiate. Bracts small or absent; bracteoles absent or present (but minute). Stamens monadelphous. Anthers dimorphic. Stigma terminal or lateral. Style incurved, glabrous. Ovary shortly stipitate, few to numerous ovules. Fruit oval to arched to slightly linear, sometimes turgid or inflated; surface smooth to pubescent or rugose; joints absent; dehiscent; exserted or enclosed in calyx. Pod length 4-20 mm. Seeds reniform to globose; 1-25 per fruit. About 75 species (Yakovlev *et al.* 1996).

Useful Taxonomic Treatments

Heyn, C.C. (1981) 'Trifolieae, Vol. 1'. (Eds. R.M. Polhill and P.H. Raven). pp. 383-385 (Royal Botanic Gardens: Kew).

Huber-Morath, A. (1970). *Ononis* L. 'Flora of Turkey and East Aegean Islands. Vol. 3'. (Ed. P.H. Davis). pp. 373-375 (Edinburgh University Press: Edinburgh).

Mouterde, P. (1966). 'Nouvelle Flore du Liban et de la Syrie. Vol. 2'. pp. 233-234. (Librairie Orientale: Beirut, Lebanon).

Townsend, C. C. (1974). 'Flora of Iraq. Vol.3'. pp. 72-73. (Ministry of Agriculture and Agrarian Reform of the Republic of Iraq: Bagdad).

Yakovlev, G. P., Sytin, A. K. and Roskov, Yu. R. (1996). 'Legumes of Northern Eurasia. A Check-list'. (Royal Botanic Gardens: Kew'.

Zohary, M., (1972). 'Flora Palaestina, Vol. 2'. pp. 113-114. (Israel Academy of Sciences and Humanities: Jerusalem, Israel).

Ecology

Typical habitats include rocky ground, scrub, macchie, deciduous woodland, coniferous woodland, fallow, meadow or pasture, cultivated land, roadside, and seashore.

Distribution

The main centre of diversity for this genus is in the western Mediterranean. Its distribution extends to northern Europe, western Asia, north Africa and south to Ethiopia.

Uses

Various members have been used in medicinal preparations. The glucoside ononin from *O. spinosa* is a diuretic (Allen and Allen 1981). *O. fruticosa* is grown as an ornamental (GRIN 1997).

Ornithopus L.

Classification (from Chamberlain 1970).

Family Leguminosae

Subfamily Papilionoideae

Tribe Coronilleae

Genus *Ornithopus* L.

Taxon Description

Ornithopus L. Sp. Pl. 743 (1753).

Type species: *O. perpusillus* L.

Synonyms: *Ornothopodium* Adans. (1763); *Arthrolobium* Reichb. (1828); *Hormolotus* Oliv. (1886).

Annual herbs. Height 10-50cm. Decumbent or prostrate. Surface glabrous to pubescent. Spines absent. Leaves imparipinnate. 6-22 leaflets per leaf. Leaflet shape often narrowly oblong; margins entire; lamina pubescent; glands absent. Stipules small and/or inconspicuous, scarious, fused with petiole; stipels absent. Flowers arranged in heads or umbels; axillary (often subtended by floral leaf). 1-5 flowers per inflorescence. Long peduncle. Corolla white, pink or yellow. Standard sub-orbicular or obovate; wing oblong; keel almost blunt; wings longer than keel. Calyx tubular or campanulate, teeth subequal. Bracts small and/or inconspicuous, scarious; bracteoles present (but small). Stamens diadelphous. Anthers monomorphic. Stigma terminal. Style incurved, glabrous. Ovary sessile, numerous ovules. Fruit linear or arched, sub-terete or compressed; surface smooth and glabrous to ridged and slightly grooved; joints present; indehiscent; exserted from calyx. Pod length 12-40mm. Seeds sub-globose, ovoid or oblong; approximately seven per fruit.

6 species (Polhill 1981).

Useful Taxonomic Treatments

Chamberlain, D. F. (1970). *Ornithopus* L. 'Flora of Turkey and East Aegean Islands. Vol. 3'. (Ed. P.H. Davis). pp.

544-545 (Edinburgh University Press: Edinburgh).

Polhill, R.M. (1981) 'Coronilleae, Vol. 1'. (Eds. R.M. Polhill and P.H. Raven). pp. 375 (Royal Botanic Gardens: Kew).

Zohary, M., (1972). 'Flora Palaestina, Vol. 2'. pp. 99. (Israel Academy of Sciences and Humanities: Jerusalem, Israel).

Ecology

Members thrive on sandy loam and slightly acidic soils, especially in maritime environments. Typical habitats include fallow, meadow or pasture and seashore.

Distribution

The mediterranean is the centre of diversity with species having been introduced to other areas. Five species are found in Europe, the mediterranean, west Asia and the Atlantic Islands. One species is found in southern South America.

Uses

Species (notably *O. compressus* and *O. sativus*) are cultivated for grazing, soil cover, hay and silage. They are often cultivated in areas where white and red clovers are not favoured due to the acidic soils, and in Australia are particularly well adapted to the deep, acid sands in Western Australia (Loi *et al.* 1997). *O. microcanthus* is indigenous to Argentina and is used as a forage on a minor scale.

Pisum L.

Classification (from Davis 1970).

Family Leguminosae

Subfamily Papilionoideae

Tribe Vicieae

Genus *Pisum* L.

Taxonomic History

Historically the taxa included have been divided into a number of species, but currently only two species (*P. sativum* and *P. fulvum*) are accepted and the bulk of the diversity is contained within *P. sativum* at the subspecies or varietal level. There has been extensive debate over the infra-specific classification of *P. sativum*. The current view is that the taxa are divided into wild and cultivated subspecies, the wild subspecies (ssp. *elatius*) containing three distinct wild forms (vars. *elatius, pumilio, brevipedunculatum*) and the cultivated subspecies (ssp. *sativum*) containing the range of domesticated forms (vars. *sativum, arvense*) as suggested by Polhill and van der Maesen, (1985).

Taxon Description

Pisum L. Sp. Pl. 727 (1753).

Type species: *P. sativum* L.

Annual or rarely perennial herbs. Height 10-200cm. Erect, climbing, sprawling or creeping. Surface glabrous. Spines absent. Stem not winged. Leaves paripinnate with terminal tendril. 2-8 leaflets per leaf. Margins entire, dentate or lobed; leaflet lamina often glabrous; glands absent. Stipules large and conspicuous, foliaceous, free from petiole. Flowers solitary or arranged in racemes; axillary. 1-3 flowers per inflorescence. Pedunculate. Corolla white, pink or purple. Standard often oblong, claw short and broad; wing falcate to oblong; keel blunt; wings longer than keel. Calyx campanulate, teeth bilabiate. Bracts small and/or inconspicuous, setaceous; bracteoles absent. Stamens monadelphous or diadelphous. Anthers monomorphic. Stigma sub-terminal. Style incurved, pubescent on adaxial surface. Ovary sessile or shortly stipitate, few ovules. Fruit linear, compressed; surface smooth and glabrous. Pod length 30-70(-150)mm. Seeds sub-globose or oblong; 3-10 per fruit.

2 species (Polhill and van der Maesen 1985)

Useful Taxonomic Treatments

Davis, P. H. (1970). 'Flora of Turkey and East Aegean Islands. Vol. 3'. pp. 370-372 (Edinburgh University Press: Edinburgh).

Meikle, R. D. (1977). 'The Flora of Cyprus. Vol. 1'. pp. 578-579. (Bentham-Moxon Trust, Royal Botanic Gardens: Kew).

Polhill, R.M. and van der Maesen, L.J.G. (1985). Taxonomy of grain legumes. In 'Grain Legume Crops'. (Eds. R.J. Summerfield and E.H. Roberts). pp. 3-36 (Collins: London).

Zohary, M., (1972). 'Flora Palaestina, Vol. 2'. pp. 221-222. (Israel Academy of Sciences and Humanities: Jerusalem, Israel).

Ecology

Members prefer well-drained soils, reasonable fertility, with a pH between 5.5 and 7, and a temperature of 12-18_C (Allen and Allen 1981). Habitats include rocky ground, scrub, fallow, meadow or pasture, cultivated land, and roadside.

Distribution

Pisum species are native to the mediterranean region and western Asia. *P. sativum* is grown extensively in North America, the former USSR, India, China and northern Europe.

Uses

P. sativum has been cultivated since prehistoric times for its edible seeds which can be eaten green, dried or preserved. Increasingly markets are being found for green pods and foliage (Duke 1981). The plants often form a component of a cereal, legume, fallow rotation in the Mediterranean. It is the fourth most important seed legume and constitutes a vital source of protein (content of about 22%) for human consumption in many areas of the Near East, the Mediterranean, Ethiopia and north west India. Species are also extensively used for forage, ground cover, production of hay, silage and green manure.

Scorpiurus L.

Classification (from Heyn 1970).

Family Leguminosae

Subfamily Papilionoideae

Tribe Coronilleae

Genus *Scorpiurus* L.

Taxonomic History

Species delimitation within this genus has presented problems because of the continuous range of forms (Heyn and Raviv 1966).

Taxon Description

Scorpiurus L. Sp. Pl. 744 (1753).

Type species: *S. sulcatus* L.

Synonyms: *Scorpioides* Tourn. ex Adans. (1763); *Scorpius* Lois. (1806).

Annual herbs. Height 5-30cm. Decumbent, prostrate, sprawling or creeping. Surface glabrous to pilose or villous. Spines absent. Leaves simple. Leaflet shape elliptic to oblong; margins entire; lamina glabrous to pilose; glands absent. Stipules small and/or inconspicuous, may be scarious, partly fused with petiole; stipels absent. Flowers solitary or arranged in umbels; axillary. 1-5(-8) flowers per inflorescence. Pedunculate. Corolla yellow to red. Standard sub-orbicular or obovate, claw long; wing obliquely obovate or oblong; keel beaked to pointed. Calyx tubular, teeth slightly bilabiate. Bracts small and/or inconspicuous, scarious; bracteoles absent. Stamens diadelphous. Anthers monomorphic. Stigma terminal. Style incurved, glabrous. Ovary sessile, numerous ovules. Great variation in fruit form. Fruit coiled or arched, terete; surface ornamentally ridged or grooved, spiny or rugose; joints present (may be reduced); dehiscent or indehiscent. Seeds globose to ovoid; many per fruit.

2 - 4 species (Polhill 1981).

Useful Taxonomic Treatments

Heyn, C.C. (1970). *Scorpiurus* L. 'Flora of Turkey and East Aegean Islands. Vol. 3'. (Ed. P.H. Davis). pp. 547-548 (Edinburgh University Press:

Polhill, R.M. (1981). 'Coronilleae, Vol. 1'. (Eds. R.M. Polhill and P.H. Raven). pp. 375 (Royal Botanic Gardens: Kew).

Heyn, C.C. and Raviv, V. (1966). Experimental taxonomic studies in the genus *Scorpiurus* (Papilionaceae). *Bull. Torrey Bot. Club* 93(4), 259-267.

Mouterde, P. (1966). 'Nouvelle Flore du Liban et de la Syrie. Vol. 2'. pp. 378. (Librairie Orientale: Beirut, Lebanon).

Ecology

Typical habitats include rocky ground and fallow, meadow or pasture.

Distribution

Found in Europe, Mediterranean, west Asia and north Africa.

Uses

S. vermiculatus is cultivated in Argentina, where it is used as a salad supplement (Allen and Allen 1981). *S. murucatus* is also grown as a vegetable (GRIN 1997).

Securigera DC.

Classification (from Lassen 1989).
Family Leguminosae
Subfamily Papilionoideae
Tribe Coronilleae
Genus *Securigera* D.C.

Taxonomic History

The genus is often quoted as monotypic (*S. securidaca*). A new delimitation by Lassen (1989) includes former members of *Coronilla*, increasing the size of the genus. In the 'Flora URSS' the genus is included in the tribe Loteae.

Taxon Description

Securigera DC. Lam. and DC., Fl. Fr. ed.3,4:609 (1805).

Type species: *S. securidaca* L.

Synonyms: *Bonaveria* Scop. (1777); *Securina* Medik. (1787); *Securilla* Gaertn. ex Steud. (1821).

Annual herbs. Height 10-50cm. Sprawling or creeping. Surface glabrous. Leaves imparipinnate. 10-16 leaflets per leaf. Leaflet shape obovate or cuneate; margins entire; lamina glabrous; glands absent. Stipules small and/or inconspicuous, scarious, partly fused with petiole; stipels absent. Flowers arranged in umbels; axillary or terminal. 4-8 flowers per inflorescence. Pedunculate. Corolla yellow. Standard sub-orbicular; wing obliquely oblong; keel curved or beaked. Calyx shortly campanulate, teeth subequal. Bracts small and/or inconspicuous; bracteoles absent. Stamens diadelphous. Anthers monomorphic. Stigma terminal. Style incurved, glabrous. Ovary sessile, numerous ovules. Fruit linear, plano-compressed; surface ridged or grooved (with thickened margins); joints absent; indehiscent. Pod length 50-100mm. Seeds compressed; 2-6 per fruit. 12 - 15 species (Yakovlev *et al.* 1996).

Useful Taxonomic Treatments

Chamberlain, D.F. P. H. (1970). *Securigera* DC. 'Flora of Turkey and East Aegean Islands. Vol. 3'. (Ed. P.H. Davis). pp. 537-538 (Edinburgh University Press: Edinburgh).

Lassen, P. (1989). A new dekimitation of the genera *Coronilla, Hippocrepis* and *Securigera. Willdenowia* 19,49-62.

Yakovlev, G. P., Sytin, A. K. and Roskov, Yu. R. (1996). 'Legumes of Northern Eurasia. A Check-list'. (Royal Botanic Gardens: Kew'.

Taxonomic Affinities

This genus closely resemble *Coronilla,* especially *C. grandiflora,* but can be distinguished by the fruit which does not break into segments.

Ecology

Typically found in marsh and damp areas, deciduous woodlands, by roadsides and in cultivated fields.

Distribution

This species is endemic to the Mediterranean and west Asia.

Uses

The forage value of this genus is reduced by the presence of coronillin, a toxic glucoside, in the leaves and flowers. However, Duke (1981) reports *S. varia* (crownvetch) is easily digested and in food value compares to alfalfa. The creeping roots of this species make it important in the USA and Europe as a weed controller, for cover by roadsides and in soil erosion control. Parts of the plants have been used medicinally for diuretic purposes (Allen and Allen 1981).

Tetragonolobus Scop.

Classification (from Heyn 1970).
Family Leguminosae
Subfamily Papilionoideae
Tribe Loteae
Genus *Tetragonolobus* Scop.

Taxonomic History

This genus formerly constituted a section of *Lotus.*

Taxon Description

Tetragonolobus Scop. Fl. Carn. ed. 2,2:87 (1772).

Type species: *T. maritimus* (L.) Roth.

Annual or perennial herbs. Height 10-35cm. Ascending or decumbent. Surface pubescent to pilose or villous.

Spines absent. Leaves trifoliate. Leaflet shape obovate or cuneate; margins ciliate; lamina pubescent to pilose; glands absent. Lower leaflets stipuloid. Stipules large and conspicuous, foliaceous, fused with petiole. Flowers often solitary; axillary. 1(-2) flowers per inflorescence. Pedunculate. Corolla pink, yellow or red. Keel beaked; standard longer or equal to wings. Calyx subulate, teeth subequal or bilabiate. Bracts conspicuous, foliaceous; bracteoles absent. Stamens diadelphous. Anthers monomorphic. Stigma lateral. Style incurved, glabrous. Ovary stipitate. Fruit narrowly linear, slightly compressed (four-cornered in cross section); surface ridged or grooved (with winged or thickened sutures); joints present; dehiscent; exserted from calyx. Pod length 20-70mm. Seeds sub-globose; many per fruit. 7 species (Yakovlev *et al.* 1996).

Taxonomic Affinities

Similar to *Lotus*, this genus differs in having trifoliate leaves with leafy stipules adnate to the petiole, flowers solitary or paired with equal calyx teeth, legumes always many-seeded and septate with thickened or winged sutures.

Useful Taxonomic Treatments

Heyn, C.C. (1970). *Tetragonolobus* Scop. 'Flora of Turkey and East Aegean Islands. Vol.3'. (Ed. P.H. Davis). pp. 531-532 (Edinburgh University Press:, Edinburgh).

Yakovlev, G. P., Sytin, A. K. and Roskov, Yu. R. (1996). 'Legumes of Northern Eurasia. A Check-list'. (Royal Botanic Gardens: Kew'.

Zohary, M., (1972). 'Flora Palaestina, Vol. 2'. pp. 95. (Israel Academy of Sciences and Humanities: Jerusalem, Israel).

Ecology

Species are very successful in meadows and wet areas. Typical habitats also include macchie, fallow, pasture and cultivated land.

Distribution

Found in central and southern Europe, the Mediterranean region and west Asia.

Uses

T. purpureus (Winged pea) is sometimes grown as a culinary herb; the edible pods are picked young, cooked whole and served as a vegetable; the seeds can be roasted and used as a substitute for coffee (Duke 1981). *T. palaestinus* has shown promise as a green manure, winter cover crop and hay crop in Israel (Allen and Allen 1981).

Trifolium L.

Classification (from Zohary and Heller 1984).

Family Leguminosae

Subfamily Papilionoideae

Tribe Trifolieae

Genus *Trifolium* L.

Sections *Lotoidea, Paramesus, Mistyllus, Vesicaria, Chronosemium, Trifolium, Trichocephalum, Involucrarium*

Taxonomic History

The genus *Trifolium* was recognised very early on by naturalists and herbalists (e.g. Turner, 1548), although it was of a much broader genus, including the current genera *Lotus* and *Melilotus*. Even Linnaeus included *Melilotus* in one of his groups, *Trifolium-Melilotus*. Post Linnaeus there have been several monographs and revisions, for example Savi (1808), Boissier (1872); Lojacono (1883), Schischkin and Bobrov (1948); Hossain (1961); Gillet (1965). However, other than Zohary and Heller (1984) there are few revisions which deal with the entire genus. The genus has been divided into five units, four groups, nine genera and seven sections by Linnaeus (1753), Moench (1794), Presl (1832) and Koch (1835) respectively. Today most of these groups are accepted as sections, although the names have been changed.

The most widely accepted classification is that proposed by Zohary and Heller (1984). They recognise eight sections, in which section *Lotoidea* is the largest and most heterogeneous. Zohary and Heller divide the section into nine subsections and thirteen series, but warn that their classification is provisional as many of the species are not adequately known.

Taxon Description

Trifolium L. Sp. Pl. 764 (1753).

Type species: *T. pratense* L.

Synonyms: *Lupinaster* Adans. (1763); *Vesicaria* Crantz (1769); *Triphylloides* Ponted. ex Moench (1794); *Lagopus* Bernh. (1800); *Pentaphyllon* Pers. (1807); *Falcatula* Brot. (1816); *Dactyphyllon* Raf. (1818); *Dactiphyllum* Raf. (1819); *Vesicastrum* Ser. (1825); *Chrysaspis* Desv. (1827); *Lotophyllum* Reichb. (1827); *Amarenus* C. Presl (1830); *Amoria* C. Presl (1830); *Calycomporhum* C. Presl (1830); *Galearia* C. Presl (1830); *Micrantheum* C. Presl (1830); *Mistyllus* C. Presl (1830); *Mystyllus* C. Presl (1830); *Paramesus* C. Presl (1830); *Fragifera* Koch (1837); *Trichocephalum* Koch (1837); *Dactyphyllum* Endl. (1840); *Amooria* Walp. (1842); *Loxospermum* Hochst. (1846); *Glycirrhizum* Bertol. (1850); *Melilotea* Bertol. (1850); *Oliganthema* Bertol. (1850); *Microphyton* Fourr. (1868); *Platystylium* Willk. (1877); *Fissicaylx* Lojac. (1883); *Eleuterosemium* Gib. and Belli (1889); *Trigantheum* Gib. and Belli (1892); *Carpohypogaea* Gib. and Belli (1893); *Medusea* Gib. and Belli (1893).

Annual or perennial herbs. Height 3-80(-100)cm. Erect, ascending or prostrate. Surface glabrous (or nearly so). Spines absent. Leaves trifoliate or rarely digitate. Usually 3 leaflets per leaf, rarely 5-9. Leaflet shape linear, filiform to obovate or cuneate; margins entire or dentate; lamina glabrous; glands absent. Stipules large and conspicuous, foliaceous, fused or free from petiole. Flowers rarely solitary or arranged in heads, spikes or umbels; axillary or terminal. 1-8 flowers per inflorescence. Sessile or pedunculate. Corolla white, pink, purple, yellow or red (may be part-coloured). Standard ovate or oblong, claw short; keel blunt; wings longer than keel. Calyx tubular or campanulate, teeth subequal or bilabiate. Bracts conspicuous to absent, scarious or foliaceous; bracteoles absent. Stamens diadelphous. Anthers monomorphic. Stigma terminal or rarely lateral. Style filiform and adaxially incurved, glabrous. Ovary sessile or stipitate, few ovules. Fruit oval or arched to linear, sometimes compressed; joints usually absent; often indehiscent; enclosed in calyx (calyx may be inflated, corolla may persist). Seeds reniform, oblong or globose to ovoid; 1-2(-9) per fruit.

About 250 species (Zohary and Heller 1984).

Taxonomic Affinities

The genus *Trifolium* is closely related to the genera *Trigonella, Medicago* and *Melilotus*, although it is the most isolated genus within the tribe Trifolieae.

Useful Taxonomic Treatments

Rechinger, K. H. (1984). 'Flora Iranica. Papilionaceae II'. pp. 275-280. (Akademische Druck-u. Verlagsanstalt: Graz).

Zohary, M. (1970). *Trifolium* L. 'Flora of Turkey and East Aegean Islands. Vol. 3'. (Ed. P.H. Davis). pp. 384-391 (Edinburgh University Press: Edinburgh).

Zohary, M., (1972). 'Flora Palaestina, Vol. 2'. pp. 158-162. (Israel Academy of Sciences and Humanities: Jerusalem, Israel).

Zohary, M. and Heller, D. (1984). 'The Genus *Trifolium*'. (The Israel Academy of Sciences and Humanities: Jerusalem, Israel).

Ecology

Within this large genus are species adapted to most habitats, ranging from sub-Arctic to tropical. Each species tends to be restricted to specific habitats. At least two-thirds of the species are limited to natural habitats. The genus is found from sea level to over 4000m.

Distribution

The genus is distributed throughout Europe, north, west and central Asia, north Africa, tropical and South Africa, North and South America (mainly in montane areas). No species are native to Australia. This is the only genus within the Trifolieae which occurs outside the eastern hemisphere and has a secondary centre in the Americas. Many clovers have been introduced outside their range as fodder plants and have since become naturalised.

The main centres of *Trifolium* diversity are Eurasia, North America and Africa. The agriculturally important species come predominantly from the Eurasian centre, especially the Mediterranean. The table below shows the distribution of the genus (Zohary and Heller 1984).

Location	% of genus
Mediterranean	52
Euro-Siberian and Irano-Tauranian	7
Africa	15
America	26

Uses

Most clovers are important nitrogen-fixers, improving the quality of pasture in both natural and cultivated grassland in temperate areas. Within the genus 10% of species are important as animal food, and thirteen species are of agricultural importance (Smartt and Simmonds 1995). *T. pratense* (red clover) is an excellent forage, rich in proteins and minerals and has been known in Europe since the fourth century. *T. repens* (white clover) is also a good forage plant, usually mixed with grasses as it complements the seasonal variations in the growth of grasses. It is often encouraged in permanent pastures. *T. ambiguum* (honey clover), *T. incarnatum* (crimson clover), *T. subterraneum*, (subterranean clover) *T. hybridum* (alsike clover), *T. nigrescens* (ball clover), *T. resupinatum* (persian clover), *T. medium* (zigzag clover) *T. michelianum* (balansa clover) and *T. vesiculosum* (arrow-leaf clover) are cultivated as forage, fodder, green manure and soil erosion control species. *T. fragiferum* (strawberry clover) is important in lawns and pastures, particularly for large irrigated areas where drainage is the limiting factor. Some species are also used in honey production, e.g. *T. ambiguum, T. hybridum, T. incarnatum* and *T. dubium*. A number of species are also important in erosion control. T. *uniflorumm* is grown as an ornamental (GRIN 1997).

Trigonella L.

Classification (from Rechinger 1984).

Family Leguminosae

Subfamily Papilionoideae

Tribe Trifolieae

Genus *Trigonella* L.

Sections *Ellipticae, Lunatae, Falcatulae, Verae, Cylindricae, Bucerates, Reflexae, Callicerates, Unicinatae, Capitatae, Biebersteinianae, Gladiatae*

Taxon Description

Trigonella L. Sp. Pl. 776 (1753).

Type species: *T. foenum-graecum* L.

Synonyms: *Kentia* Adans. (1763); *Fenu-graecum* Adans. (1763); *Buceras* Hall. ex Allioni (1785); *Melissitus* Medik. (1787); *Trifoliastrum* Moench (1794); *Pocockia* Ser. (1825); *Botryolotus* Jaub. and Spach (1842); *Nephomedia* Kostel (1844); *Grammocarpus* Schur (1853); *Aporanthus* Bromf. (1856); *Folliculligera* Pasq. (1867); *Falcatula* Fourr. (1868).

Annual or rarely perennial herbs. Height 3-60cm. Erect to decumbent. Surface glabrous or pilose to villous. Spines absent. Leaves trifoliate. Leaflet shape obovate, cuneate, linear or filiform to lanceolate (rarely); margins entire, dentate or lobed; lamina glabrous and pilose; glands absent or rarely present. Stipules large and conspicuous, foliaceous, fused with petiole. Flowers solitary or arranged in heads, racemes or umbels; axillary. 1-20 flowers per inflorescence. Pedunculate. Corolla white, blue or yellow. Standard sub-orbicular, oblong or obovate, claw free from staminal tube; wing oblong; keel blunt; wings longer than keel. Calyx tubular or campanulate, teeth subequal or bilabiate. Bracts small or absent, scarious; bracteoles absent. Stamens diadelphous. Anthers monomorphic. Stigma terminal. Style sometimes linear or filiform, glabrous. Ovary sessile or shortly stipitate, numerous ovules. Fruit arched or linear, terete, compressed, turgid or rarely inflated; surface smooth or pubescent; joints absent; dehiscent or rarely indehiscent; exserted from calyx. Pod length 5-100mm. Seeds oblong or globose and sometimes compressed; 1-many per fruit. About 80 species (Heyn 1981).

Taxonomic Affinities

Melilotoides is closely related to *Trigonella*.

Useful Taxonomic Treatments

Huber-Morath, A. (1970). *Trigonella* L. 'Flora of Turkey and East Aegean Islands. Vol. 3'. (Ed. P.H. Davis). pp. 452-456 (Edinburgh University Press: Edinburgh).

Heyn, C.C. (1981) 'Trifolieae, Vol. 1'. (Eds. R.M. Polhill and P.H. Raven). pp. 383-385 (Royal Botanic Gardens: Kew).

Rechinger, K. H. (1984). 'Flora Iranica. Papilionaceae II'. pp. 389-391. (Akademische Druck-u. Verlagsanstalt: Graz).

Townsend, C. C. (1974). 'Flora of Iraq. Vol.3'. pp. 88-89. (Ministry of Agriculture and Agrarian Reform of the Republic of Iraq: Bagdad).

Zohary, M., (1972). 'Flora Palaestina, Vol. 2'. pp. 123-124. (Israel Academy of Sciences and Humanities: Jerusalem, Israel).

Ecology

Species tend to prefer loam soils which are well-drained and receive 50-150 cm of rainfall per annum. They are found on rocky ground, steppe, scrub, macchie, coniferous woodland, fallow, meadow or pasture, cultivated land, roadsides, and seashores.

Distribution

Native to the Mediterranean and distributed to central Asia, north Africa (and as isolated representatives in South Africa and Australia). The main centre of diversity is in the eastern Mediterranean.

Uses

The seeds of *T. foenum-graecum* (Fenugreek, Greek Clover or Greek Hay) are an edible condiment and were widely grown in ancient times throughout the Mediterranean and southern Asia. Today it is a minor pulse crop in the Mediterranean, south-west Asia, Ethiopia and northern India, cultivated for its spicy seeds. It is sometimes grown as a fodder crop, for a source of yellow dye, for food flavouring, and is used both medicinally and in cosmetics. All species have soil-binding and improvement properties. Several species are cultivated as a fodder crops (e.g. *T. hamosa, T. laciniata, T. arabica* and *T. suavissima*). *T. caerulea* is uesd as a food flavouring (GRIN 1997) and *T. corniulatus* is used as a vegetable (GRIN. 1997).

Vavilovia A. Fed.

Classification (from Davis 1970).

Family Leguminosae

Subfamily Papilionoideae

Tribe Vicieae

Genus V*avilovia* A.Fed.

Taxonomic History

Vavilovia was formerly classified as a member of *Pisum* (*P. formosum* and *P. aucheri*), but was separated by Federov (1939) into a mono-specific genus, mainly on the basis of its perennial life-form.

Taxon Description

Vavilovia A. Fed. Trud. Biol. Inst. Arm. Fil. A N SSSR 1:52 (1939).

Type species: *V. formosa* (Steven) A. Fed.

Synonyms: *Alophotropis* (Jaub. and Spach) Grossheim (1949); *Pisum formosum* (Steven) Alef. *Vicia aucheri* Boiss.

Perennial herbs. Height 5-15cm. Sprawling or creeping. Surface glabrous. Spines absent. Stem not winged. Leaves paripinnate, with terminal mucro. 2 leaflets per leaf. Leaflet shape obovate or cuneate to ovate or cordate; margins entire; lamina glabrous; glands absent. Stipules small and/or inconspicuous, foliaceous, free from petiole. Flowers usually solitary; axillary. Pedunculate. Corolla pink or purple (keel may be white). Standard often oblong, claw short and broad; wing falcate to oblong; keel blunt; wings longer than keel. Calyx campanulate, teeth subequal. Bracts small and/or inconspicuous; bracteoles absent. Stamens monadelphous or diadelphous. Anthers monomorphic. Stigma sub-terminal. Style linear, pubescent on adaxial surface. Fruit linear, compressed; surface smooth and glabrous; dehiscent. Pod length 20-35mm. Seeds globose; 3-5 per fruit.

One species, *V. formosa*, (Davis 1970).

Taxonomic Affinities

This monotypic genus is closely allied to *Pisum*.

Ecology

Typically found in high-mountain areas on shale or rocky ground.

Distribution

Found in Turkey, Lebanon, Iraq, Iran and Caucasia.

Uses

Vavilovia is of scientific interest as a source of a perennial gene that could be potentially be transferred into closely related *Pisum* material..

Vicia L.

Classification (from Kupicha 1976 and Maxted 1993).

Family Leguminosae

Subfamily Papilionoideae

Tribe Vicieae

Genus *Vicia* L.

Subgenus *Vicilla*

Sections *Vicilla, Cassubicea, Perditae, Cracca, Variegatae, Pedunculatae, Americanae, Subvillosae, Volutae, Panduratae, Ervum, Ervoides, Ervilia, Lentopsis, Trigonellopsis, Australes, Mediocinctae*

Subgenus *Vicia*

Sections *Atossa, Microcarinae, Hyperchusa, Peregrinae, Wiggersia, Vicia, Bithynicae, Narbonensis, Faba.*

Taxonomic History

The precise delimitation of *Vicia, Lens* and *Lathyrus* has varied considerably, with systems often recognising the genera *Ervum* L. (containing members of *Vicia* and *Lens*) and *Orobus* L. (containing members of *Lathyrus* and *Vicia*). The taxonomic history of the genus is extensive, with 20 major classifications of the subgenus *Vicia* having been produced since the work of Linnaeus (Maxted 1993). The genus tends to be divided into three or four groups (*Cracca, Vicia, Ervum* and Faba) from sectional to subgeneric rank. The infrageneric structure of *Vicia* was not organised on a world-wide basis until the work of Kupicha (1976). She recognised two subgenera - *Vicia* and *Vicilla* and twenty-two sections. Subgenus *Vicia* is the smaller taxon, contains less variation, but nearly all the economic species. This subgenus was revised by Maxted (1993) and the taxa were split into nine sections.

Taxon Description

Vicia L. Sp. Pl. 734 (1753).

Type species: *V. sativa* L.

Synonyms: *Ervum* L. (1753); *Faba* Adans. (1763); *Arachus* Medik. (1787); *Bona* Medik. (1787); *Cracca* iv. Medik. (1787); *Viciodes* Moench (1794); *Wiggersia* Gaertn. (1801); *Ervilia* Link (1822); *Orobella* C. Presel (1837); *Coppoleria* Todaro (1845); *Endiusa* Alef. (1859); *Parallosa* Alef. (1859); *Sellunia* Alef. (1859); *Swantia* Alef. (1859); *Hypechusa* Alef. (1860); *Abacosa* Alef. (1861); *Atossa* Alef. (1861); *Cujunia* Alef. (1861); *Tuamina* Alef. (1861); *Endusia* Benth. and Hook f. (1865); *Vicilla* Schur (1866); *Rhynchium* Dulac (1867).

Annual, biennial or perennial herbs. Height 5-150cm. Often climbing. Surface glabrous to pilose or villous. Spines absent. Stems not winged. Leaves paripinnate or occasionally imparipinnate, with terminal tendril, mucro or rarely leaflet. 2-many leaflets per leaf. Leaflet shape linear, filiform or lanceolate; margins entire or dentate; lamina pubescent; glands absent. Stipules large and conspicuous to small and/or inconspicuous, foliaceous, free from petiole; stipels absent. Flowers solitary arranged in racemes; axillary. 1-40 flowers per inflorescence. Sessile or pedunculate. Corolla white, yellow, purple or blue. Standard obovate or oblong, claw broad; wing obliquely obovate or oblong; keel blunt; wings longer than keel. Calyx tubular to campanulate, teeth subequal or bilabiate. Bracts small and/or inconspicuous (caducous); bracteoles absent. Stamens monadelphous or diadelphous. Anthers monomorphic. Stigma terminal. Style glabrous or pubescent complete or on adaxial surface (rarely). Ovary sub-sessile or shortly stipitate, numerous ovules or rarely few. Fruit linear, compressed; joints absent. Pod length 6-70(-200)mm. Seeds sub-globose or rarely compressed; 1-many per fruit.

About 140 species (Kupicha 1976).

Taxonomic Affinities

Vicia is closely related to *Lens* and *Lathyrus*. It is separated by having a pubescent style (rather than on the upper side only) and generally has more leaflets. Members do not have winged stems, a characteristic of many *Lathyrus* species.

Useful Taxonomic Treatments

Ball, P.W. (1968). 'Flora Europaea. Vol. 2'. (Eds. Tutin, T.G., Heywood, V.H., Burges, N.A., Moore, D.M., Valentine, D.H., Walters, S.M., and Webb, D.A.). pp. 129-136 (Cambridge University Press: Cambridge).

Chrtkova-Zertova, A., Van der Maesen, J. G. and Rechinger, K. H. (1979). 'Flora Iranica. Papilionaceae I - Vicieae'. pp. 17-21. (Akademische Druck-u. Verlagsanstalt: Graz).

Davis, P. H. and Plitmann, U. (1970). *Vicia* L. 'Flora of Turkey and East Aegean Islands. Vol. 3'. (Ed. Davis, P. H). pp. 274-279. (Edinburgh University Press: Edinburgh).

Kupicha, F.K. (1976). The Infragenic Structure of *Vicia*. *Notes Roy. Bot. Gard. Edinburgh*, 34, 287-326.

Maxted, N. (1993). A phenetic investigation of *Vicia* L. subgenus *Vicia* (Leguminosae, Vicieae). Bot. J. Linn. Soc. 111, 155-182.

Ecology

Species of *Vicia* thrive in many habitats and from sea level to over 3500m. They prefer well-drained, sandy soils and upland loams. Typical habitats include rocky ground, scrub, meadows or pasture, cultivated land, waste areas, roadsides, by rivers and in open woodland.

Distribution

Distributed chiefly in Europe and Asia, *Vicia* now also extends to temperate North and South America and tropical east Africa. The centre of diversity of this genus is in the Mediterranean and Irano-Tauranian regions

Uses

Vicia species are among the most important legumes grown in the world. *V. faba* (Broad or Faba Bean) has been cultivated since prehistoric times for human consumption and as a fodder. It is important in much of the north-temperate zone and at higher altitudes of some subtropical areas. China is the greatest producer (60% of the world's crop). FAO estimated that in 1981 annual world production was in excess of 4 million tonnes (Smartt 1990). It is useful because the plants are erect, have easily threshed pods and contain large seeds with a high protein content (20-25%). Selection under domestication has produced three sizes of faba bean, the smaller tending to be used in livestock foods and the larger for human consumption. Consumption of the bean can occasionally cause favism, a haemolytic anaemia. The main product is the seeds, but limited use can be made of immature pods and the tips of plants for human consumption.

Closely linked to the cultivation of *V. faba* is the evolution and domestication of *V. narbonensis* (Narbon vetch) as a grain and forage crop. *V. sativa* subsp. *sativa*, the Common vetch, is widely grown as a forage crop and as green manure in the Mediterranean. Other members of the subgenus *Vicia* have high potential as forages of the future: *V. hyaeniscyamus*, *V. noeana*, *V. pannonica* (Hungarian vetch) and *V. sativa* ssp. *amphicarpa*, ssp. *macrocarpa*, ssp. *sativa* (Maxted 1995). *V. villosa* (winter vetch) is particular interesting as a forage of adverse soil and climate conditions (Duke 1981). The bittervetch, *V. ervilia*, is grown as a minor crop in the Mediterranean basin and the Near East, especially in Turkey. Its seeds are bitter and toxic to humans and some animals, but the toxins can be removed by soaking. Its is used primarily as a winter green manure and the seeds can be used as animal feed.

APPENDIX 2. BRACKETED KEY TO TEMPERATE FORAGE LEGUME GENERA.

1(0). Leaves simple 2
Leaves compound 9

2(1). Stipules present 3
Stipules absent 8

3(2). Stipules conspicuous 4
Stipules small and inconspicuous 5

4(3). Fruit turgid or inflated .... *Astragalus*
Fruit laterally compressed .... *Coronilla*

5(3). Fruit curved or coiled; inflorescence axillary 6
Fruit straight; inflorescence terminal .... *Anthyllis*

6(5). Calyx tubular; calyx teeth bilabiate; fruit terete .... *Scorpiurus*
Calyx campanulate; calyx teeth subequal; fruit laterally compressed 7

7(6). Bracts scarious; ovules numerous; seed oblong .... *Coronilla*
Bracts foliaceous; ovules few; seed reniform .... *Hymenocarpus*

8(2). Inflorescence axillary; fruit dehiscent; fruit exserted from calyx .... *Cytisopsis*
Inflorescence terminal; fruit indehiscent; fruit enclosed in calyx .... *Anthyllis*

9(1). Leaves with terminal leaflet; style glabrous 10
Leaves usually with terminal tendril or mucro; style pubescent 67

10(9). Anthers monomorphic 11
Anthers dimorphic 66

11(10). Leaflet arrangement digitate 12
Leaflet arrangement trifoliate 14
Leaflet arrangement pinnate 37

12(11). Stipules present; fruit indehiscent 13
Stipules absent; fruit dehiscent .... *Cytisopsis*

13(12). Stamens monadelphous; stipules small and inconspicuous; stem or leaf surface pubescent; stipules scarious; fruit septa present .... *Hammatolobium*
Stamens diadelphous; stipules conspicuous; stem or leaf surface glabrous; stipules foliaceous; fruit septa absent .... *Trifolium*

14(11). Stigma terminal 15
Stigma lateral 34

15(14). Stamens monadelphous 16
Stamens diadelphous 18

16(15). Stem or leaf surface glabrous; margins of leaf or leaflet not entire .... *Melilotus*
Stem or leaf surface pubescent or pilose; margins of leaf or leaflet entire 17

17(16). Inflorescence axillary; standard orbicular .... *Hammatolobium*
Inflorescence terminal; standard ovate .... *Anthyllis*

18(15). Ovules few 19
Ovules numerous 26

19(18). Stem or leaf surface glabrous 20
Stem or leaf surface pubescent or pilose 22

20(19). Fruit turgid or inflated .... *Astragalus*
Fruit laterally compressed 21

21(20). Fruit exserted from calyx .... *Melilotus*
Fruit enclosed in calyx .... *Trifolium*

22(19). Stipules conspicuous 23
Stipules small and inconspicuous 24

23(22). Wing obovate .... *Cicer*
Wing oblong .... *Astragalus*

24(22). Margins of leaf or leaflet entire 25
Margins of leaf or leaflet not entire .... *Cicer*

25(24). Inflorescence axillary .... *Ebenus*
Inflorescence terminal .... *Anthyllis*

26(18). Inflorescence axillary 27
Inflorescence terminal 33

27(26). Style filiform 28
Style linear 29
Style subulate 30
Style incurved 31

28(27). Stipules fused with petiole; wing oblong .... *Trigonella*
Stipules free from petiole; wing obovate .... *Cicer*

29(27). Fruit septa present .... *Astragalus*
Fruit septa absent .... *Trigonella*

30(27). Leaves trifoliate; seed reniform .... *Medicago*
Leaves imparipinnate or upper trifoliate and lower simple; seed oblong .... *Coronilla*

31(27). Margins of leaf or leaflet entire 32
Margins of leaf or leaflet not entire .... *Cicer*

32(31). Bracts scarious .... *Coronilla*
Bracts foliaceous .... *Lotus*

33(26). Stipules conspicuous .... *Astragalus*
Stipules small and inconspicuous .... *Anthyllis*

34(14). Fruit exserted from calyx 35
Fruit enclosed in calyx .... *Trifolium*

35(34). Bracts scarious; style subulate .... *Medicago*
Bracts foliaceous; style incurved 36

36(35). Margins of leaf or leaflet entire; stipules small and inconspicuous; calyx .... *Lotus*
Margins of leaf or leaflet not entire; stipules conspicuous; calyx subulate .... *Tetragonolobus*

37(11). Stipules conspicuous 38
Stipules small and inconspicuous 44

38(37). Fruit turgid or inflated 39
Fruit laterally compressed 40

39(38). Wing obovate .... *Cicer*
Wing oblong .... *Astragalus*

40(38). Fruit septa present 41
Fruit septa absent 73

41(40). Bracteoles present .... *Hedysarum*
Bracteoles absent .... *Coronilla*

42(40). Stipules fused with petiole .... *Trifolium*
Stipules free from petiole 43

43(42). Stem winged .... *Lathyrus*
Stem without wings .... *Vicia*

44(37). Inflorescence axillary 45
Inflorescence terminal 64

45(44). Fruit turgid or inflated 46
Fruit terete 48
Fruit laterally compressed 52

46(45). Margins of leaf or leaflet entire 47
Margins of leaf or leaflet not entire .... *Cicer*

47(46). Keel blunt .... *Dorycnium*
Keel beaked or pointed .... *Lotus*

48(45). Bracts scarious 49
Bracts foliaceous 50

49(48). Bracteoles present .... *Ornithopus*
Bracteoles absent .... *Hippocrepis*

50(48). Stamens monadelphous; fruit indehiscent .... *Hammatolobium*
Stamens diadelphous; fruit dehiscent 51

51(50). Keel blunt .... *Dorycnium*
Keel beaked or pointed .... *Lotus*

52(45). Fruit septa present 53
Fruit septa absent 60

53(52). Bracts scarious 54
Bracts setaceous .... *Hedysarum*
Bracts foliaceous 57

54(53). Bracteoles present 55
Bracteoles absent 56

55(54). Style filiform .... *Hedysarum*
Style incurved .... *Ornithopus*

56(54). Seed oblong .... *Coronilla*
Seed arcuate .... *Hippocrepis*

57(53). Stamens monadelphous .... *Hammatolobium*
Stamens diadelphous 58

58(57). Bracteoles present; fruit enclosed in calyx; style filiform .... *Ebenus*
Bracteoles absent; fruit exserted from calyx; style incurved 59

59(58). Fruit dehiscent; ovules numerous; lower leaflets stipuloid .... *Lotus*
Fruit indehiscent; ovules few; lower leaflets not stipuloid .... *Hymenocarpus*

60(52). Fruit arched 61
Fruit circular .... *Onobrychis*
Fruit linear 62

61(60). Bracts scarious; fruit indehiscent; ovules few; calyx tubular .... *Onobrychis*
Bracts foliaceous; fruit dehiscent; ovules numerous; calyx campanulate .... *Lotus*

62(60). Keel blunt .... *Vicia*
Keel curved .... *Secuigera*
Keel beaked 63
Keel pointed .... *Lotus*

63(62). Fruit dehiscent; wing obovate .... *Lotus*
Fruit indehiscent; wing oblong .... *Securigera*

64(44). Fruit dehiscent .... *Dorycnium*
Fruit indehiscent 65

65(64). Stem or leaf surface glabrous; standard orbicular .... *Securigera*

Stem or leaf surface pubescent or pilose; standard ovate .... *Anthyllis*

66(10). Stipules conspicuous; glands on leaf lamina present; ovary stipitate .... *Ononis*

Stipules small and inconspicuous; glands on leaf lamina absent; ovary sessile 67

67(66). Leaves digitate or trifololiate; 1, 3 or rarely 5 leaflets per leaf. Stipules free petiole. Flowers solitary raceme or umbels .... *Lotononis*

Leaves palmate; 5-15 leaflets per leaf. Stipules adanate to base of petiole. Flowers raceme or spike .... *Lupinus*

68(9). Leaves glandular; style filiform, incurved .... *Cicer*

Leaves never glandular; style compressed with dilated apex 69

69(68). Stipules considerably bigger than the leaves; stems terete; leaves with terminal tendrils; style folded longitudinally .... *Pisum*

Stipules smaller than the leaves; stems angled or winged; leaves with or without terminal tendrils: style not folded longitudinally 70

70(69). Style pubescent all over or only on abaxial face (rarely); stem not winged .... *Vicia*

Style pubescent only on adaxial surface; stem may be winged 71

71(70). Dwarf perennials; leaflets 1-paired; style with reflexed margins .... *Vavilovia*

Annuals or perennials; leaflets 1-many paired; style without reflexed margins 72

72(71). Legumes usually more than 2-seeded; more than 15mm long; seeds more or less globose; annuals or perennials .... *Lathyrus*

Legumes 1-2 seeded; usually less than 15mm long; seeds compressed; annuals .... *Lens*

73(40). Leaves trifoliate; no terminal tendril or mucro .... *Trifolium*

Leaves imparipinnate or paripinnate; terminal tendril or mucro 74

74(73). Few leaflets; winged stem; style pubescent on upper side only .... *Lathyrus*

Many leaflets; stem not winged; style pubescent all over or on lower side only .... *Vicia*

APPENDIX 3. TABULAR KEY TO TEMPERATE FORAGE LEGUME GENERA.

Note. Character numbers and character state codes are explained in Appendix 1.

Astragalus	11a 19a 60a 47a
Coronilla	11a 19a 60a 47c
Scorpiurus	11a 19a 60b 25a 47b 35b 34a
Coronilla	11a 19a 60b 25a 47c 35a 34b 37a 61b 53c
Hymenocarpus	11a 19a 60b 25a 47c 35a 34b 37c 61a 53a
Anthyllis	11a 19a 60b 25b
Cytisopsis	11a 19b 25a 50a 51a
Anthyllis	11a 19b 25b 50b 51b
Hammatolobium	11b 43a 40a 62a 19a 50b 39a 60b 9b 20a 49a
Trifolium	11b 43a 40a 62a 19a 50b 39b 60a 9a 20b 49b
Lupinus	11b 43a 40a 62a 19a 50a
Cytisopsis	11b 43a 40a 62a 19b 50a
Melilotus	11b 43a 40a 62b 41a 39a 9a 16b
Hammatolobium	11b 43a 40a 62b 41a 39a 9b 16a 25a 28c
Anthyllis	11b 43a 40a 62b 41a 39a 9b 16a 25b 28d
Astragalus	11b 43a 40a 62b 41a 39b 61a 9a 47a
Melilotus	11b 43a 40a 62b 41a 39b 61a 9a 47c 51a
Trifolium	11b 43a 40a 62b 41a 39b 61a 9a 47c 51b
Cicer	11b 43a 40a 62b 41a 39b 61a 9b 60a 30b
Astragalus	11b 43a 40a 62b 41a 39b 61a 9b 60a 30d
Ebenus	11b 43a 40a 62b 41a 39b 61a 9b 60b 16a 25a
Anthyllis	11b 43a 40a 62b 41a 39b 61a 9b 60b 16a 25b
Cicer	11b 43a 40a 62b 41a 39b 61a 9b 60b 16b
Trigonella	11b 43a 40a 62b 41a 39b 61b 25a 42a 21a 30d
Cicer	11b 43a 40a 62b 41a 39b 61b 25a 42a 21b 30b
Astragalus	11b 43a 40a 62b 41a 39b 61b 25a 42b 49a
Trigonella	11b 43a 40a 62b 41a 39b 61b 25a 42b 49b
Medicago	11b 43a 40a 62b 41a 39b 61b 25a 42D 53a
Coronilla	11b 43a 40a 62b 41a 39b 61b 25a 42D 53c
Coronilla	11b 43a 40a 62b 41a 39b 61b 25a 42E 16a 37a
Lotus	11b 43a 40a 62b 41a 39b 61b 25a 42E 16a 37c
Cicer	11b 43a 40a 62b 41a 39b 61b 25a 42E 16b
Astragalus	11b 43a 40a 62b 41a 39b 61b 25b 60a
Anthyllis	11b 43a 40a 62b 41a 39b 61b 25b 60b
Medicago	11b 43a 40a 62b 41b 51a 37a 42d
Lotus	11b 43a 40a 62b 41b 51a 37c 42e 16a 60b 34b
Tetragonolobus	11b 43a 40a 62b 41b 51a 37c 42e 16b 60a 34c
Trifolium	11b 43a 40a 62b 41b 51b
Cicer	11b 43a 40a 62c 60a 47a 30b
Astragalus	11b 43a 40a 62c 60a 47a 30d
Hedysarum	11b 43a 40a 62c 60a 47c 49a 38a
Coronilla	11b 43a 40a 62c 60a 47c 49a 38b
Vicia	11b 43a 40a 62c 60a 47c 49b
Lathyrus	11b 43a 40a 62c 60a 47c 49b
Trifolium	11b 43a 40a 62c 60a 47c 49b
Dorycnium	11b 43a 40a 62c 60b 25a 47a 16a 31b

Lotus	11b 43a 40a 62c 60b 25a 47a 16a 31d
Lotus	11b 43a 40a 62c 60b 25a 47a 16a 31e
Cicer	11b 43a 40a 62c 60b 25a 47a 16b
Ornithopus	11b 43a 40a 62c 60b 25a 47b 37a 38a
Hippocrepis	11b 43a 40a 62c 60b 25a 47b 37a 38b
Hammatolobium	11b 43a 40a 62c 60b 25a 47b 37c 39a 50b
Dorycnium	11b 43a 40a 62c 60b 25a 47b 37c 39b 50a 31b
Lotus	11b 43a 40a 62c 60b 25a 47b 37c 39b 50a 31d
Lotus	11b 43a 40a 62c 60b 25a 47b 37c 39b 50a 31e
Hedysarum	11b 43a 40a 62c 60b 25a 47c 49a 37a 38a 42a
Ornithopus	11b 43a 40a 62c 60b 25a 47c 49a 37a 38a 42e
Coronilla	11b 43a 40a 62c 60b 25a 47c 49a 37a 38b 53c
Hippocrepis	11b 43a 40a 62c 60b 25a 47c 49a 37a 38b 53f
Hedysarum	11b 43a 40a 62c 60b 25a 47c 49a 37b
Hammatolobium	11b 43a 40a 62c 60b 25a 47c 49a 37c 39a
Ebenus	11b 43a 40a 62c 60b 25a 47c 49a 37c 39b 38a 51b
Lotus	11b 43a 40a 62c 60b 25a 47c 49a 37c 39b 38b 51a
Hymenocarpus	11b 43a 40a 62c 60b 25a 47c 49a 37c 39b 38b 51a
Onobrychis	11b 43a 40a 62c 60b 25a 47c 49b 46b 37a 50b 61a
Lotus	11b 43a 40a 62c 60b 25a 47c 49b 46b 37c 50a 61b
Onobrychis	11b 43a 40a 62c 60b 25a 47c 49b 46c
Vicia	11b 43a 40a 62c 60b 25a 47c 49b 46e 31b
Secuigera	11b 43a 40a 62c 60b 25a 47c 49b 46e 31c
Lotus	11b 43a 40a 62c 60b 25a 47c 49b 46e 31d 50a 30b
Securigera	11b 43a 40a 62c 60b 25a 47c 49b 46e 31d 50b 30d
Lotus	11b 43a 40a 62c 60b 25a 47c 49b 46e 31e
Dorycnium	11b 43a 40a 62c 60b 25b 50a
Securigera	11b 43a 40a 62c 60b 25b 50b 9a 28c
Anthyllis	11b 43a 40a 62c 60b 25b 50b 9b 28d
Ononis	11b 43a 40b 60a 18a 45b
Lotononis	11b 43a 40b 60b 18b 45a
Vicia	11b 43b 13a 44a
Lathyrus	11b 43b 13a 44b 58a
Lens	11b 43b 13a 44b 58b 35a 9b
Pisum	11b 43b 13a 44b 58b 35b 9a
Cicer	11b 43b 13a 44c
Vicia	11b 43b 13b 44a
Lathyrus	11b 43b 13b 44b 60a 58a
Lens	11b 43b 13b 44b 60a 58b
Vavilovia	11b 43b 13b 44b 60b
Cicer	11b 43b 13b 44c

APPENDIX 4. GENERIC CHARACTER SET.

1. citation.
2. type species.
3. synonyms.
4. tribe: a, Hedysareae; b, Loteae; c, Coronilleae; d, Vicieae; e, Cicereae; f, Trifolieae; g, Lotononideae; h, Galegeae.
5. life span: a, annual; b, biennial; c, perennial.
6. habit: a, plant a herb <without persistent stem and definite wood structure>; b, plant a shrub <low, woody>.
7. height range / cm.
8. growth habit: a, plant erect <without climbing>; b, plant climbing; c, plant ascending <indirectly upwards>; d, plant decumbent <lying on ground but with leaf ascending>; e, plant prostrate <lying flat>; f, plant sprawling or creeping.
9. stem or leaf surface: a, glabrous; b, pubescent or pilose.
10. spines <on stems or leaves>: a, present; b, absent.
11. leaves <type>: a, simple; b, compound.
12. leaf apex: a, paripinnate; b, imparipinnate.
13. leaves with terminal: a, tendril (more than 5mm); b, mucro (less than 5mm).
14. leaflet number per leaf (range).
15. leaflet shape: a, linear or filiform <long and narrow>; b, lanceolate <wide at base, tapering at apex>; c, ovate or cordate <egg or heart shaped>; d, elliptic <widest in middle>; e, oblong <broad with sides nearly parallel>; f, obovate or cuneate <with apex broader than base>.
16. margins of leaf or leaflet: a, entire; b, not entire.
17. leaflet lamina: a, glabrous; b, pubescent or pilose.
18. glands on leaf lamina: a, present; b, absent.
19. stipules: a, present; b, absent.
20. stipules <form>: a, scarious <membrane-like>; b, foliaceous; c, tubercule-like.
21. stipules <fusion to petiole>: a, fused with petiole; b, free from petiole.
22. stipels: a, present; b, absent.
23. flowers arrangement: a, solitary; b, inflorescent.
24. inflorescence arranged in: a, heads; b, racemes; c, spikes; d, umbels.
25. inflorescence position: a, axillary; b, terminal.
26. flower number per inflorescence (range).
27. corolla colour: a, white; b, pink; c, purple; d, blue; e, yellow; f, red.
28. standard shape: a, obovate <terminal half broader than basal>; b, obcordate <lobed and wider at apex>; c, orbicular <circular>; d, ovate <egg-shaped>; f, acuminate <sides concave, to point>; g, oblong <longer than broad, sides parallel>;
29. claw description.
30. wing shape: a, falcate <sickle-shaped>; b, obovate; c, ovate; d, oblong; e, obtuse <blunt>.
31. keel shape: a, oblique <slanting>; b, blunt; c, curved; d, beaked; e, pointed.
32. relative sizes of petals.
33. peduncle: a, present <inflorescence pedunculate>; b, absent <inflorescence sessile>.
34. calyx shape: a, tubular; b, campanulate; c, subulate.
35. calyx teeth: a, subequal; b, bilabiate.
36. bracts: a, present; b, absent.
37. bracts type: a, scarious <membrane-like>; b, setaceous <bristled>; c, foliaceous.
38. bracteoles: a, present; b, absent.
39. stamens: a, monadelphous; b, diadelphous.
40. anthers: a, monomorphic; b, dimorphic.
41. stigma: a, terminal; b, lateral.

42. style shape: a, filiform <thread-like>; b, linear; c, spatulate <spoon-shaped>; d, subulate <taper from base to apex>; f, incurved.
43. style hairiness: a, glabrous; b, pubescent.
44. style pubescence: a, complete; b, on adaxial surface; c, on abaxial surface.
45. ovary: a, sessile; b, stipitate.
46. fruit shape: a, oval; b, arched; c, circular; d, coiled; e, linear.
47. fruit transverse: a, turgid or inflated; b, terete <cylindrical>; c, laterally compressed.
48. fruit surface: a, smooth and glabrous; b, not smooth.
49. fruit septa: a, present; b, absent.
50. fruit dehiscence: a, dehiscent; b, indehiscent.
51. fruit exertion: a, exserted from calyx; b, enclosed in calyx.
52. pod length range / mm.
53. seed shape: a, reniform; b, compressed; c, oblong; d, globose; e, ovoid; f, arcuate.
54. seeds number per fruit (range).
55. flowering time.
56. altitude range / m asl.
57. typical habitats include: a, rocky ground; b, steppes; c, scrub; d, macchie; e, deciduous woodland; f, coniferous woodland; g, fallow, meadow or pasture; h, cultivated land; i, roadside; j, marsh or damp areas; k, seashore.
58. stem: a, winged; b, without wings.
59. lower leaflets: a, stipuloid; b, not stipuloid.
60. stipules: a, conspicuous; b, small and inconspicuous.
61. ovules <number>: a, few <up to 3>; b, numerous <more than 3>.
62. leaflet arrangement: a, digitate; b, trifoliate; c, pinnate.

APPENDIX 5. GLOSSARY (DEFINITIONS FROM HICKEY AND KING 1988 AND TOOTHILL 1984).

Abaxial	The side of an organ facing away from the main axis ie. the lower surface; dorsal
Acuminate	An acute apex with concave sides, tapering to a protracted point
Adaxial	The side of an organ towards the axis ie. the upper surface; ventral
Annual	A plant which lasts for one year from seed to maturity and death
Arched	Curved or bowed (as arcuate)
Arcuate	As arched
Ascending	Rising up obliquely or indirectly
Axillary	In an axil (the upper angle that a petiole or peduncle makes with the stem)
Beak	A long, prominent point
Biennial	A plant that takes two years to complete its life cycle
Bilabiate	Two-lipped
Blunt	Without a sharp edge or point
Bracteoles	A leaf-like organ subtending an inflorescence that is itself subtended by a bract
Bracts	A leaf-like organ subtending an inflorescence
Calyx	The collective term for the sepals, constituting the outer whorl of the perianth
Campanulate	Bell-shaped
Circular	Having form of a circle
Claw	The long narrow petiole-like base of petals or sepals
Climbing	Growing upwards with support by tendrils or twining with a surface
Coiled	Twisted into circular or spiral shape; arranged in concentric rings
Complete	Entire, having all parts
Compound(leaf)	A leaf of two or more leaflets
Compressed	Flattened, especially laterally
Corolla	Collective term for the petals, constituting the inner whorl of the perianth
Cordate	Heart-shaped with the lobe at the base
Cuneate	Wedge-shaped or triangular with the narrow end at the point of attachment
Curved	Surface with no part straight
Decumbent	Lying on the ground but with end ascending
Dehiscent	The splitting open of a plant organs such as anthers or pods
Dentate	Toothed; with sharp notches perpendicular to the margin
Diadelphous	An androecium comprised of two sets of stamens
Digitate	Hand-like. A compound leaf with all the leaves arising from one point
Dimorphic	Occurring in two forms
Elliptic	Oval in outline, widest in the middle
Enclosed	Surrounded, contained
Entire margin	With an edge completely continuous; not indented at all
Exserted	Sticking out; not included
Falcate	Sickle-shaped
Filament	The stalk of a stamen with the anther at the apex
Filiform	Thread-like
Foliaceous	Leaf-like
Glabrous	Not hairy
Gland	A secreting part or appendage
Glaucous	Surface with a bloom or whitish-grey substance that rubs-off
Globose	Rounded
Head	A short, dense spike
Herb	A plant without persistent stem or definite wood structure
Keel	The two lower fused petals of a papilionaceous flower which forms a boat-like structure

Imparipinnate	A terminal leaflet is present and the total number of leaflets is an odd number
Incurve	Bend into a curve, curve inwards
Inflorescence	A flowering system consisting of more than one flower; a cluster
Indehiscent	Not regularly opening, as seed pod or anther
Lamina	The bladelike portion of a leaf, usually flattened; a blade or expanded portion
Lanceolate	Longer than broad; widens above the base and tapers to the apex
Lateral	On or at the side
Linear	Long and narrow, the sides parallel or almost so
Lobed	Divided into curved or rounded parts
Monadelphous	Stamens united into one group by connation of their filaments
Monomorphic	Occurring in a single form
Mucro	A fine small point arising abruptly, usually at the tip
Obcordate	Deeply lobed at the apex
Oblique	Slanting; unequal-sided
Oblong	Broad with the sides nearly parallel
Obovate	The terminal half broader than the basal
Obtuse	Blunt, rounded
Orbicular	Circular or disc-shaped
Ovary	The swollen basal part of the carpel, containing the ovules
Oval	Egg-shaped or ellipsoidal
Ovate	Egg-shaped
Ovoid	Oval in flat outline
Ovules	The egg-containing unit of the ovary which becomes the seed after fertilisation
Paripinnate	No terminal leaflet present and the total number of leaflets is an even number
Peduncle	Stalk of a flower cluster
Perennial	A plant that lives for many years
Petiole	The stalk that attaches the leaf lamina to the stem
Pilose	Long soft hairs
Pinnate	A compound leaf in which the leaflets are arranged in two rows, either side of the midrib
Pitted	With little depressions or cavities
Pointed	With a sharp tip or end
Prostrate	Lying flat
Pubescent	Covered in short, soft hairs
Raceme	A elongated, indeterminate inflorescence in which the flowers are formed on individual pedicels on the main axis
Reniform	Kidney-shaped
Rugose	Wrinkled
Scarious	Dry, thin and membranous, often translucent
Septa	A partition or cross wall
Sessile	Not stalked
Setaceous	Bearing bristles
Setose	Covered in bristles
Shrub	A low, woody plant
Simple (leaves)	Not compounded into leaflets
Spatulate	Spoon-shape
Spike	An inflorescence in which the flowers are sessile and borne on an elongated axis
Spine	A modified leaf or part of a leaf forming a sharp pointed structure; a sharp woody body arising from the wood of the stem
Sprawling	Spreading out or straggling
Stamens	The unit of the androecium and typically composed of anther and filament
Standard	The upper, broad, often erect petal of a papilionaceous flower
Stigma	The receptive part of the carpel that receives the pollen
Stipels	Stipules of a leaflet

Stipitate	Born on a short stalk
Stipules	A pair of structures at the base of a leaf or along a petiole. They may be leaf-like, spines, glands or scales
Stipuloid	Resembling stipules
Style	The part of the pistil between the ovary and the stigma
Subequal	More or less the same
Subulate	Shaped like an awl ie. narrow, tapering from base to apex, more or less flattened
Tendril	A threadlike process or extension by which a plant grasps an object. It may be a modified stem or leaf
Terete	Circular in transverse section; imperfectly cylindrical
Terminal	At the tip, apical or distal end
Trifoliate	A compound leaf with three leaflets arising from the same point
Tubercule	A small tuber or rounded protruding body
Tubular	Shaped as a long hollow cylinder
Turgid	Swollen from fullness
Umbel	An inflorescence in which the flowers are borne on undivided pedicels originating from a common node on the main axis. The whole inflorescence is flat topped, resembling an umbrella
Villous	Long soft hairs, shaggy
Wing	The lateral petals of a papilionaceous flower / The membranous outgrowth of certain fruits /
A flange	running down a stem or stalk.

CHAPTER 18

Conservation and utilisation of agricultural diversity in Greece. Current status and perspectives

S. Samaras, N. Stavropoulos, A. Mattheou
N.AG.RE.F, Greek Gene Bank, 570 01 Thermi- Thessaloniki, Greece.

1. INTRODUCTION

The Greek Gene Bank (GGB) was established in 1981 by the Ministry of Agriculture at the Agricultural Research Center of Makedonia and Thraki (ARCMT) with a mandate to protect, evaluate, document and eventually utilize the plant genetic resources of the country.

Although systematic germplasm collections started in Greece early this century when most major crop breeding institutes of the country were established, those early germplasm collections have now been lost, primarily due to the limited scientific knowledge of effective long-term seed storage at that time, the lack of technical and financial support, and above all the belief that plant genetic resources were secure and would always remain intact and available to breeders.

Systematic efforts of genetic resource conservation in Greece resumed in 1981 with the establishment, by the Ministry of Agriculture, of the Greek Gene Bank in the Agricultural Research Centre of Makedonia and Thraki. This new establishment received significant support in its early years of operation by UNDP, FAO, IBPGR and ECP/GR, for the construction and provision of the necessary facilities and equipment, for the training of its scientific staff and for urgent germplasm rescue expeditions of priority crops.

The Greek Gene Bank (GGB) belongs today to the National Agricultural Research Foundation and operates as the scientific and administrative organ of the Ministry of Agriculture for the coordination and implementation of national policies in the sector of Plant Genetic Resources. To this end it collaborates closely with all crop plant institutes of NAGREF and has developed links with the major relevant University departments in the country and certain non-governmental organisations operating in this field or on related issues.

GGB has carried out a number of collecting missions over the last 15 years and today it conserves safely, in its active and base collections, orthodox seeds of approximately 7220 accessions belonging to 66 genera and 169 species of crop plants or economic plants. Clonal material (fruit trees and grapevines) and germplasm of aromatic and medicinal plants are maintained as field collections in respective institutes. Minor specific collections are maintained by a number of institutes, university faculties or other relevant organisations collaborating with the Ministry of Agriculture, NAGREF and the GGB in the framework of the National Program on Genetic Resources, introduced in 1990 by the Presidential Decree 80/1990 but not yet fully effective.

The GGB coordinates and implements national policies for plant genetic resources (*ex-situ* and *in-situ* on farm conservation and access to PGR through collections and germplasm availability or exchanges. It also regulates EU 1467/94 and 2078/92, matters related to the GPA, CBD and IU etc.) and collaborates closely with all plant institutes, agricultural departments of national universities, and various non-government conservation organizations to fulfill its mandate.

S. J. Bennett and P. S. Cocks (eds.), Genetic Resources of Mediterranean Pasture and Forage Legumes, 227–230.

2. COLLECTION AND *EX-SITU* CONSERVATION

The collection depicts the rich genetic diversity of the country. Major collections include *Beta, Brassica, Allium, Triticum, Hordeum, Aegilops, Nicotiana, Gossypium, Phaseolus, Cicer, Dactylis, Medicago, Trifolium, Zea, Prunus* and *Vitis*. During the early 1980's the collection priorities had to be in accordance with the FAO mandate. The collection is weak in wild and cultivated vegetables, medicinal and aromatic plants and ornamentals. These crops have been targeted for collection in the coming years. Complimentary collections are needed to fill the existing gaps in species representation and in areas of high genetic diversity.

The GGB collection contains 1895 accessions of forage crops and pastures. The greatest collections are of *Trifolium* (363), *Cicer* (177), *Dactylis* (150), *Medicago* (117), *Phaseolus* (387), *Vicia* (202), *Lens* (97) and *Lupinus* (80).

To date a first rough exploration of most of the country has been completed. The plans for the immediate future emphasise the completion of the existing gaps of the collection (vegetables, legumes etc.) and on detailed explorations and germplasm collections in areas already identified as containing high species and germplasm diversity.

3. STORAGE FACILITIES

Longterm storage facilities (base collection) at the GGB has a capacity of $40m^3$ and can hold approximately 10.000 samples. Seed is stored in sealed galvanised tin cans or in aluminum foil packages placed on fixed shelves at -18 to -21°C. Control of air humidity in the base collection is unnecessary due to the air-tight and moist-proof sealed packaging. Under these conditions seed viability of most orthodox species is maintained for more than 20 years.

Short to medium-term storage facilities (active or working collection) have a capacity of $40m^3$ and can also hold approximately 10.000 samples. Seed is stored in non-sealed packaging (paper or cloth bags), placed on fixed shelves at 0 to + 5°C and at 30 % relative air humidity. Under these conditions seed moisture is maintained at 5-6 % and in equilibrium with the humidity of the surrounding air, allowing reliable seed conservation for at least 10 years.

Grapevine clonal plant material is conserved in the field collections of the GGB and at the Grapevine Institute of Athens. A field collection of *Prunus* germplasm is maintained by the Pomology Institute in Naoussa. A rich working and breeders collection of forages and pulses exists in the Forage Crops and Pastures Institute in Larissa. Greece is rich in medicinal and aromatic plant species, and an excellent herb collection is maintained both in the field and in seed form by the Aromatic and Medicinal Plants Department at ARCMT.

Future goals of the GGB include development of a cryopreservation facility and a tissue culture unit for efficient clonal germplasm storage and pre-breeding.

4. *IN-SITU* AND ON FARM CONSERVATION

Endeavours to protect forage crops *in-situ* are undertaken by the Institute of Forage Crops and Pulses in Larissa. In addition, the GGB has identified through numerous explorations, surveys and collecting expeditions certain areas rich in indigenous wild relatives of crop plants. For example, wild progenitors of cultivated cereals, such as *Triticum boeoticum, Aegilops speltoides, Aegilops* spp., *Haynaldia* spp. are found in the counties of Mesti, Rodopi, Thermopyles, Fthiotis, Prespa, Florina and Kipourio, Grevena. The area of Mani in Lakonia county is exceptionally rich in medicinal and aromatic plants and cereals.

The GGB has identified areas where traditional agricultural systems still survive. These areas should be preserved through a system of field conservation or on-farm conservation with local, national and international support. This will ensure the conservation of endangered local landraces associated with the traditional agricultural practices within the rural landscape. Of the many places identified the main areas are mountainous central Greece and the Greek islands (the plateaus of Lassithi, Crete and Englouvi, Lefkas), the Aegean and Kalavryta area in Peloponnese. These areas are particularly rich in pulses and merit special protection, having exceptional aesthetic, natural or cultural value and therefore should receive first priority.

5. DOCUMENTATION

All germplasm is documented in a database using DBASE 4

Almost all material is currently documented only for passport data using the characters of the standard collection form of FAO / IBPGR. Only a limited number of accessions have undergone regeneration, characterisation and some preliminary evaluation, yielding data which will evantually be used to create species- specific characterisation and evaluation databases.

6. GERMPLASM ENHANCEMENT, EVALUATION AND OTHER ACTIVITIES

The GGB carries out germplasm regeneration, multiplication, characterisation, evaluation and pre-breeding on a limited number of wild and cultivated species. Efforts are concentrated on cereals, vegetables and pulses. Students use the facility for training in seed conservation techniques.

International links are maintained with a number of organizations involved in genetic resources conservation. GGB participates in many European programs and collaborations. Crop breeding institutes and University departments use the GGB for long-term conservation of their genetic stocks.

7. USE AND EXCHANGE OF THE COLLECTION

Germplasm collection and exchange of PGR is regulated by Presidential Decree 80/1990. According to its provision, permission to collect is joinly granted by the Directorates of the Agricultural Research and Environmental Protection of the Ministry of Agriculture, following approval by the Greek Gene Bank, the positive opinion of the appropriate major crop research institute of the country, and of the local authorities in the areas were the collection is to take place, in the case of endangered species.

The collecting team has to be accompanied by a collector of the Greek Gene Bank or the appropriate crop breeding institute, to facilitate the collection and the contacts with local people. Half of the collected samples are stored in the Greek Gene Bank. It has been informally introduced as a move towards benefit sharing that the cost of participation of the local collector is covered by the visiting organisation.

With regard to the general issue of access to PGR's in Greece, we expect through the implementation of the multilateral or bilateral agreements envisaged by the international undertaking currently under revision, to have a legal and globally respected system of germplasm availability either through direct collections or germplasm donations and exchanges.

8. LEGAL FRAMEWORK AND NATIONAL PROGRAM

Innovative legislation was proposed by Presidential Decree No 80/1990 providing a legal framework for the establishment of a national system for plant genetic resources. The GGB is the coordinating scientific and administrative body for this project. However funding must be made available to elaborate and implement the program. Other conservation bodies, such as

Universities, botanical gardens, natural history and agricultural museums, environmental protection groups and heritage seed networks could be linked through this framework and could contribute to the protection, study and utilization of the country's genetic resources. This could form the backbone of a Greek Plant Conservation Strategy.

9. PERSPECTIVES

Until recently the only active European body in this field was the European Cooperative Program on Plant Genetic Resources (ECP/GR) operating under FAO and IBPGR guidelines and funds, and EU Regulation 1467/94 covering primarily relevant research.

The major advance this year was the implementation of the Global Plan of Action and its beneficial implications to PGR matters. In order to participate actively in the GPA, all nations are obliged to introduce or reactivate their national PGR programs. The agreements for germplasm accessibility will create political interest, but also, and most importantly, better international scientific collaboration between participating nations or groups of nations.

CHAPTER 19

Conserving plant genetic resources in Cyprus

A. Della
Agricultural Research Institute, Nicosia, Cyprus

1. INTRODUCTION

The Mediterranean region is extremely important as a centre of origin and genetic diversity of many cultivated plants, the most important being the cereals and legumes. Other crop species such as beet, flax, olive, and some vegetables are also of importance in the region (Anon. 1979).

Genetic variability of many crops, and their wild relatives has been collected in Cyprus and utilised by local and foreign scientists. Vavilov (Schreiber 1934) visited Cyprus in 1927 and made rich collections from a number of areas. Cyprus ,medics have been widely collected and utilised (Crawford 1963; 1970), and Bennett (1973) reported a remarkable amount of genetic variation in durum wheat varieties of Cyprus. Hadjichristodoulou and Della (1976) showed that large amounts of adaptive genetic diversity has been conserved in commercial, field crop cultivars.

Cyprus in 1979, in co-operation with the International Board for Plant Genetic Resources (now the International Plant Genetic Resource Institute, IPGRI), launched its programme for collecting, conserving and utilising its plant genetic resources.

2. ECOGEOGRAPHY OF CYPRUS

Cyprus is the third largest island in the Mediterranean basin, with an area of 9,251 km^2. It is composed of four distinct topographical areas:

1. The Troodos range in the central western part of the island, with the highest point at Chionistra, Olympus (1.951m);
2. The Kyrenia or Pentadactylos range towards the extreme north of the island, a narrow ridge stretching from west to east, at an average of 8km from the north coast. The highest peak is 1,024m;
3. The Mesaoria or central plain, stretching between the Troodos and Kyrenia ranges, and from the Morphou Bay in the west to Famagusta Bay in the east; and,
4. The coasts, generally low and usually without slopes or sea cliffs, often rocky and stony, but with sandy bays in between. Many areas of the coastal belt are fertile and tilled almost to the edge of the sea.

The climate of Cyprus is an intense mediterranean one, with wet, changeable weather in winter from November to March, and hot, dry summers from May to September. These are separated by short spring and autumn seasons of rapidly changing weather.

The geographical attributes of Cyprus; its extremely irregular coastline, the great variation in temperature and rainfall, its location bridging three continents, along with 9,000 years of history and civilisation, and the fact that it suffered no glacial epoch, have resulted in a flora of great richness and diversity (Della 1987).

S. J. Bennett and P. S. Cocks (eds.), Genetic Resources of Mediterranean Pasture and Forage Legumes, 231–234.

3. INDIGENOUS PLANT GENETIC RESOURCES

The vegetation of Cyprus is composed of typical mediterranean types; coniferous forest, maquis, garique and batha vegetation, while more localised communities occur around the salt marshes, sand dunes, stone walls and mountain streams. A total of 1651 taxa have been recorded in Cyprus, both native and naturalised. 122 taxa have been recorded as endemics (Della 1998). Most of these occur on the igneous rocks of the Troodos mountains, or on the calcareous rocks of the Kyrenia range.

The most common conifer on the island is *Pinus brutia* Tenore (Calabrian pine) while *Pinus nigra* ssp. *pallasiana* (D. Don) Holmboe (Black pine) forms the forest to the top of Chionistra. The endemic *Cedrus brevifolia* (Hook f.) Henry (Cyprus cedar) dominates an isolated patch in the Paphos forest. The ecosystem under *Pinus nigra* is very important because it includes many endemic plants.

A number of native aromatic, medicinal and other useful plants are being exploited in their wild state, e.g. *Origanum* species., *Salvia fruticosa* Mill. (sage), *Thymus capitatus* (L.) Hoffsgg. Et Link (thyme), *Capparis spinosa* L. var. *canescens* Cosson, (wild caper), *Pistacia* species and *Laurus nobilis* L. (laurel). Leaves of *Rhus coriaria* L. (Sumach), an industrial plant rich in tanins, are collected every year and exported. The carob tree, *Ceretonia siliqua* L., a native of Cyprus, is also cultivated for its fruit.

Wild progenitors of ornamental species such as *Tulipa, Narcissus, Anemone, Cyclamen, Orchis, Ophrys* and *Fritillaria* species are present in Cyprus. Wild relatives of cereals, such as *Hordeum, Aegilops* and *Avena* species, leguminous forages such as *Vicia, Medicago, Trifolium* and *Lathyrus* species, and grasses occur in abundance in Cyprus.

4. LANDRACES AND OLD CULTIVARS

A considerable amount of variation is still present in fruit trees and nuts in Cyprus. Local varieties of loquats, citrus, hazelnuts, figs, pomegranate, walnuts, apricots, and plums are still grown. The local olive tree is the main variety grown. There is tremendous variation present in local almond germplasm. Local varieties of artichoke, water bottle, loofach and sweet pepper are still grown in Cyprus.

High yielding, uniform cereal varieties have been introduced in recent years, almost replacing completely the local germplasm. Relicts of the old durum wheat varieties are only found in semi-mountain vine areas, where the improved varieties are not competitive and mechanisation can not be practised. Barley local varieties have disappeared from farmer fields (Della 1995; 1996).

Local varieties of pulses are still the main varieties grown. However, improved chickpea, cowpea and common bean varieties as well as selected local faba beans were recommended for cultivation by the Agricultural Research Institute and have been released. The area under cultivation is increasing rapidly, subjecting local germplasm to genetic erosion or even extinction. Cultivation of lentil, chickpea and ochrus vetch is rare. The main reasons for this are the shortage and high labour costs and the lower prices of imported pulses. With regard to forage legumes, the cultivation of bittervetch, *Vicia ervilia* (L.) Willd. and of chickling vetch (*Lathyrus sativus* L.) is also rare and local germplasm has almost disappeared. *Vicia sativa* L. is still grown on a large scale.

5. *IN-SITU* CONSERVATION - NATIONAL LEGISLATION

The state forests are managed and protected by the Forestry department in accordance with existing forest law. Important ecosystems which were declared National Forest Parks or Nature Reserves, as well as very old trees are protected by law. Plants within the state forests are also reasonably protected. However, no direct measures have been taken yet for in-situ conservation of the wild relatives of crop species. There are no measures of on-farm conservation of local varieties either.

Cyprus has ratified the Washington convention on the international trade in endangered species of wild fauna and flora, the Bern convention on the conservation of European wildlife and natural habitats as well as the convention concerning the protection of world cultural and natural heritage and the convention on biodiversity.

Nineteen taxa, eighteen of which are endemics, proposed in the Bern convention in 1979 (reviewed in Strasburg in 1992) are also protected by law.

Legislation for the protection of nature was recently prepared by the Cyprus government in accordance with European Union regulations and directives. It includes inter alia legislation for the protection of the Cyprus Agricultural Research Institute (CYPARI) genebank and of the National herbarium. It also incorporated regulations for the national implementation of commitments undertaken by the ratification of the above conventions (Della 1996).

6. *EX-SITU* COLLECTIONS - CYPARI GENEBANK

Cyprus agreed in 1975 to co-operate with the International Board for Plant Genetic Resources (IBPGR) through the Agricultural Research Institute (ARI). Cyprus has participated in the European Networks (ECP/GR) since 1980, and in IPGRI, formerly IBPGR, and the West Asian and north African network (WANANET) since 1992. Cyprus endorsed the international undertaking on plant genetic resources, established by the United Nations and the Food and Agriculture Organisation (FAO), in 1983 and in 1985 became a member of the commission of plant genetic resources.

In the framework of co-operation with IPGRI, ARI organised a number of collecting missions. Germplasm collected is conserved at the Cyprus Agricultural Research Institute's (CYPARI) genebank, at 0-4°C in hermetically sealed laminated foil packets. The CYPARI genebank which contains mainly local material, was established in 1985. The collection, consists of approximately 12,000 samples of mainly cereals, food and forage legumes. Duplicate collections are kept at the Germplasm Institute, Bari, Italy, at the International Centre for Agricultural Research in Dry Areas (ICARDA), Syria, at the Greek National Genebank, Thessaloniki, and at Agriculture Western Australia (AgWA), Perth, Australia.

Field genebanks of olives, citrus, loquats, figs, pomegranates, hazelnuts, walnuts, and pistachios have been established. These collections are at stations belonging to the Ministry of Agriculture, Natural Resources and Environment. There is no botanical garden yet in Cyprus. However there are some botanical collections at the Forestry Department and the Department of Agriculture (Della 1996).

7. NATIONAL HERBARIUM

A herbarium was initiated in 1966 at the Agricultural Research Institute. It's aim was to serve as a reference collection for the identification of unknown plants. The herbarium has been used to provide information on weeds, fodder and host plants, wild relatives of crops, endemic and rare plants, and plants which cause an allergic reaction.

Duplicates of the ARI herbarium were sent to the herbarium at the Royal Botanic Gardens, Kew, where unknown specimens were identified. A great number of them are cited in the Flora of Cyprus (Meikle 1985; 1997) . Among them several plants were recorded for the first time, while others contributed to the study of the distribution of the taxa of Cyprus.

In 1990 the ARI herbarium was established as the National Herbarium in which all historical collections will also be maintained. Already a number of such collections have been donated to the ARI herbarium.

It is estimated that more than 10,000 named specimens exist in the ARI herbarium, while a large number of specimens collected during recent years have still to be identified and named. Special emphasis has been placed on the identification of endemic and rare plants.

8. CONCLUSION

Following ratification of the conventions signed by the Cyprus government, it is important to prepare legislation according to the European directives detailing implementation of the commitments that have been undertaken. Without this future protection of natural resources in Cyprus can not be guaranteed. There are areas, at present, where protective measures are reasonably effective, such as the State forests and in particular the National forest parks and nature reserves. Individual plants within the State forests are also reasonably protected. However, the protection of taxa whose habitat is outside these areas, even those which are protected by law, the current measures are inadequate.

The selection of representative areas of different ecosystems and their protection is urgently required to prevent further extinction of taxa. Necessary action must also be taken to conserve selected sites for *in situ* conservation of endangered taxa. Genetic resources of cereals and legumes, as well as their wild relatives, are well represented and conserved in the CYPARI genebank. *Ex situ* conservation of vegetable species and fruit trees, however, are under-represented and are a high priority for collection.

REFERENCES

Anon. (1973). IBPGR Mediterranean programme. *Plant Genetic Resources Newsletter* **38**, 23.

Bennett, E. (1973). Wheats of the Mediterranean basin. In 'Survey of crop genetic resources in their centres of diversity'. (Ed. O. H. Frankel) pp.1-8. (FAO: Rome).

Crawford, E. J. (1963). Early Cyprus barrel medicDepartment of Agriculture, South Australia. Leaflet No. 3725.

Crawford, E. J. (1970). Variability in a large Mediterranean collection of introduced lines of *Medicago* truncatula Gaertn. In 'Proceedings of the XI International Grassland Congress.' pp. 188-192.

Della, A. (1987). The plant world of Cyprus. *Cyprus To-day* **25**, 2-13.

Della, A. (1995). History and geography combine to make Cyprus a floristic wonder. *Diversity* **2**, 92-94.

Della, A. (1996). Cporus country report, ICPPGR, for the state of the world's plant genetic resources for food and agriculture. FAO International Technical Conference on Plant Genetic Resources. June 17-32. (Leipzip: Germany).

Della, A. (1998). The Cyprus flora in checklist format: native or naturalised, clutivated, endemics, rarities, additions. (IPGRI through WANA: Nicosia, Cyprus).

Hadjichristodoulou, A and Della, A. (1976). Genetic diversity in Cyprus. *Plant Genetic Resources Newsletter* **32**, 8-15.

Meikle, R.D. (1977). 'Flora of Cyprus. Vol. 1'. (Royal Botanic Gardens: Kew)

Meikle, R.D. (1985). 'Flora of Cyprus. Vol. 2'. (Royal Botanic Gardens: Kew)

CHAPTER 20

Observations on the distribution and ecology of annual *Medicago* species in northern Tunisia

Aziza Zoghlami and H. Hassen
Institut National de la Recherche Agronomique de Tunisie (INRAT), 2080-Ariana, Tunisia

Abstract. An ecogeographical survey was conducted on the distribution of forage and pasture legumes, in three geographical areas in Tunisia, in 1994.

275 accessions from 17 species of *Medicago* were collected, but only 8 of them were used in this preliminary study. *M. polymorpha* was the most widespread and was found at 81% of the sites. *M. rugosa, M. minima, M. italica* and *M. rigidula* were found to be rare.

Statistical analysis showed that the presence of the genus was not affected by climate. Soil factors such as electrical conductivity, and nitrogen and phosphorus content were more important. *M. orbicularis* and *M. intertexta* revealed the same requirements for low levels of salt and nitrogen. *M. truncatula* showed a preference for relatively low rainfall areas and high pH sites. *M. aculeata* occurred particularly at high rainfall areas and *M. murex* at low pH sites.

Key-words: Ecology, distribution, *Medicago*, northern Tunisia.

1. INTRODUCTION

In 1994, a collection mission was conducted in northern and coastal Tunisia. The aim of the mission was to complete the sampling of genetic variability of natural populations of forage and pasture legumes in Tunisia, and to determine the ecogeographical distribution of *Medicago* species in relation to soil and climatic conditions.

2. METHOD OF COLLECTION

The collection mission was conducted in June 1994 (5th-20th), jointly with the International Centre for Agricultural Research in the Dry Areas (ICARDA) and the Centre for Legumes in Mediterranean Agriculture (CLIMA). It covered the north of the line Kef-Sousse, including the northern and eastern coast where the rainfall varies from 200 to 1500 mm.

During the mission, 902 accessions from 16 genera and 71 species of pasture and forage legumes were collected from 91 sites. These were distributed over 3 bioclimatological areas. At each site, data such as slope, aspect, altitude and associated species were collected. Soil samples were also taken at a depth of 30 cm, and shipped to ICARDA for chemical and physical analysis. Meteorological data including rainfall and temperature (minimum and maximum) were collected from both the Meteorological National Institute and the Bioclimatological Laboratory of the Institut National de la Recherche Agronomique de Tunisie (INRAT).

The data were analysed to determine the effect of ecogeographical factors on the species distribution (Tables 1 and 2) and to extract the most discriminating variables in relation to the presence of species using a factor analysis.

S. J. Bennett and P. S. Cocks (eds.), Genetic Resources of Mediterranean Pasture and Forage Legumes, 235–238.

Table 1. Mean climatic variables for sites where Medicago species were recorded

	Fr(%)	Altitude	Rainfall	TMin	TMax
Genus		166,4	686,9	6,3*	32,1
M. polymorpha	81	142,3	609,4**		32,5*
M. ciliaris	53	161,2	794,9	6,1*	31,8
M. orbicularis	32	239,4*	858,0+	6,1+	32,4
M. murex	21	212,9	943,6*	5,8*	32,4
M. aculeata	20	192,7	558,0***	6,7	31,6
M. truncatula	18	136,4	765,9	6,2	31,9
M. intertexta	15	135,6	565,8*	6,8	31,4
M. littoralis	10	142,9	695,7	6,8	32,0
Mean		(166,2)	(283,6)	(3,9)	(1,8)

+: significant at 10%, *: significant at 5%, **: significant at 1%, ***: significant at 0,01%.

Table 2. Means of soil variables for sites where Medicago species were collected

	pH	EC	CaCO3	N	K	P	OM
Genus	NS	**	NS	**	NS	*	NS
M. polymorpha	7*	0,5	16,6	1342*	0,6+	13,5	2,5
M. ciliaris	6,9	0,8	20,2	1532,4	0,8	13,3	2,8
M. orbicularis	6,8+	0,3***	14,5	1156,5**	0,6	17,3	2,3
M. murex	6,4*	1,1	17	1728	0,6	12,1	3,3
M. aculeata	7,7	0,6	17,7	1477	0,8	25,7	2,7
M. truncatula	7,9*	1,0	30,6*	1487,5	0,7	9,2*	2,6
M. intertexta	7	0,2***	11,1	964,2*	0,4**	4,8***	4,5
Mean	7,7	0,7	17,9	1536,2	0,2	13,0	2,8

+: significant at 10%, *: significant at 5%, **: significant at 1%, ***: significant at 0,01%.

3. FREQUENCY OF SPECIES

17 species of *Medicago* have been collected, but only eight species had adequate and significant frequency distributions to be statistically analysed.

M. polymorpha was collected at 81% of the sites, and was the most widespread species. Less frequent, but still common were *M. ciliaris* (53%) and *M. orbicularis* (32%). *M. murex, M. aculeata, M. truncatula, M. intertexta* and *M. littoralis* occurred at only 10-20% of sites. *M. rugosa, M. minima, M. italica* and *M. rigidula* were rare (frequency <10%).

In 1992, a collection was made in the Centre of Tunisia, where rainfall is between 85 and 490 mm per year. Hassen *et al.* (1994) recorded that *M littoralis, M. truncatula, M. minima* and *M. laciniata* were dominant in that area (freq > 50%). The occurrence of *M. polymorpha* and *M. truncatula,* observed in Tunisia is similar to that reported in Algeria (Abdelguerfi *et al.* 1988) and in Morocco (Bounejemate *et al.* 1992). It was noticed that *M. ciliaris* is more common in Tunisia.

M. aculeata, M. intertexta and *M. truncatula* were each found to prefer a particular niche. The two first species were restricted to humid and sub-humid areas, whereas *M. truncatula* is limited to semi-arid ones. The rest of species are common to the three areas covered by the collection.

4. ECOLOGY AND ENVIRONMENTAL ADAPTATION OF *MEDICAGO* SPECIES

4.1 Influence of climatic factors

The distribution of most of the collected species of *Medicago* were affected by one or more climatic factors. For example the distribution of *M. polymorpha* and *M. orbicularis* are influenced by low minimum temperatures, and *M. truncatula* and *M. littoralis* are influenced by low rainfall (Table 1). Altitude and maximum temperature had a strong influence on the occurrence of *M. murex* and *M. orbicularis*, *M. murex* preferring medium altitude sites (with a mean altitude of 240m compared to 143m, the mean altitude of all the sites), and *M. orbicularis* preferring a relatively high maximum temperature (32.5°C).

4.2 Influence of soil factors

The soil factors appeared to have a greater affect on the occurrence of the *Medicago* species than the climatic factors (Table 2). Electrical conductivity, nitrogen and phosphorus content all had an important influence on the distribution of the genus. pH also seemed to affect the distribution of all the medic species except for *M. aculeata, M. intertexta* and *M. littoralis* which were indeterminate. The organic matter and the calcium carbonate did not have an influence on the distribution of the *Medicago* species. *M. orbicularis* and *M. littoralis* showed a preference for soils with low salinity. *M. truncatula* preferred soils with a high calcium carbonate content, high pH and a relatively low phosphorus content. This is similar to the requirements of *M. truncatula* reported by Prosperi *et al.* (1989) in Corsica and by Zoghlami *et al.* (1996) in central Tunisia. *M. murex* showed a specific requirement for low pH sites. This is thought to be related to the exceptional nodulation ability of this species where rhizobial numbers are low (Honieson and Ewing 1986).

5. DISCUSSION

The comparison of means provides useful information on the effect of each of the environmental variables on the distribution of the 8 species.

In north Tunisia, climatic factors were found to have no effect on the distribution *Medicago*, compared to the soil factors. However a few species had specific requirements, such as *M. murex* for altitude and *M. truncatula, M. littoralis, M. ciliaris* and *M. aculeata* for rainfall. M. polymorpha was the most widespread species showing a wide adaptation to climate, in particular altitude and rainfall. *M. orbicularis* and *M. intertexta* seem to have same requirements for a low levels of salt and nitrogen content.

Phosphorus content seems to affect only the occurrence of *M. intertexa* and *M. truncatula* which showed a preference for low phosphorus soils. In central Tunisia (Zoglami *et al.* 1996) noted that the level of phosphorus was lower in the north, and no requirement was observed for all the analysed species.

In general the area of north Tunisia which was studied, the main selection criteria are rainfall and minimum temperature, with regard to climate, and pH and nitrogen for the edaphic factors.

Following the results from this study, *M. murex* is being considered as a species for acid soils, and *M. aculeata* as a species for humid and cold areas.

REFERENCES

Abdelguerfi, A., Chapot, J. Y and Conesa, A. P. (1988). Contribution à l'étude de la répartition des luzernes annuelles spontanées en Algérie selon certains facteurs du milieu. *Fourrages* **113**, 89-106.

Bounejemate, M., Beale, P. E. and Robson, A. D. (1992). Annual medicago species in Morocco. I. Species and their abundance. *Aust. J. Agric. Res* **43**. 739-749.

Hassen, H., Zoghlami, A and Sassi, S. (1994). Contribution à l'étude de quelques espèces spontanées de légumineuses pastorales en Tunisie centrale: répartition géographique et relation avec le milieu environnant. *Annales de l'INRAT* **67** (1, 2), 203-222.

Howieson, J. G., and Ewing, M. A. (1986). Acid tolerance in Rhizobium *Meliloti-Medicago* symbiosis. *Aust.J. Agric. Res.* **37**, 55-64.

Prosperi, J. M., Gensollen, V., Olivieri, I. and Mansat, P. (1989). Observations sur la répartition et l'écologie des luzernes annuelles et de trèfle souterrain en Corse. pp. 295-296. (XVI Congrès International des herbages: Nice- France).

Zoghlami, A., Hassen, H., Seklani, H., Robertson, L. D and Saklini, A. K. (1996). Distribution des luzernes annuelles en Tunisie centrale en fonction des facteurs édaphiques et climatiques. *Fourrages* **145**, 5-16.

Index of legumes species

Current Plant Science and Biotechnology in Agriculture

1. H.J. Evans, P.J. Bottomley and W.E. Newton (eds.): *Nitrogen Fixation Research Progress.* Proceedings of the 6th International Symposium on Nitrogen Fixation (Corvallis, Oregon, 1985). 1985 ISBN 90-247-3255-7
2. R.H. Zimmerman, R.J. Griesbach, F.A. Hammerschlag and R.H. Lawson (eds.): *Tissue Culture as a Plant Production System for Horticultural Crops.* Proceedings of a Conference (Beltsville, Maryland, 1985). 1986 ISBN 90-247-3378-2
3. D.P.S. Verma and N. Brisson (eds.): *Molecular Genetics of Plant-microbe Interactions.* Proceedings of the 3rd International Symposium on this subject (Montréal, Québec, 1986). 1987 ISBN 90-247-3426-6
4. E.L. Civerolo, A. Collmer, R.E. Davis and A.G. Gillaspie (eds.): *Plant Pathogenic Bacteria.* Proceedings of the 6th International Conference on this subject (College Park, Maryland, 1985). 1987 ISBN 90-247-3476-2
5. R.J. Summerfield (ed.): *World Crops: Cool Season Food Legumes.* A Global Perspective of the Problems and Prospects for Crop Improvement in Pea, Lentil, Faba Bean and Chickpea. Proceedings of the International Food Legume Research Conference (Spokane, Washington, 1986). 1988 ISBN 90-247-3641-2
6. P. Gepts (ed.): *Genetic Resources of* Phaseolus *Beans.* Their Maintenance, Domestication, Evolution, and Utilization. 1988 ISBN 90-247-3685-4
7. K.J. Puite, J.J.M. Dons, H.J. Huizing, A.J. Kool, M. Koorneef and F.A. Krens (eds.): *Progress in Plant Protoplast Research.* Proceedings of the 7th International Protoplast Symposium (Wageningen, The Netherlands, 1987). 1988 ISBN 90-247-3688-9
8. R.S. Sangwan and B.S. Sangwan-Norreel (eds.): *The Impact of Biotechnology in Agriculture.* Proceedings of the International Conference The Meeting Point between Fundamental and Applied in vitro Culture Research (Amiens, France, 1989). 1990. ISBN 0-7923-0741-0
9. H.J.J. Nijkamp, L.H.W. van der Plas and J. van Aartrijk (eds.): *Progress in Plant Cellular and Molecular Biology.* Proceedings of the 8th International Congress on Plant Tissue and Cell Culture (Amsterdam, The Netherlands, 1990). 1990 ISBN 0-7923-0873-5
10. H. Hennecke and D.P.S. Verma (eds.): *Advances in Molecular Genetics of Plant–Microbe Interactions.* Volume 1. 1991 ISBN 0-7923-1082-9
11. J. Harding, F. Singh and J.N.M. Mol (eds.): *Genetics and Breeding of Ornamental Species.* 1991 ISBN 0-7923-1094-2
12. J. Prakash and R.L.M. Pierik (eds.): *Horticulture – New Technologies and Applications.* Proceedings of the International Seminar on New Frontiers in Horticulture (Bangalore, India, 1990). 1991 ISBN 0-7923-1279-1
13. C.M. Karssen, L.C. van Loon and D. Vreugdenhil (eds.): *Progress in Plant Growth Regulation.* Proceedings of the 14th International Conference on Plant Growth Substances (Amsterdam, The Netherlands, 1991). 1992 ISBN 0-7923-1617-7
14. E.W. Nester and D.P.S. Verma (eds.): *Advances in Molecular Genetics of Plant–Microbe Interactions.* Volume 2. 1993 ISBN 0-7923-2045-X

Current Plant Science and Biotechnology in Agriculture

15. C.B. You, Z.L. Chen and Y. Ding (eds.): *Biotechnology in Agriculture.* Proceedings of the First Asia-Pacific Conference on Agricultural Biotechnology (Beijing, China, 1992). 1993 ISBN 0-7923-2168-5
16. J.C. Pech, A. Latché and C. Balagué (eds.): *Cellular and Molecular Aspects of the Plant Hormone Ethylene.* 1993 ISBN 0-7923-2169-3
17. R. Palacios, J. Mora and W.E. Newton (eds.): *New Horizons in Nitrogen Fixation.* Proceedings of the 9th International Congress on Nitrogen Fixation (Cancún, Mexico, 1992). 1993 ISBN 0-7923-2207-X
18. Th. Jacobs and J.E. Parlevliet (eds.): *Durability of Disease Resistance.* 1993 ISBN 0-7923-2314-9
19. F.J. Muehlbauer and W.J. Kaiser (eds.): *Expanding the Production and Use of Cool Season Food Legumes.* A Global Perspective of Peristent Constraints and of Opportunities and Strategies for Further Increasing the Productivity and Use of Pea, Lentil, Faba Bean, Chickpea, and Grasspea in Different Farming Systems. Proceedings of the Second International Food Legume Research Conference (Cairo, Egypt, 1992). 1994 ISBN 0-7923-2535-4
20. T.A. Thorpe (ed.): In Vitro *Embryogenesis in Plants.* 1995 ISBN 0-7923-3149-4
21. M.J. Daniels, J.A. Downie and A.E. Osbourn (eds.): *Advances in Molecular Genetics of Plant-Microbe Interactions.* Volume 3. 1994 ISBN 0-7923-3207-5
22. M. Terzi, R. Cella and A. Falavigna (eds.): *Current Issues in Plant Molecular and Cellular Biology.* Proceedings of the VIIIth International Congress on Plant Tissue and Cell Culture (Florence, Italy, 1994). 1995 ISBN 0-7923-3322-5
23. S.M. Jain, S.K. Sopory and R.E. Veilleux (eds.): In Vitro *Haploid Production in Higher Plants.* Volume 1: Fundamental Aspects and Methods. 1996 ISBN 0-7923-3577-5
24. S.M. Jain, S.K. Sopory and R.E. Veilleux (eds.): In Vitro *Haploid Production in Higher Plants.* Volume 2: Applications. 1996 ISBN 0-7923-3578-3
25. S.M. Jain, S.K. Sopory and R.E. Veilleux (eds.): In Vitro *Haploid Production in Higher Plants.* Volume 3: Important Selected Plants. 1996 ISBN 0-7923-3579-1
26. S.M. Jain, S.K. Sopory and R.E. Veilleux (eds.): In Vitro *Haploid Production in Higher Plants.* Volume 4: Cereals. 1996 ISBN 0-7923-3978-9
27. I.A. Tikhonovich, N.A. Provorov, V.I. Romanov and W.E. Newton (eds.): *Nitrogen Fixation: Fundamentals and Applications.* 1995 ISBN 0-7923-3707-7
28. N.L. Taylor and K.H. Quesenberry: *Red Clover Science.* 1996 ISBN 0-7923-3887-1
29. S.M. Jain, S.K. Sopory and R.E. Veilleux (eds.): In Vitro *Haploid Production in Higher Plants.* Volume 5: Oil, Ornamental and Miscellaneous Plants. 1996 ISBN 0-7923-3979-7
30. R.H. Ellis, M. Black, A.J. Murdoch and T.D. Hong (eds.): *Basic and Applied Aspects of Seed Biology.* 1997 ISBN 0-7923-4363-8
31. C. Elmerich, A. Kondorosi and W.E. Newton (eds.): *Biological Nitrogen Fixation for the 21st Century.* 1998 ISBN 0-7923-4834-6

Current Plant Science and Biotechnology in Agriculture

32. S.M. Jain, D.S. Brar and B.S. Ahloowlia (eds.): *Somaclonal Variation and Induced Mutations in Crop Improvement.* 1998 ISBN 0-7923-4862-1
33. S.J. Bennett and P.S. Cocks (eds.): *Genetic Resources of Mediterranean Pasture and Forage Legumes.* 1999 ISBN 0-7923-5522-9

KLUWER ACADEMIC PUBLISHERS – DORDRECHT / BOSTON / LONDON

GPSR Compliance

The European Union's (EU) General Product Safety Regulation (GPSR) is a set of rules that requires consumer products to be safe and our obligations to ensure this.

If you have any concerns about our products, you can contact us on ProductSafety@springernature.com

In case Publisher is established outside the EU, the EU authorized representative is:

Springer Nature Customer Service Center GmbH
Europaplatz 3
69115 Heidelberg, Germany

Batch number: 10755292

Printed by Printforce, the Netherlands